The Human Genome in Health and Disease

A Story of Four Letters

To my parents

The Human Genome in Health and Disease

A Story of Four Letters

Tore Samuelsson

 CRC Press
Taylor & Francis Group

A GARLAND SCIENCE BOOK

CRC Press
Taylor & Francis Group
6000 Broken Sound Parkway NW, Suite 300
Boca Raton, FL 33487-2742

© 2019 by Taylor & Francis Group, LLC
CRC Press is an imprint of Taylor & Francis Group, an Informa business

No claim to original U.S. Government works

Printed on acid-free paper

International Standard Book Number-13: 978-0-367-07633-7 (Hardback)
978-0-8153-4591-6 (Paperback)

Library of Congress Cataloging-in-Publication Data

Names: Samuelsson, Tore, 1951- author.
Title: The human genome in health and disease : a story of four letters / Tore Samuelsson.
Description: Boca Raton : Taylor & Francis, 2019. | Includes bibliographical references and index.
Identifiers: LCCN 2018040829| ISBN 9780815345916 (pbk. : alk. paper) | ISBN 9780367076337 (hardback : alk. paper) | ISBN 9780429021732 (e-ISBN)
Subjects: | MESH: Genome, Human | Disease--genetics
Classification: LCC QH447 | NLM QU 460 | DDC 611/.0181663--dc23
LC record available at https://lccn.loc.gov/2018040829

Visit the Taylor & Francis Web site at
http://www.taylorandfrancis.com

and the CRC Press Web site at
http://www.crcpress.com

eResource material is available for this title at https://www.crcpress.com/
9780367076337

Contents

Preface

The DNA molecule contains genetic information in the form of sequences of nucleotides abbreviated A, T, C, and G. Therefore, genetic information is essentially a very long sequence of these four letters. In a single human individual, the genetic content amounts to about 3 billion letters. Furthermore, information in DNA is used to direct the production of proteins and RNAs. Sequences of nucleotides in DNA and RNA, as well as sequences of amino acids in proteins, are all examples of *molecular sequences*. This book attempts to present an accessible account of the molecular sequence information contained in the human genome and how it is being used to direct the production of RNA and protein. It addresses the question of what important biological signals are found in the linear sequence of nucleotides in DNA and shows how specific DNA sequences have distinct functions. Such functional sequence elements, typically in a small size range like 3–20 nucleotides, may be classified into a number of different functional categories. Examples are the three-nucleotide sequences that specify amino acids, or short DNA sequences that are targets for proteins that regulate transcription.

To provide biological motivation, the importance of the DNA sequence for biological function is illustrated with a variety of inherited disorders or with other genetic disorders such as cancer. With such examples, different functional elements of a gene, as well as different aspects of genetic information transfer within the cell, are introduced. For instance, a point mutation in a coding sequence of a globin gene gives rise to sickle cell anemia, a splicing mutation results in a form of hemophilia, and a mutation in an untranslated region of an mRNA leads to an iron metabolism disorder.

When discussing the functional consequences of DNA sequence, it is also of interest to consider the molecular impact of mutations and sequence variation. For instance, changes in DNA sequence may affect the recognition by a nucleic acid, such as in the case of a codon being read by a tRNA anticodon. Alternatively, a DNA variant will affect the recognition by a protein, such as a protein regulating transcription. There are indeed throughout the book several examples showing in structural detail the interaction of a nucleic acid with another nucleic acid or with a protein, illustrating the consequences of mutations at the level of molecular interactions.

In essence, therefore, this book has a *molecular sequence perspective*, and recurring themes are *functional DNA sequence elements*, illustration of *functional impact* with genetic disorders, and *molecular interactions* affected by sequence variation.

Why is it important to know about the different DNA sequence elements of the human genome and their functional impact? There are many medical areas where such knowledge is critical, as shown in this book. Examples include pharmacology, gene therapy, and the diagnosis of rare inherited disorders and cancer.

The book is organized such that a major part (Chapters 6 through 12) discusses specific functional sequence elements of the human genome. The preceding chapters offer an introduction to basic concepts with regard to the human genome. Chapters 2 and 3 introduce human genes and their relationship to human disease using sickle cell anemia as an example. Proteins and their structure are introduced (Chapter 2), and the focus is then on DNA structure, the flow of genetic information from DNA to protein, as well as the basic principles of translating an mRNA using the genetic code (Chapter 3). Human genome sequencing and the organization and overall content of the human genome are topics of Chapter 4. Individual genetic variation, the phenotypic effects of mutations, as well as cancer are then discussed in the following chapter.

Chapters 6 through 12 provide details about functional sequence elements of the genome and their role in the flow of genetic information. The importance of these elements is illustrated with inherited disorders or cancer. First, the role of the coding regions of mRNA molecules is discussed in Chapter 6. Chapter 7 covers repeat expansions in coding regions. Then the role of mRNA untranslated regions (Chapter 8) and the process of splicing to form mature mRNA (Chapter 9) is considered. The complex but important area of transcriptional control is then discussed in Chapter 10. The noncoding RNAs are the subject of Chapter 11. The analysis of DNA and protein sequences requires a substantial amount of computing. Therefore, in Chapter 12, computational methods used with biological sequences are briefly reviewed.

Chapters 13 through 15 provide additional biological motivation as they further illustrate why careful studies of the relationship between sequence and function are essential. These chapters deal with important medical applications of human genome sequencing. Thus, experimental methods to diagnose errors in DNA sequences are described, including successful efforts to reveal causative mutations in rare inherited disorders (Chapter 13). Furthermore, there is recent progress in the area of gene therapy. Some of the most important methods are covered, as well as a few cases of successful therapy for inherited disease (Chapter 14). Finally, some of the applications of human genome sequencing raise a number of ethical concerns, and these are discussed in the final chapter.

When it comes to level of difficulty, I have attempted to present the material at a basic level to make it understandable for a reader without previous studies of genetics and molecular biology. However, the reader will benefit from a basic knowledge about chemistry and biochemistry. Following the principle that a picture says more than a thousand words, the book is richly illustrated. Furthermore, a web supplement to the book that includes scientific updates and answers to selected chapter questions is available at http://toresamuelsson.se/hg.

Why did this book come about? We currently see a dramatic development in terms of human genome sequencing. Research laboratories around the world generate a wealth of genomic sequence data. Sequencing is becoming widely used in the clinic to analyze a variety of genetic disorders, including cancer. In addition, you can order your own genome sequence from a company ("direct-to-consumer" sequencing). With this wealth of genetic information, it becomes increasingly important to know what the human genome sequence is all about, how the sequence should be understood in terms of biological function, and how particular variants in the genome should be interpreted. I wanted to write a book providing an introduction to these topics. In the early days of my scientific career, I worked as a molecular biologist. Eventually, I turned to bioinformatics with a focus on molecular sequences. For a long time I have been intrigued by the digital nature of genetic information, that the genome sequence can be handled with computers as a long string of letters, and that computing with molecular sequences may be used to address a variety of biological problems. There are already very good books dealing with the human genome, but this book has a focus on molecular sequences and it examines in a systematic manner the functional role of DNA sequence elements as illustrated with human genetic disorders.

ACKNOWLEDGMENTS

Illustrations of the book were typically created with CorelDRAW version X4. All images of protein or nucleic acid structures were made with the UCSF Chimera package and structures available in the Protein Data Bank (www. rcsb.org; Berman HM, Westbrook J, Feng Z, Gilliland G, Bhat TN, Weissig H, Shindyalov IN, Bourne PE. 2000. The Protein Data Bank. *Nucleic Acids Res* 28:235–242). Chimera is developed by the Resource for Biocomputing, Visualization, and Informatics at the University of California, San Francisco (supported by NIGMS P41-GM103311). Chimera is described in Pettersen EF, Goddard TD, Huang CC, Couch GS, Greenblatt DM, Meng EC, Ferrin TE. 2004. UCSF Chimera—A visualization system for exploratory research and analysis. *J Comput Chem* 25(13):1605–1612. I am indebted to the KEGG/GenomeNet Support Team for permission to use a metabolic pathway image.

I am grateful to a number of people for their help in the context of this book. Jan Korbel, European Molecular Biology Laboratory, Heidelberg, and Evan Eichler, University of Washington, Seattle, helped out with comments about human structural variation. Johanna Rommens at the Hospital for Sick Children, Toronto, Canada, provided information about the Cystic Fibrosis Mutation Database, and Catherine Porcher, University of Oxford, United Kingdom, about the ELK/GATA sequence logo. Oxford Nanopore Technology allowed the use of a photograph of the MinION apparatus. Aravinda Chakravarti and Sumantra Chatterjee, Johns Hopkins University School of Medicine, Baltimore, Maryland, informed me on the role of transcription factor binding sites with regard to Hirschsprung disease. Eric Ottesen, Iowa State University, Ames, provided information regarding SMN2 exon 8 and nusinersen. I am grateful to Stefan Mundlos, Max Planck Institute for Molecular Genetics, Berlin, Germany, for providing clinical photographs in the context of developmental disorders resulting from genomic rearrangements in the EPHA4 locus. Sally Heywood, Cardiff University, United Kingdom, provided statistics of the collection of mutations in the Human Gene Mutation Database (HGMD). Mark Johnson, reporter at the Milwaukee *Journal Sentinel*, helped out with questions regarding the Nicholas Volker case. The National Center for Biotechnology Information (NCBI) User Services provided help regarding human single-nucleotide variants that are part of the NCBI dbSNP. I am grateful to Retta Beery for permission to use a photograph of her family. Niclas Juth, Karolinska Institute, Sweden, commented on the ethical topics of Chapter 15.

Colleagues at my department were also very helpful. Gunnar Hansson helped out with the parts of the book dealing with cystic fibrosis, and the noncoding RNA chapter was examined by Chandrasekhar Kanduri. I am also grateful to Per Elias for discussions on a variety of topics. Erik Larsson carefully examined specific parts of the manuscript and had very useful suggestions regarding the contents of Chapter 10.

A number of anonymous reviewers recruited by the publisher provided important comments to the manuscript. Furthermore, I am indebted to staff at Garland Science/CRC Press. Elizabeth Owen, Senior Editor at Garland Science, carefully read all of the text and offered very helpful comments in an early phase of the book writing. I am also grateful for the work and support of Developmental Editor Jordan Wearing and Senior Editor Chuck Crumly at CRC Press.

Finally, I'm very grateful for a stipend from The Royal Society of Arts and Sciences in Gothenburg, Sweden, that allowed me to work on this book during one month in 2017 at Hotel Chevillon, Grez-sur-Loing, France.

Tore Samuelsson

Introduction

1

DNA, short for deoxyribonucleic acid, is a universal carrier of hereditary information. In all life forms—viruses, bacteria, fungi, plants, and animals—it carries important instructions for the design of the organism. And not only does it carry information—it is also a molecule designed so that it may be accurately copied to the next generation. DNA is built from simple units, referred to as nucleotides, that are joined to form very long molecules. Each nucleotide contains any of four different nitrogenous bases: adenine, thymine, cytosine, or guanine, abbreviated A, T, C, and G, respectively. It is the sequence of these bases that forms the actual genetic message. Thus, the information in DNA may be expressed as a long sequence of the letters A, T, C, and G—for an example, see **Figure 1.1**.

We refer to the complete genetic material of an organism as its **genome**. The human genome is an astounding three billion letters. An important milestone was reached in biomedical research in 2001 when, for the first time, a draft of the human genome was presented and the complete sequence of letters could be read. A small fraction of the human genome is shown in Figure 1.1. Consider the whole genome printed as a physical book. A total of 6,400 bases are in Figure 1.1. You would need in the order of 500,000 pages like this to cover the full human genome. That would correspond to more than 1,600 books, each with 300 pages. For more on printing the human genome on paper, see **Figure 1.2**.

The issues addressed by this textbook are related to the three billion letter sequence of the human genome. How are we to make sense of and understand this vast information? What different biological signals are contained in the DNA? How important are different regions of the sequence? Are some regions more important than others? What are the effects in the event the sequence of letters in DNA is changed? In molecular biology laboratories, scientists have carried out experiments to address these questions. In addition, as changes or mutations in DNA are natural components of evolution, nature has by itself carried out experiments during billions of years that may guide us in understanding the relationship between genetic information and biological function. For instance, mutations in DNA can

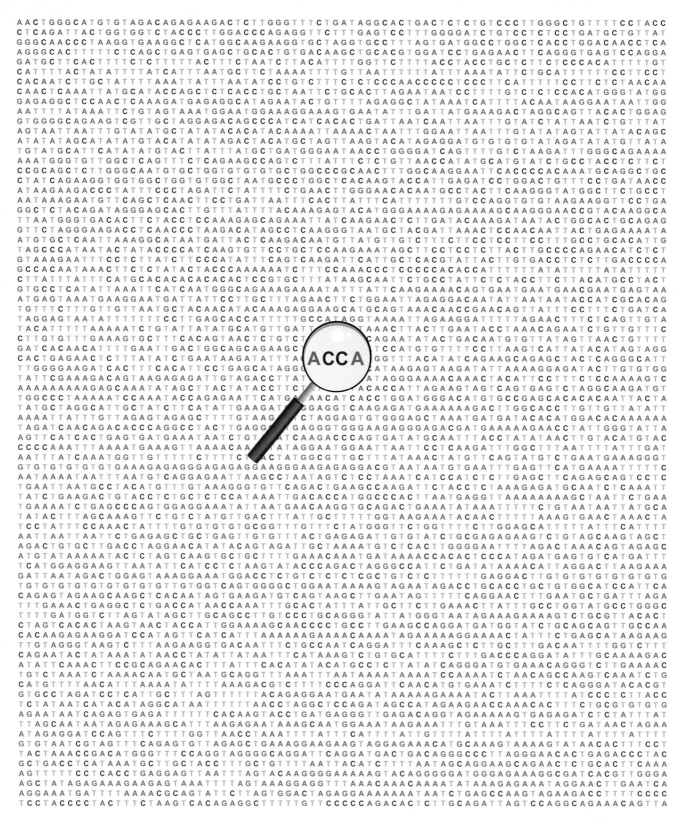

Figure 1.1 Portion of the human genome. Letters A, T, C, and G represent the DNA bases adenine, thymine, cytosine, and guanine, respectively. This page has 6,400 bases. It would take 500,000 pages like this to cover the full human genome. This would correspond to more than 1,600 books assuming one book contains 300 pages. The magnifying glass is to indicate that the object of research concerning the human genome is to associate sequence elements with biological functions. Scientists use experimental methods, as well as computational methods, to do this—a work in progress.

give rise to specific inherited diseases as well as cancer. What are the changes in the DNA sequence that cause such deleterious effects?

To answer these questions, we need to understand the organization of the human genome, as well as the different functional sequence elements in that genome. The flow of genetic information is crucial. Hence, DNA specifies what RNA molecules are to be made. One subclass of these RNAs is subject to processing to form messenger RNA (mRNA) molecules. These mRNA molecules in turn act as templates for the production of proteins. Another abundant class of RNA molecules has functions other than to specify proteins. Throughout the elaborate flow of genetic information that includes copying of DNA sequences to RNA, RNA processing, as well as the synthesis of protein using mRNA, specific nucleotide sequences have distinct functions.

In essence, this book explores the information in the human genome and all of the important biological signals that are present. It illustrates various functions of DNA sequences. Examples include protein coding sequences and sequences that regulate the flow of genetic information. For all of the different sequence elements, the relationship between sequence and function is illustrated with disorders of a genetic background.

Figure 1.2 Human genome printed on paper. Scientists at the University of Leicester printed the whole human genome on paper. It resulted in 130 book volumes that would take 95 years to read. (Published under CC BY-SA 2.0.)

A Molecular Disorder

2

The theme of this book is the information contained within the human genome as outlined in Chapter 1. As a first element of information, we consider regions in the genome that specify the proteins to be made. Proteins are molecules built from amino acids, and the sequence of amino acids is determined by the sequence of nucleotides in the genome. Regions specifying proteins make up only a minute portion of the entire genome but are nevertheless significant.

We first consider how amino acid sequences are related to inherited disorders. As an example, we discuss the disorder **sickle cell anemia**. It is caused by a mutation that gives rise to a replacement of the amino acid glutamic acid to valine in the protein hemoglobin. There are multiple reasons why we discuss this particular disorder in some detail in the first book chapters. Early studies of sickle cell anemia were based on nonmolecular clinical observations, and it seemed likely that the disease is inherited. But as research progressed, we eventually obtained a detailed molecular understanding of sickle cell anemia from the changes in DNA to detailed structural information about the hemoglobin protein. Sickle cell anemia is historically significant as it was the first disease to be characterized where a genetic change is associated with a well-defined change in a protein molecule. This finding gave an early clue as to the power of molecular medicine. In addition, very few inherited disorders have been so thoroughly examined as sickle cell anemia, and information about the disease is still being collected today. There is a significant medical impact from this knowledge, which is why we return to this disorder also in other contexts such as gene therapy (Chapter 14).

THE DISCOVERY OF SICKLE CELL ANEMIA

Abotutuo. Chwechweechwe. Nwiiwii. Nuiduidui. These are all names of a disease common in Western Africa—a disease we now know also as sickle cell anemia. Its history in Africa may be tracked as far back as the seventeenth century. Classification of the disease was difficult on this continent, because the symptoms were closely related to those of other diseases in tropical areas. It was not until the early twentieth century that sickle cell anemia was first described in a medical publication. The affected individual was Walter Clement Noel.

Noel was born in 1884 on a large estate on Grenada. At this time, this island was a British colony. Noel was from a wealthy black family. He suffered from sickle cell anemia but was still able to attend school, and he completed his undergraduate studies in 1904. The same year, he sailed to New York. During this week-long journey, he developed a leg ulcer, a common complication of sickle cell anemia. When he arrived in New York, he immediately sought medical attention. His ulcer was treated with iodine—this chemical had the effect of killing the bacteria of the ulcer. Noel was thus cured and then traveled to Chicago, Illinois, where he was to study dentistry at the Chicago College of Dental Surgery. Noel was an unusual student at this time in the sense that it was uncommon for students of African descent to reach higher studies. In November 1904, his disease unfortunately got more severe as he developed respiratory problems. These problems persisted for more than 1 month. He finally sought medical attention at the Presbyterian Hospital in Chicago and was examined by an intern, Ernest Irons. Among other tests, Irons examined a blood sample from Noel. Under the microscope he could see that the sample contained, as he phrased it, "many pear-shaped and elongated forms—some small." Iron discussed these findings with his supervisor James Brian Herrick. Despite further investigations, Herrick and Irons could not reveal the cause of these unusual cells.

Noel eventually recovered from his respiratory problems and returned to continue his studies at the dentistry school. However, he experienced additional illnesses during a period of more than two years of studies. Thus, he once was hospitalized for bronchitis and once for painful muscular crises and gallstones. He was then under the care of Irons who kept dutiful notes. When Irons was done with his training, he gave these notes to Herrick. Herrick then took care of Noel for two years.

Herrick was at a national meeting in 1910 and there presented the case of Noel—incidentally without giving credit to Irons. He published a report the same year. To describe the blood cells of the disease, Herrick used the term "sickle shaped cells" (**Figure 2.1**).

Despite his illnesses, Noel graduated from dental school. He then went back to Grenada where he set up a general dentistry practice in St. George's, the capital of Grenada. Not much is known of him from then on. However, in 1916, he overexerted himself. He attended a horse race on Grenada a long way from home and traveled home, all on the same day. As a result of this, he developed a serious respiratory infection. Only a few weeks after the horse race, he died from pneumonia at 32 years old. Noel is buried in a churchyard with a view of the Caribbean Sea. He is next to his sister Jane, who also died young of respiratory problems.

Only three months after Herrick published the case of Noel in 1910, a second case of the same disorder was described. Blood samples from a 25-year-old woman, Ellen Anthony, a resident of Virginia, showed the same strange shape of red blood cells as was observed in Noel. As more cases were identified in the 1920s, it was noted that all individuals with the disease were of African origin. The disease was eventually to be named *sickle cell anemia* (SCA).

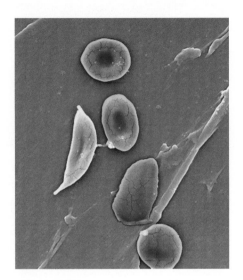

Figure 2.1 Sickle or crescent shape of red blood cells characteristic of sickle cell anemia. A sickle cell is shown (to the left) along with normal red blood cells. (Published under CC BY 3.0.)

A RECESSIVE INHERITED DISORDER

Significant advances in understanding sickle cell anemia were made by the end of the 1940s—both with respect to genetic inheritance and as to the molecular basis of the disease. Already in 1923, John Huck studied families with sickle cell anemia and noted that the disease was probably inherited, although his studies did not provide any firm evidence of this theory. Studies were complicated by the fact that some individuals had symptoms related to sickle cell anemia but had a much milder form of the disease that did not shorten their lives. These individuals were said to have the **sickle cell trait**. It was often difficult to distinguish between these two categories of patients. But in 1949, James Neel carried out a careful examination of families affected by sickle cell anemia and was able to conclude that the disorder is indeed hereditary.

How is sickle cell anemia inherited? We turn to basic principles of genetics and inheritance, first elucidated by Gregor Mendel in the nineteenth century. The genetic makeup of an individual is referred to as the **genotype**, whereas the **phenotype** is the collection of observable characteristics. The phenotype is determined by the genotype and/or environmental factors. Human individuals—like all other animals—have two copies of each gene, one of paternal and one of maternal origin. A gene may have two or more variants—in the language of genetics, such variants are referred to as **alleles**. If two individuals have the same allele, they are said to be **homozygous** for that allele, and if they have two different alleles, they are **heterozygous**. From the perspective of an allele, one basic principle of inheritance is outlined in **Figure 2.2**. In this example, two different alleles "A" and "a" are considered, and the two parents are heterozygous, since they both have the allele configuration (genotype) Aa. During the formation of sperm and egg cells that occurs during **meiosis**, each cell ends up with any one of the two alleles. During fertilization, alleles are combined, and a child of the two parents may in this case have any of the genotypes AA, Aa, and aa.

For the discussion of sickle cell anemia, the two different alleles we consider are the sickle cell variant and the normal form not associated with disease. Individuals with two copies of the sickle gene develop sickle cell anemia, whereas patients with one copy of the sickle gene and one normal copy of the hemoglobin gene have milder symptoms and express the sickle cell trait. This is illustrated by the pedigree (family tree) with members affected by the disorder in **Figure 2.3**.

The inheritance of sickle cell anemia follows the rules of a **recessive** disorder. In such a disorder, two copies of the disease allele are required to develop the disease. The rules of inheritance also inform us on probabilities on inheriting disorders as explained in the diagram in **Figure 2.4**,

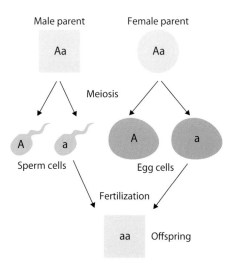

Figure 2.2 Inheritance of alleles. Two different alleles (gene variants) "A" and "a" are considered. In this example, both parents have the same setup of these alleles. The sperm and egg cells formed during meiosis have only one copy of each allele. The probability that a certain sperm or egg cell has a specific allele is about 50%. The fertilized egg will have two copies of each autosomal chromosome, and in the example shown here, "a" from the father is combined with the same allele from the mother. However, other outcomes are possible given the parent genotypes. Thus, offspring may be of three different genotypes: AA, Aa, or aa.

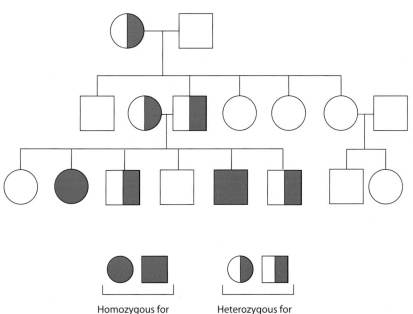

Figure 2.3 Sickle cell anemia is a recessive inherited disorder. The pedigree illustrates inheritance of sickle cell anemia. Males are represented by squares, females by circles. The disease gene is indicated in red. In recessive disorders, two copies of the affected allele are required to develop the disease.

Homozygous for sickle cell variant (express sickle-cell anemia)

Heterozygous for sickle cell variant (express sickle-cell trait)

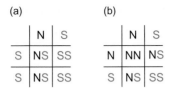

(a)   **(b)**

	N	S
S	NS	SS
S	NS	SS

	N	S
N	NN	NS
S	NS	SS

Figure 2.4 Probabilities related to sickle cell anemia. Squares have been arranged to discover all possible genotypes that occur in children, given the genotypes of their parents. From this diagram, we may also infer the probability of each offspring genotype. The sickle cell allele is represented by "S" and the normal allele as "N." The top row has the genotype of one parent, and the leftmost column the genotype of the other parent. The other boxes are obtained by copying the parent letters across or down. In this way, we get the predicted frequencies of all the potential genotypes. (a) This is a case where one parent has sickle cell anemia (genotype SS) and the other parent is heterozygous. As the SS genotype occurs in two of the four cells, we estimate that the probability that a child has sickle cell anemia is 50%. In (b) both parents are heterozygous. Here, the SS genotype occurs in one of the four cells, and we estimate that the probability is 25% that a child is affected by the disease. The example shows sickle cell anemia, a recessive disorder, but the method of inferring probabilities may be used to examine just about any case of recessive or dominant inheritance.

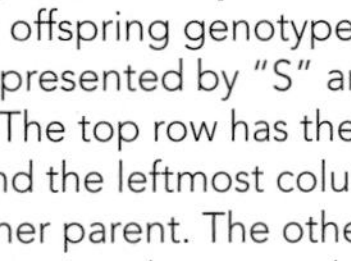

Figure 2.5 Chemical structure of an amino acid. All amino acids contain a carbon atom to which is connected an amino group, a carboxyl group, a hydrogen atom, and a side chain (R) specific to the amino acid.

known as a Punnett square. For instance, if one parent is homozygous for the sickle variant and the other parent is heterozygous, the probability that a child of these parents develops sickle cell anemia is 50%. If both parents are heterozygous, the probability is instead 25%.

As opposed to recessive disorders, in **dominant** disorders, only one disease gene is sufficient to develop the disease. Recessive and dominant diseases may be distinguished because they have different inheritance patterns (for an example of a dominant disorder and its inheritance, see Chapter 7).

SICKLE CELL ANEMIA AND MALARIA

Sickle cell anemia affects many hundreds of thousands around the world. In particular, it is common among individuals of Sub-Saharan African descent. In the United States, it occurs among about 1 out of every 365 black American births, and about 1 in 13 black American babies is born with sickle cell trait. In Africa, sickle cell disease is even more common—in some parts of Sub-Saharan Africa, up to 1 in 30 of all newborns are affected by the disease.

Why is sickle cell anemia such a common inherited disorder in parts of Africa? It was demonstrated in the 1950s that people with the sickle cell trait are more resistant against malaria caused by the protozoan parasite *Plasmodium falciparum*. Therefore, in areas of malaria, the sickle cell gene has a selective advantage. It is still not clear why the sickle cell gene protects against malaria.

CHARACTERIZING SICKLE CELL ANEMIA FURTHER: THE ROLE OF HEMOGLOBIN

What is actually the cause of sickle cell anemia? Research eventually zoomed in on **hemoglobin**, a protein.

Proteins are important molecules in biology, as they carry out many critical functions. For instance, they act as enzymes, transporters, and receptors mediating hormonal response. By the turn of the twentieth century, it was shown that proteins are composed of **amino acids**. All amino acids have a carboxyl group and an amino group in addition to a side chain specific to the amino acid (**Figure 2.5**). The chemical structures of selected amino acids are shown in **Figure 2.6** (see Appendix Figure A.1 for the structure of all 20 amino acids). In proteins, the amino acids are joined through **peptide bonds** (**Figure 2.7**).

Hemoglobin belongs to the family of transporter proteins. It is an important protein responsible for transporting oxygen from the lungs to the tissues; in addition, it carries carbon dioxide back to the lungs. The oxygen-binding properties of hemoglobin had already been discovered in the nineteenth century. Could the amino acid composition of hemoglobin somehow

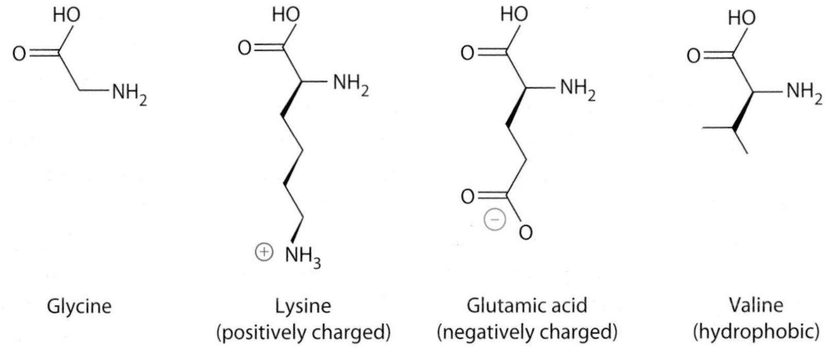

Figure 2.6 Chemical structures of selected amino acids. Glycine is the smallest amino acid with a hydrogen as its side chain. Lysine and glutamic acid are amino acids that are charged because of their amino and carboxyl groups, respectively. Valine belongs to a group of nonpolar amino acids.

Glycine   Lysine (positively charged)   Glutamic acid (negatively charged)   Valine (hydrophobic)

be changed in sickle cell anemia? In one classic experiment in 1949, Linus Pauling compared hemoglobin from healthy individuals to that of sickle cell patients. It turned out that sickle cell globin had a different electrophoretic mobility—that is, when the proteins were subjected to an electric field, the sickle cell hemoglobin moved as a positively charged protein, in contrast to the normal hemoglobin (**Figure 2.8**). Pauling suggested that the difference was due to a change in the number of charged amino acids in the sickle cell version of the protein. The amino acids glutamic acid and aspartic acid are negatively charged because they carry a carboxyl group, and lysine and arginine are positively charged as they have an amino group (see Appendix Figure A.1). It seemed likely that sickle cell hemoglobin was affected in one or more of these amino acids. In the context of this finding, Pauling also used the term "molecular disease." This expression referred to the fact that for the first time the molecular background to a disease was identified.

MAPPING A DELETERIOUS CHANGE IN HEMOGLOBIN

Further details as to the molecular basis of the disease were elucidated in the 1950s. Vernon Ingram did one crucial experiment in 1956. A protein may be cut up into smaller pieces—peptides—using an enzyme that is able to break up peptide bonds. One such enzyme is trypsin that cuts on the carboxyl-terminal side of lysine and arginine residues. Ingram used this enzyme to fragment hemoglobin and was able to separate the resulting peptides with a newly developed technique of two-dimensional separation by electrophoresis and chromatography. When comparing normal hemoglobin to the sickle cell variant, it turned out that one single peptide was different (**Figure 2.9a**). The peptide in sickle cell hemoglobin was more positively charged than that of normal hemoglobin. Analysis of its amino acid composition showed that it had less glutamic acid and more of valine, suggesting that in sickle cell anemia, glutamic acid had been replaced by valine. The exact sequence of amino acids in the peptide was soon after determined. In the sickle cell protein, the peptide had valine instead of glutamic acid in one position (**Figure 2.9b**).

The analysis by Pauling and Ingram of sickle cell hemoglobin in the 1940s and 1950s was an important milestone in molecular biology. For the first time, a genetic inherited disorder was explained in terms of a specific amino acid substitution in a protein. We now know many more examples of inherited disorders associated with amino acid replacements as will be apparent in forthcoming chapters.

A PROTEIN FOLDS INTO A THREE-DIMENSIONAL SHAPE BASED ON ITS AMINO ACID SEQUENCE

To understand properly the molecular basis of sickle cell anemia, we also need to know the structural consequences of the amino acid substitution. Proteins are built from one or more **polypeptide** chains—each such chain is a string of amino acids joined by peptide bonds. The sequence of amino acids in each of the polypeptide chains will determine the three-dimensional shape of the protein. We commonly refer to four different levels of protein structure: **primary**, **secondary**, **tertiary**, and **quaternary** (**Figure 2.10**). The primary structure refers to the amino acid sequence of the polypeptide chain (**Figure 2.10a**). Noncovalent interactions between amino acids in the same polypeptide chain give rise to structures known

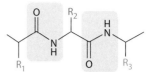

Figure 2.7 Amino acids are connected through peptide bonds. A unit with CONH represents the peptide bond. R1, R2, and R3 represent amino acid side chains.

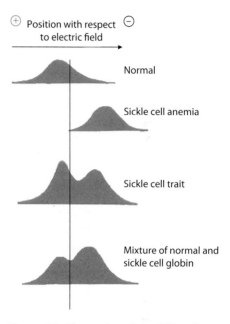

Figure 2.8 Electrophoretic mobility of normal hemoglobin and its sickle cell variant as studied by Linus Pauling. Pauling used a method of electrophoresis where the protein samples to be separated are in solution. By applying a current, the proteins will move in the solution dependent on their charge. The position with respect to the electric field is on the x-axis, and the peak height (y-axis) is proportional to protein concentration. The vertical line represents the position of an electrically neutral molecule. The results of Pauling's experiment demonstrated that the sickle cell hemoglobin moved as a positively charged molecule, whereas the normal protein migrated as a negatively charged molecule. The four different samples are (1) normal hemoglobin, (2) sickle cell hemoglobin, (3) hemoglobin prepared from individuals with the sickle cell trait (such individuals have approximately equal proportions of sickle and normal protein), and (4) a synthetic mixture with equal amounts of normal and sickle cell protein.

Figure 2.9 Two-dimensional separation of hemoglobin peptides. (a) Normal hemoglobin and sickle cell globin were treated with trypsin to generate a number of peptides. These peptides were separated in two dimensions using paper electrophoresis and paper chromatography. The patterns obtained are identical except for one peptide (N and S, for normal and sickle cell hemoglobin, respectively). (b) Determination of the peptide sequences showed that glutamic acid in position 6 of the normal protein had been replaced by valine in the sickle cell protein.

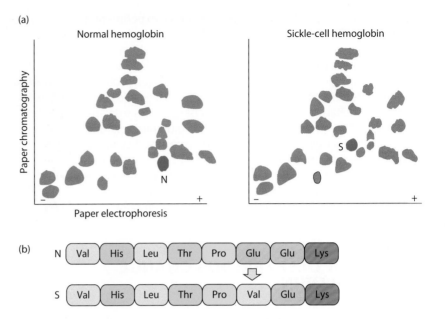

as α-**helices** and β-**sheets**. These are secondary structure elements (**Figure 2.10b**). The tertiary structure of a protein refers to the global folding of the entire polypeptide chain of a protein. This structure may contain α-helices and/or β-sheets, as well as loop regions in between (**Figure 2.10c**). In the event the protein has at least two polypeptide chains, also known as **subunits**, we also need to consider the number and arrangement of these polypeptide chains—the quaternary structure (**Figure 2.10d**).

There are three major experimental methods whereby the three-dimensional structure of a protein may be inferred. In **x-ray crystallography**, the protein is in the form of a crystal. The crystal is irradiated with x-rays, and the diffraction pattern obtained is used to calculate the electron density of the crystal. The electron density is finally used to elucidate the coordinates of the different atoms in the protein. Most available protein structures have been determined with x-ray crystallography. Second, **nuclear magnetic resonance** (**NMR**) is a method for protein structure determination where the protein is in solution. Third, **cryo-electron microscopy** is a more recent method used for structural studies of large proteins or complexes involving proteins and other large molecules. Structure may also be predicted from the sequence of amino acids using computational tools, although this is much less reliable than the experimental methods.

SICKLE CELL DISEASE MAY NOW BE UNDERSTOOD IN THE CONTEXT OF PROTEIN STRUCTURE

In 1959, the structure of hemoglobin was elucidated by Max Perutz using the technique of x-ray crystallography (**Figure 2.11**). The hemoglobin molecule is built from two different polypeptides, α-globin and β-globin. In one molecule of hemoglobin, there are two α-chains and two β-chains. It is the β-chain that is affected in sickle cell anemia.

Hemoglobin and a related protein myoglobin were in fact the first protein structures ever to be presented. The structure of hemoglobin, as well as that of the sickle cell variant, have been further refined in later work.

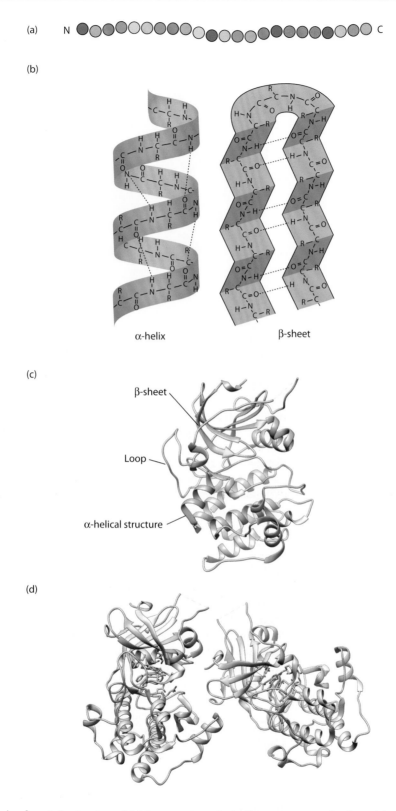

Figure 2.10 The four levels of protein structure. (a) Primary structure is the linear sequence of amino acids along the polypeptide chain. (b) Protein secondary structure elements. An α-helix is a common helical structure in proteins in which every backbone N-H group donates a hydrogen bond to the backbone C = O group of the amino acid located three or four residues earlier along the protein sequence. Another common secondary structure element in proteins is the β-sheet . It consists of strands connected laterally by backbone hydrogen bonds. (c) The tertiary structure of a protein is its global fold, including α-helices (blue), β-sheets (orange), and loop regions without ordered structure. The protein shown is a subunit of the enzyme Akt2 with both α-helical and β-sheet structures. (d) The quaternary structure of Akt2 is shown— a dimer of two identical polypeptide chains.

Figure 2.11 Three-dimensional structure of hemoglobin. Hemoglobin is built from two α-chains (orange) and two β-chains (cyan). Each of the four subunits contains heme groups (shown in ball-and-stick representation). (Adapted from Paoli M et al. 1996. *J Mol Biol* 256:775. PDB ID: 1GZX.)

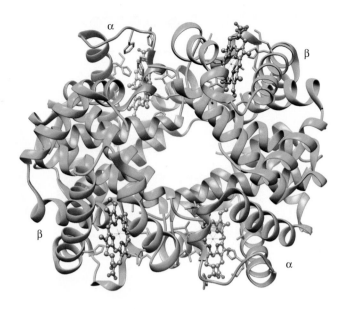

The replacement in sickle cell anemia of glutamic acid—a negatively charged amino acid (see Figure 2.6 and **Figure 2.12a**)—with valine has a significant effect on the properties of hemoglobin. Valine is a hydrophobic amino acid and interacts with a pair of hydrophobic amino acids—phenylalanine in position 85 and leucine in position 88 of the protein chain—that are located in the β-chain of a neighboring globin molecule (**Figure 2.12b**). As a result, an aggregation process is initiated where globin molecules form fibers. These fibers extend through the red blood cells such that the cells are distorted. Because of this alteration in shape and because sickle cells tend to stick to the wall of blood vessels, normal blood flow is prevented. Pain may arise when blood flow is restricted. Sickled cells have a shorter life span, giving rise to a lowered amount of red blood cells and poor oxygen transport (anemia).

A MOLECULAR SEQUENCE: PROTEINS CAN BE DEPICTED AS A STRING OF AMINO ACID SYMBOLS

Whereas Ingram focused on the amino acid sequence of selected peptides, the complete amino acid sequences of the human α- and β-globins were elucidated in the early 1960s. The sequence of the β-chain is shown in **Figure 2.13**, where the amino acids are shown with a background color based on their physical and chemical properties. A polypeptide chain built from amino acids has a distinct polarity, where one end of the polypeptide has a free amino-terminal group (the N-terminus) and the other end has a free carboxyl terminal group (C-terminus). Amino acid sequences are, as a rule, displayed with the N-terminus to the left and the C-terminus to the right. The amino acids are typically abbreviated with one letter code (see Figure 2.13) or three letter codes. The reader is referred to Figure A.1 for details of these codes.

Amino acid sequences of proteins may be compared using a computational procedure of **alignment**. Such an alignment is shown in **Figure 2.14**, where α- and β-chains of five different vertebrates are compared. From this alignment, we observe that all the globin sequences are similar, reflecting the fact that they have been well conserved during evolution. We also identify features that distinguish the α- and β-chains,

(a)

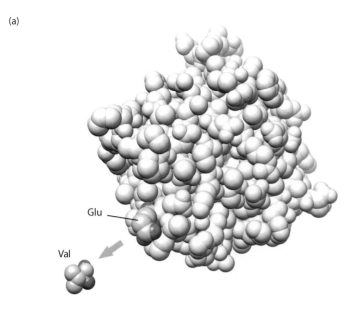

(b)

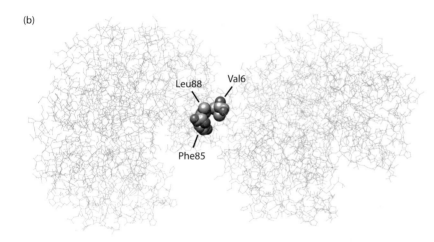

Figure 2.12 Sickle cell hemoglobin molecules interact with each other in an abnormal manner. (a) Space-filling model of normal β-chain of hemoglobin showing surface location of glutamic acid (colored with side-chain carboxyl oxygen atoms in red) in position 6 of the polypeptide chain. For comparison is shown the structure of valine, the amino acid replacing glutamic acid in sickle cell anemia. (b) Two molecules of sickle cell hemoglobin showing interaction between hydrophobic amino acids in neighboring subunits. The β-chains are in green and α-chains in red (wireframe representation). Highlighted in space-fill representation is an interaction between valine 6 in one polypeptide with a pair of two other hydrophobic amino acids—phenylalanine 85 and leucine 88—in a neighboring polypeptide. This interaction initiates aggregation of globin molecules into harmful fibers. (Adapted from Tame J, Vallone B. 2000. *Acta Crystallogr* 56:805–811. PBD ID: 1A3N. Harrington DJ et al. 1997. *J Mol Biol* 272:398–407. PDB ID: 2HBS.)

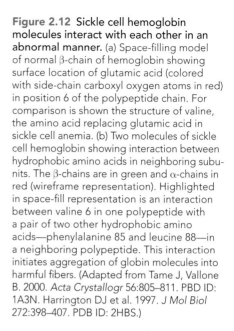

Figure 2.13 Amino acid sequence of the β-chain of hemoglobin. Amino acids are colored according to their physical and chemical properties. For instance, the basic (positively charged) amino acids arginine (R) and lysine (K) have a red background. Cleavage sites for trypsin as used in the experiment of Figure 2.9 are indicated with arrows.

such as an insertion of five amino acids in the β-chains, as compared to the α-chains. However, there is a lot of additional information that we may extract from amino acid sequences and alignments as further discussed in Chapter 12.

Amino acids in proteins are our first instance of a **molecular sequence**. There will be more of these as we enter into the world of nucleic acids—DNA and RNA—in the next chapter. We then address the important problem of how, at a molecular level, glutamic acid was changed into valine in sickle cell anemia.

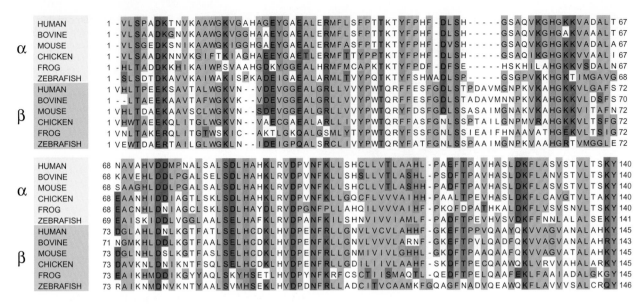

Figure 2.14 Alignment of vertebrate hemoglobin amino acid sequences. Coloring of amino acids is as in Figure 2.13.

SUMMARY

- Sickle cell anemia is a recessive inherited disorder. For the disease to be expressed, individuals must carry two copies of the sickle cell allele. Individuals with only one copy have the sickle cell trait with less severe symptoms.
- Protein structure may be described at the four levels: primary, secondary, tertiary, and quaternary.
- Hemoglobin is built from two chains of α-globin and two chains of β-globin.
- In sickle cell anemia, the amino acid glutamic acid of β-globin is replaced by the hydrophobic amino acid valine, resulting in an aggregation process.
- Sickle cell anemia was the first disease to be characterized where a genetic change is associated with a well-defined change in a protein molecule.
- The primary structure of a protein—the amino acid sequence—may be depicted as a string of amino acid symbols. Amino acid sequences may be compared using a computational procedure of alignment.

QUESTIONS

1. What is characteristic of recessive and dominant inheritance, respectively?
2. What is the difference between *sickle cell trait* and *sickle cell anemia*?
3. Explain the concepts *genotype* and *phenotype*.
4. Explain what is meant when we say that an individual is *homozygous* for a specific genetic variant (allele).
5. Use a Punnett square to show the offspring when one parent is homozygous for the sickle cell allele and one parent has two normal globin alleles.
6. Why is the sickle cell gene variant (allele) fairly common in specific populations?
7. Draw the chemical structure of an *amino acid*. Also draw a dipeptide (two amino acids joined with a *peptide bond*).
8. What are the acidic and basic amino acids, respectively?
9. Give examples of hydrophobic amino acids.

10. What were the experiments carried out in the 1940s and 1950s to elucidate the molecular basis of sickle cell anemia?
11. Explain what is meant by *primary, secondary, tertiary,* and *quaternary* structures of proteins.
12. What are the methods used to infer the three-dimensional structure of proteins and other large molecules?
13. In sickle cell anemia, one amino acid is replaced with another in the β-globin subunit of hemoglobin. What are the consequences in terms of amino acid interactions and protein structure? What are the physiological consequences?
14. What is meant by a *molecular sequence*?

FURTHER READING

General on sickle cell anemia and its history

Bjorklund R. 2011. *Sickle cell anemia.* Marshall Cavendish Benchmark, New York.

Orkin SH, Higgs DR. 2010. Medicine. Sickle cell disease at 100 years. *Science* 329(5989):291–292.

Serjeant GR. 2001. The emerging understanding of sickle cell disease. *Br J Haematol* 112(1):3–18.

Serjeant GR. 2010. One hundred years of sickle cell disease. *Br J Haematol* 151(5):425–429.

Early work on human hemoglobin and myoglobin

Baglioni C. 1961. An improved method for the fingerprinting of human hemoglobin. *Biochim Biophys Acta* 48:392–396.

Braunitzer G, Hilschmann N, Rudloff V et al. 1961. The hæmoglobin particles. *Nature* 190:480–482.

Braunitzer G, Gehring-Mueller R, Hilschmann N et al. 1961. The structure of normal adult human hemoglobins. *Hoppe Seylers Z Physiol Chem* 325:283–286.

Ingram VM. 1959. Abnormal human haemoglobins. III. The chemical difference between normal and sickle cell haemoglobins. *Biochim Biophys Acta* 36:402–411.

Kendrew JC, Bodo G, Dintzis HM et al. 1958. A three-dimensional model of the myoglobin molecule obtained by x-ray analysis. *Nature* 181(4610):662–666.

Perutz MF, Rossmann MG, Cullis AF et al. 1960. Structure of haemoglobin: A three-dimensional Fourier synthesis at 5.5-A. resolution, obtained by x-ray analysis. *Nature* 185(4711):416–422.

Sickle cell anemia: Early genetic studies

Neel JV. 1949. The inheritance of sickle cell anemia. *Science* 110(2846):64–66.

Sickle cell anemia: Early paper by Pauling

Pauling L, Itano HA, Singer SJ, Wells IC. 1949. Sickle cell anemia: A molecular disease. *Science* 110(2865):543–548.

A Code of Life

3

In Chapter 2 we presented an example of an inherited disorder, sickle cell anemia. Through the work of Ingram and others, it was learned that this disease is the result of a specific change in the hemoglobin β-chain—a substitution of the amino acid glutamic acid to valine. Now we examine further the cause of this amino acid replacement. The answer is to be found in the molecule known as **DNA**.

EARLY DAYS OF DNA RESEARCH

DNA, short for deoxyribonucleic acid, was known long before the discovery of the defective sickle cell hemoglobin. It was isolated in pure form by Friedrich Miescher in 1869. His starting material were bandages collected from a surgical clinic. From the pus of those bandages, Miescher isolated white blood cells. From these cells, in turn, he purified nuclei, and by adding acetic acid to the nuclei, a precipitate was formed. Miescher characterized this precipitate that he called *nuclein* and that we now know as DNA.

By the end of the 1920s, Phoebus Levene showed that DNA is composed of four different nitrogenous bases—adenine (A), thymine (T), cytosine (C), and guanine (G) (**Figure 3.1**)—as well as the sugar deoxyribose and phosphate. He also showed that these components are linked together in the order phosphate-sugar-base to form repeating units. The unit is referred to as a **nucleotide** (**Figure 3.2**). However, Levene was not correct about the structure of DNA, as he thought it had only four nucleotides per molecule. We now know there is a lot more.

DNA IS THE CARRIER OF GENETIC INFORMATION

Scientists believed for a long time that proteins were the hereditary material, and for this reason DNA was not extensively examined as a potential carrier of genetic information. But in the 1940s, a famous experiment was carried out by Oswald Avery and his colleagues Colin MacLeod and

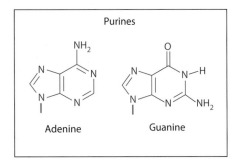

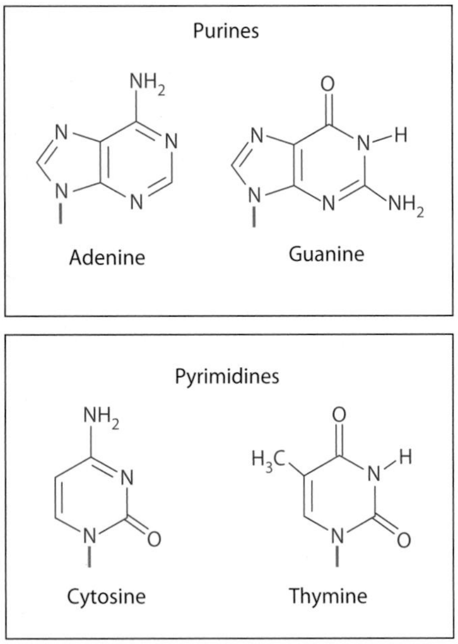

Figure 3.1 **Chemical structures of bases in DNA.** Bases are classified into purines and pyrimidines. The sites of attachment to deoxyribose are highlighted with green lines.

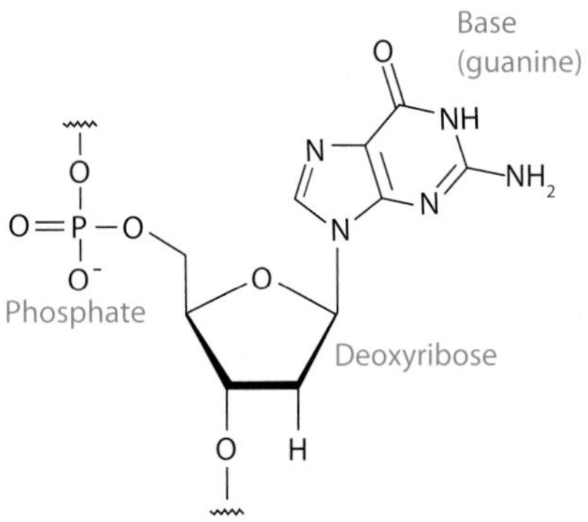

Figure 3.2 Chemical structure of a DNA nucleotide. DNA is a polymer of nucleotides. Each nucleotide is composed of a sugar, deoxyribose, phosphate group, and nitrogenous base (here guanine). The curly lines represent the sites of attachment to other nucleotides of the DNA molecule.

Maclyn McCarty. *Streptococcus pneumoniae* is a dangerous bacterium as it gives rise to pneumonia in humans and other animals. In experiments with mice, it had previously been shown that a nonvirulent strain of this bacterium could be changed—or "transformed" in the language of microbiology—into a virulent strain by adding killed cells of virulent *S. pneumoniae*. The important component in the dead cells was first thought to be protein material, but Avery and coworkers discovered that it was in fact DNA. Other experiments later verified DNA as the genetic material. We now also know that DNA is a universal carrier of genetic information, present not only in bacteria, but in all three domains of life—**eukaryotes**, **bacteria**, and **archaea**.[1]

THREE-DIMENSIONAL STRUCTURE OF DNA

DNA is a molecule built from repeating nucleotide units, but what exactly is its three-dimensional structure? Early studies of DNA indicated that it had a relatively small size. But these studies used conditions where the DNA was severely degraded. Instead, Avery showed in 1944 that DNA, in this case from *S. pneumoniae*, was a high molecular weight molecule. Its molecular weight was estimated to about 500,000, corresponding to a number of nucleotides larger than 1,000.

A major milestone in the history of DNA was in 1953 when James Watson and Francis Crick presented a model for its three-dimensional structure. This model presented basic features of DNA and provided in addition an explanation of how DNA is copied in preparation for cell division. For this work, Watson and Crick, together with Maurice Wilkins, received the Nobel Prize in Physiology or Medicine in 1962.

A crucial experimental result that allowed the elucidation of DNA structure was an x-ray crystallographic pattern of DNA obtained by Rosalind Franklin (**Figure 3.3**). The X pattern in the middle suggested a helical shape. In addition, the dimensions of the helix could be calculated from the x-ray pattern. Thus, the height of a turn should be 34 angstroms (1 angstrom = 10^{-10} meters). The distance between the stacked bases should be 3.4 angstroms. Therefore, there should be 10 nucleotides to every helical repeat. It was also deduced that the phosphate groups were on the outside of the helical structure.

Watson and Crick arrived at a model of DNA where the molecule is composed of two chains forming a helical structure (**Figure 3.4**). Each chain is a polymer built from nucleotides (**Figure 3.5**). The two chains of DNA are held together by pairing between the bases, such that A always pairs with T and G always with C (**Figure 3.6**). It is the sequence of the bases in DNA that constitutes the genetic message.

POLARITY OF DNA

Every DNA strand has a polarity. Thus, one end of a DNA strand is referred to as its 5′ end, as it typically has a phosphate attached to the fifth carbon of deoxyribose. The other end of the DNA strand has a free OH group at carbon

---

[1]Eukaryotes contain a membrane-surrounded nucleus, as well as other characteristic components such as mitochondria. Eukaryotes are often multicellular and include animals, plants, and fungi. Bacteria and archaea do not contain a nucleus and are not multicellular. In addition to the three domains of life, there are viruses. These are not free-living organisms but are dependent on a host organism for propagation. All kinds of organisms may be invaded by viruses.

number 3; consequently, this end of the DNA strand is named 3' (see Figure 3.5). The two chains of a double-stranded DNA run in opposite directions. The polarity of a DNA chain means that a DNA sequence being read from one end is not the same as the same sequence read from the other end. Different proteins and nucleic acids that interact with DNA are able to identify this polarity of the DNA molecule. As an example, this double-stranded DNA

$$5'\text{- GTAACCTA -}3'$$
$$| \ | \ | \ | \ | \ | \ | \ |$$
$$3'\text{- CATTGGAT -}5'$$

is not the same molecule as this one, where the bases occur in reverse order:

$$5'\text{- ATCCAATG -}3'$$
$$| \ | \ | \ | \ | \ | \ | \ |$$
$$3'\text{- TAGGTTAC -}5'$$

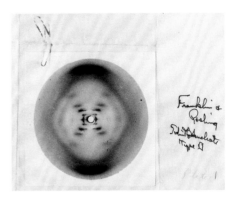

Figure 3.3 Famous x-ray crystallographic pattern of DNA. This pattern was obtained by Rosalind Franklin and Raymond Gosling on May 2, 1952. It was used by James Watson and Francis Crick to obtain a model of the three-dimensional structure of DNA.

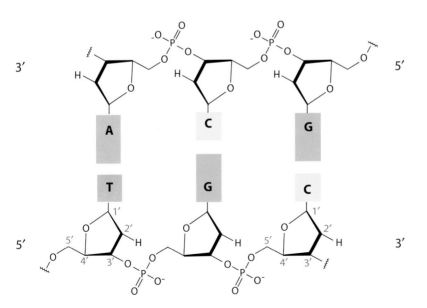

Figure 3.4 Three-dimensional structure of DNA. A double-stranded helical DNA with the sugar-phosphate backbones of the two strands in gray. The bases (involved in pairing) are shown in ball-and-stick representation. The view of DNA is based on the PDB entry 7BNA in which the base sequence of both strands is 5' CGCGAATTCGCG 3'.

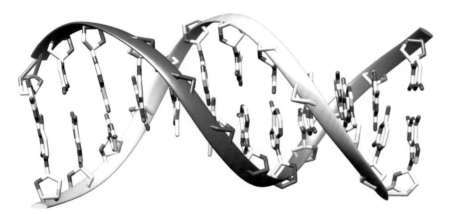

Figure 3.5 Chemical structure of a double-stranded DNA. DNA has a backbone with alternating phosphate and deoxyribose units. A phosphate group is attached to one deoxyribose at carbon 5 and to another deoxyribose at carbon 3. Attached to deoxyribose at carbon 1 is a nitrogenous base, adenine, thymine, cytosine, or guanine.

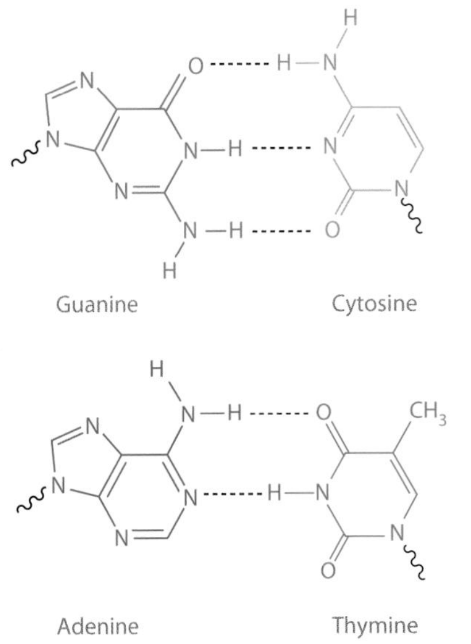

Figure 3.6 Base-pairing in DNA. Bases are held together in DNA by hydrogen bonding. Guanine pairs with cytosine with three hydrogen bonds and adenine with thymine with two hydrogen bonds.

A MECHANISM OF REPLICATION WAS SUGGESTED BY THE STRUCTURE OF DNA

The model of Watson and Crick also suggested a simple mechanism whereby the information in DNA is copied to a new molecule whenever chromosomes are duplicated before cell division. During copying, DNA would unzip to form two separate strands, and each strand would then be able to serve as a template for formation of a new strand (**Figure 3.7**). Further experimental studies showed that the process of copying, commonly called **DNA replication**, is indeed based on this mechanism involving base-pairing.

RELATIONSHIP OF DNA AND PROTEIN

In Chapter 2, it was shown that the inherited disease sickle cell anemia is the result of a change of one single amino acid in one of the chains of the hemoglobin protein, a transporter of oxygen. This change has the unfortunate effect of making globin chains stick to each other, in turn forming harmful fibers. Exactly why and how is the amino acid changed? The answer is to be found in the context of the DNA molecule and its relationship to amino acid sequences in proteins.

Even before the characterization of sickle cell hemoglobin and the discovery of its altered amino acid composition, it had been realized that there was a distinct relationship of genes—that we now know are contained within DNA—and proteins. For instance, in the early 1940s, Beadle and Tatum examined mutations in the bread mold *Neurospora crassa*. They could introduce mutations in genes by x-ray irradiation and discovered that many of these mutations changed the activity of an enzyme (and enzymes are proteins). Each mutation was such that it affected the activity of a single enzyme. This led to the **one gene–one enzyme hypothesis**. With Ingram's discovery described in Chapter 2, this hypothesis could be generalized to a **one gene–one protein hypothesis**.

In the 1950s it became increasingly clear that DNA contains information for the production of proteins and that it is the sequence of nucleotides in DNA

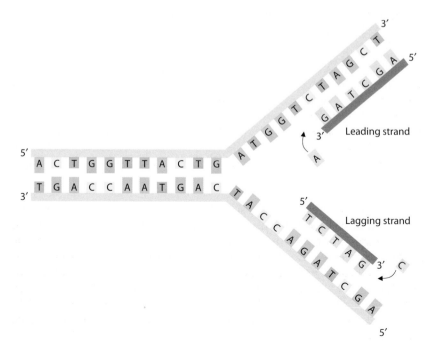

Figure 3.7 Principle of DNA replication. During replication, the two strands of DNA are separated, and each of these strands will act as templates for the synthesis of novel strands (shown in darker gray). DNA synthesis is always in the 5′ to 3′ direction and occurs on one of the strands (the **leading** strand) in the same direction as the growing replication fork. The other strand (**lagging** strand) is synthesized in short, separated segments. These segments are then joined by a DNA ligase enzyme.

that specifies the sequence of amino acids in proteins. But what exactly is the relationship between nucleotide and amino acid sequences? By the end of the 1950s, researchers wanted to determine the details of the **genetic code**.

BASIC FEATURES OF THE GENETIC CODE

In the work to elucidate the genetic code, a starting point was a number of theoretical considerations. A certain nucleotide or a sequence of nucleotides specifies a distinct amino acid. How many nucleotides specify a certain amino acid? Remember that there are 20 different amino acids in a protein, but only 4 different nucleotides. Therefore, a single nucleotide is clearly insufficient to encode an amino acid. What about a sequence of two nucleotides? There are 16 (4^2) possible combinations of two letters, and for this reason a sequence of two nucleotides is not quite sufficient to code for the 20 different amino acids. Three nucleotides specifying an amino acid is yet another possibility. In this case, there are $4^3 = 64$ possible triplets. Genetic experiments were carried out with bacteriophages, viruses that infect bacteria. Mutations in phages could be introduced so as to add one base or delete one base in DNA. It was discovered that in triple mutants, for instance when three extra bases were inserted, normal (wild-type) behavior of the phage was restored. The results of such experiments suggested that the genetic code is based on three-letter words.

The organization of information was next considered. Is the code overlapping, like what is shown in **Figure 3.8a**? This organization of the code would be such that one substitution mutation is likely to change two different amino acids. However, genetic experiments showed that this was never the case. Another possibility is that there is a comma or punctuation in the code. For instance, one base could be the punctuation symbol (**Figure 3.8b**). Again, results from genetic experiments made this punctuation system unlikely. Instead, it was concluded that the information in DNA is being read as three-letter words arranged in a linear order with no symbol in between (**Figure 3.8c**). Now to the remaining and significant problem, what three-letter words—also known as **codons**—correspond to what amino acids?

MOLECULES INVOLVED IN PROTEIN PRODUCTION

During the 1950s, it was realized that DNA does not directly act as a template for the production of protein—there is a missing molecular link between DNA and protein. This missing link was identified in the early 1960s and was named **messenger RNA** or **mRNA**. The mRNA is a class of **RNA, ribonucleic acid**. RNAs are structurally different from DNA in three respects. First, an RNA molecule is typically single stranded. Second, the base thymine (T) in DNA is replaced with **uracil** (U) in RNA. Third, the sugar unit in RNA is **ribose** instead of deoxyribose (**Figure 3.9**).

Another important element of protein production was predicted by Francis Crick in 1958. He suggested that adaptor molecules exist that are composed of nucleotides and able to recognize sequences by "pairing" to mRNA. Furthermore, the adaptor molecules should be able to carry the appropriate amino acid. Crick's prediction was correct, an adaptor molecule called **transfer RNA (tRNA)** with these properties was soon identified. The nucleotide sequence of a tRNA was first revealed in 1965 when Robert W. Holley determined the sequence of an alanine-specific tRNA from yeast (**Figure 3.10**). Incidentally, this was the first nucleic acid to be sequenced. When examining the nucleotide sequence of this tRNA, it was

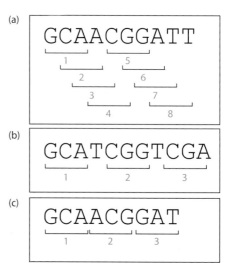

Figure 3.8 Three possible modes of reading codons in mRNA contemplated before the genetic code was elucidated. (a) An overlapping code. (b) A punctuation code. (c) Codons are read in a nonoverlapping mode without a punctuation symbol.

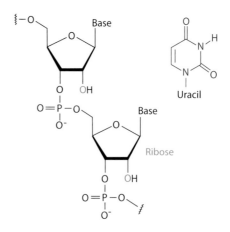

Figure 3.9 Chemistry of RNA, ribonucleic acid. As compared to DNA, RNA has ribose instead of deoxyribose and uracil instead of thymine. The difference between ribose and deoxyribose is that an OH group is present at carbon 2 of the sugar moiety (oxygen in red), whereas a hydrogen is at this position in DNA. The difference between uracil and thymine is that uracil is lacking the methyl group present in thymine.

Figure 3.10 Nucleotide sequence of alanine tRNA from the yeast *Saccharomyces cerevisiae* as determined by Robert Holley in 1965. A cloverleaf shape of the tRNA is formed through internal base-pairing in the molecule. In addition to the regular RNA bases A, U, C, and G, tRNAs contain a number of modified nucleotides. In this tRNA they are 1-methylguanosine (m1G), 5-methyluridine (T), N2,N2-dimethylguanosine (m2G), dihydrouridine (D), inosine (I), 1-methylinosine (m1I), and pseudouridine (Ψ). The dashed lines indicate interactions between bases that take place in the three-dimensional structure of the tRNA (tertiary base pairs).

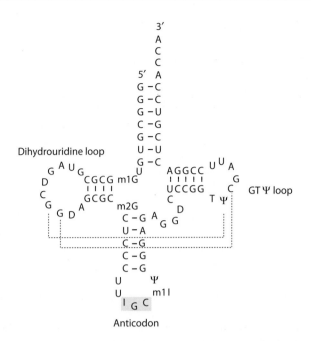

noted that it could fold into a cloverleaf-type shape. This shape is based on internal base-pairing in the molecule, making use of base pairs similar to those found in the double-stranded DNA. Later, when more tRNA sequences became known, these tRNAs could also be folded into a similar type of structure. All tRNAs contain elements critical for their adaptor function. First, they have an **anticodon**, which is complementary to the mRNA codon. Second, at their 3′ end is attached an amino acid that corresponds to the codon being read by the anticodon. In the 1970s, the three-dimensional structure of a tRNA was elucidated for the first time (**Figure 3.11**).

Crick also predicted the existence of ribonucleic-protein complexes that make possible the assembly of amino acids into proteins, following the instructions in the mRNA. The "ribonucleic-protein complexes" later became known as **ribosomes**. We now know that they are particles 20–30 nm in diameter,[2] with several RNA and protein subunits.

Figure 3.11 Three-dimensional structure of phenylalanine tRNA from the yeast *S. cerevisiae*. The structure has features not predicted by the secondary structure tRNA model in Figure 3.10. For instance, there are interactions between the dihydrouridine loop and the GTΨ loop (Figure 3.10), giving rise to an L-shaped structure. Functionally important elements in the tRNA structure are the anticodon loop (with anticodon GAA) and the 3′ end with a CCA sequence, which is the attachment site for the amino acid. (Adapted from Shi H, Moore PB. 2000. *RNA* 6:1091–1105. PDB ID: 1EHZ.)

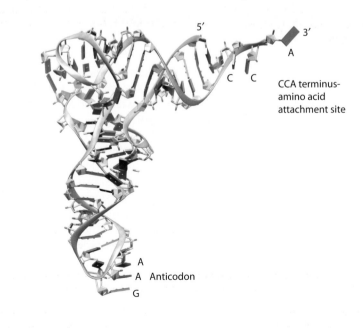

_____

[2] One nanometer is 10^{-9} m.

MAKING SENSE OF THE CODONS

Marshall Nirenberg conducted a number of experiments that provided the first clues to the genetic code. He designed a cell-free system where he could study the production of protein as dependent on added synthetic RNAs acting as mRNAs. In 1961, a critical experiment was carried out by Nirenberg and Heinrich Matthaei. A set of 20 reactions was carried out, each with one of the 20 amino acids radioactively labeled. A bacterial extract provided necessary components for protein synthesis, such as ribosomes and tRNAs. After allowing protein synthesis to occur, protein material was isolated from each reaction and radioactivity determined in this protein. When a synthetic polymer containing only U, poly(U), was added to the system, only one of the 20 reactions gave rise to a radioactive product. This was one where the amino acid phenylalanine was radioactive (**Figure 3.12a**). The conclusion from this experiment was therefore that the three-letter word UUU is specifying the amino acid phenylalanine. One of the 64 words of the genetic code had been cracked.

Using the same experimental system, it could be shown that poly(A) directed the production of a protein containing only the amino acid lysine. Similarly, poly(C) gave rise to an incorporation of proline. Nirenberg now knew the meaning of two more words in the genetic code, AAA and CCC. Working out the rest of the code required a substantial amount of work.

For the elucidation of a majority of codons of the genetic code, Nirenberg developed another method. He showed that a synthetic RNA molecule containing a specific sequence of only three nucleotides could with the help of the ribosome specifically bind a tRNA specific to the corresponding amino acid. **Figure 3.12b** shows how a cysteine tRNA specifically binds to the ribosome in the presence of the triplet UGU. In this manner, UGU was assigned as a codon specifying cysteine.

The codons that could not be assigned using Nirenberg's experiments were sorted out through the work of Gobind Khorana. His strategy was to make synthetic RNA polymers of a more complex composition as compared to the polymers poly(U), poly(A), and poly(C). These could be used in a cell-free protein production system. For instance, a polymer of a repeating unit with the sequence UG specified a protein composed of valine and cysteine. This result showed that UGU and GUG specified these two amino acids. Nirenberg had already shown that UGU codes for cysteine. Therefore, GUG must be valine.

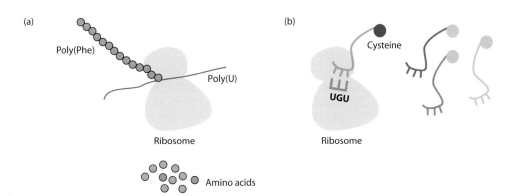

Figure 3.12 Experiments to elucidate the genetic code. (a) An experiment using a cell-free protein synthesizing system is shown, which demonstrates that UUU encodes the amino acid phenylalanine. Protein synthesis was allowed to occur with the help of a cell-free bacterial system that provided components for protein synthesis such as ribosomes and tRNAs. In a pool of amino acids, phenylalanine was radioactively labeled (orange), and the mRNA was a synthetic poly(U) RNA molecule. Radioactive protein was formed, showing that poly(U) gave rise to the protein poly(Phe). (b) Another strategy for elucidating the genetic code was to use specific binding of synthetic triplets to the corresponding tRNA in the presence of ribosome. Here is shown how a cysteine tRNA specifically binds to the triplet UGU.

Figure 3.13 The genetic code. Amino acids are shown with three- and one-letter abbreviations. Codons UAA, UAG, and UGA specify termination of protein synthesis. Families of four codons that correspond to the same amino acid and where the difference is in the third position of the codon are shaded.

UUU	Phe	F	UCU	Ser	S	UAU	Tyr	Y	UGU	Cys	C
UUC	Phe	F	UCC	Ser	S	UAC	Tyr	Y	UGU	Cys	C
UUA	Leu	L	UCA	Ser	S	UAA	Stop		UGA	Stop	
UUG	Leu	L	UCG	Ser	S	UAG	Stop		UGG	Trp	W
CUU	Leu	L	CCU	Pro	P	CAU	His	H	CGU	Arg	R
CUC	Leu	L	CCC	Pro	P	CAC	His	H	CGC	Arg	R
CUA	Leu	L	CCA	Pro	P	CAA	Gln	Q	CGA	Arg	R
CUG	Leu	L	CCG	Pro	P	CAG	Gln	Q	CGG	Arg	R
AUU	Ile	I	ACU	Thr	T	AAU	Asn	N	AGU	Ser	S
AUC	Ile	I	ACC	Thr	T	AAC	Asn	N	AGC	Ser	S
AUA	Ile	I	ACA	Thr	T	AAA	Lys	K	AGC	Arg	R
AUG	Met	M	ACG	Thr	T	AAG	Lys	K	AGG	Arg	R
GUU	Val	V	GCU	Ala	A	GAU	Asp	D	GGU	Gly	G
GUC	Val	V	GCC	Ala	A	GAC	Asp	D	GGC	Gly	G
GUA	Val	V	GCA	Ala	A	GAA	Glu	E	GGA	Gly	G
GUG	Val	V	GCG	Ala	A	GAG	Glu	E	GGG	Gly	G

Figure 3.14 The translation of human β-globin mRNA to protein. The nucleotide sequence of a human β-globin mRNA is shown together with the translation product as elucidated with the genetic code. Translation is initiated at an AUG codon and terminated at UAA.

```
AUGGUGCAUCUGACUCCUGAGGAGAAGUCUGCCGUUACUGCCCUGUGGGGCAAGGUGAAC
 M   V   H   L   T   P   E   E   K   S   A   V   T   A   L   W   G   K   V   N

GUGGAUGAAGUUGGUGGUGAGGCCCUGGGCAGGCUGCUGGUGGUCUACCCUUGGACCCAG
 V   D   E   V   G   G   E   A   L   G   R   L   L   V   V   Y   P   W   T   Q

AGGUUCUUUGAGUCCUUUGGGGAUCUGUCCACUCCUGAUGCUGUUAUGGGCAACCCUAAG
 R   F   F   E   S   F   G   D   L   S   T   P   D   A   V   M   G   N   P   K

GUGAAGGCUCAUGGCAAGAAAGUGCUCGGUGCCUUUAGUGAUGGCCUGGCUCACCUGGAC
 V   K   A   H   G   K   K   V   L   G   A   F   S   D   G   L   A   H   L   D

AACCUCAAGGGCACCUUUGCCACACUGAGUGAGCUGCACUGUGACAAGCUGCACGUGGAU
 N   L   K   G   T   F   A   T   L   S   E   L   H   C   D   K   L   H   V   D

CCUGAGAACUUCAGGCUCCUGGGCAACGUGCUGGUCUGUGUGCUGGCCCAUCACUUUGGC
 P   E   N   F   R   L   L   G   N   V   L   V   C   V   L   A   H   H   F   G

AAAGAAUUCACCCCACCAGUGCAGGCUGCCUAUCAGAAAGUGGUGGCUGGUGUGGCUAAU
 K   E   F   T   P   P   V   Q   A   A   Y   Q   K   V   V   A   G   V   A   N

GCCCUGGCCCACAAGUAUCACUAA
 A   L   A   H   K   Y   H STOP
```

By the mid-1960s, the meanings of all of the codons of the genetic code had been elucidated. In 1968 Nirenberg and Khorana were awarded the Nobel Prize in Physiology or Medicine for their work on the code. They shared the prize with Holley for his sequencing of tRNA. The genetic code is shown in **Figure 3.13**. Out of the 64 codons, there are three codons—UAA, UAG, and UGA—that do not represent amino acids. Instead, they specify a stop in protein synthesis. The codon AUG that specifies methionine also has a special role as it also marks a starting point for protein synthesis. **Figure 3.14** shows the sequence of nucleotides in the β-globin mRNA and the corresponding protein sequence as specified by the genetic code.

THE GENETIC CODE IS BY AND LARGE UNIVERSAL

The genetic code was clarified in the 1960s using a cell extract of the bacterium *Escherichia coli*. Is the genetic code the same in other living organisms? It would later turn out that the genetic code, by and large, is universal—it is the same in bacteria, yeast, animals, and plants. There are only a few genetic systems where there are minor deviations from the table shown in Figure 3.13. Bacteria known as mycoplasmas are one example. Deviations from the standard code also occur in the organelles **mitochondria** and **chloroplasts**. Mitochondria are organelles found in all animals, plants, and fungi and are specialized for energy metabolism, designed to extract energy out of carbohydrate oxidation. Chloroplasts are organelles in plants that use light energy to produce metabolic energy. Both mitochondria and chloroplasts were originally bacteria that came to join another type of cell early in the evolution of life. Because of their bacterial origin, mitochondria and chloroplasts have their own genetic system and machinery for making proteins. The genetic code in many mitochondria is different from the universal scheme in that UGA is not a stop codon but instead encodes the amino acid tryptophan.

The genetic code has 61 codons that specify amino acids (see Figure 3.13). As there are only 20 amino acids, some amino acids have more than one codon—the code is said to be **degenerate**. For instance, there are four different codons, GGA, GGG, GGU, and GGC, that specify glycine. This organization of the code will be important when we discuss the effect of mutations, changes in the sequence of nucleotides in DNA, and how likely they are to cause disorders.

Crick suggested in the 1960s that the third position of a codon, referred to as the **wobble** position, is read in a less strict manner as compared to the other positions. Thus, a single tRNA anticodon is able to decode more than one codon. For instance, the anticodon 5′ GGC 3′ of a glycine tRNA can read not only the codon GCC but also GCU.

The genetic code features 20 different amino acids. However, in the late 1980s it was discovered that there is yet another amino acid, selenocysteine, which is encoded in DNA. The codon UGA is typically a stop word during the production of proteins, but when it is in a specific sequence context, it will be recognized by a specific tRNA. This tRNA has an anticodon able to pair with UGA, and at the same time it carries the amino acid selenocysteine to be incorporated into protein.

When the genetic code was clarified in the early 1960s, we did not have access to any DNA sequences. The situation is now different, as we for instance know the complete sequence of the human genome. We also know the amino acid sequence of many proteins through advances in protein sequencing technology. Thus, we are able to directly compare coding sequences to the corresponding amino acid sequences. In this manner, the genetic code as deduced in the 1960s has been verified many times over.

A SECOND GENETIC CODE

Every tRNA carries an amino acid consistent with the anticodon in the molecule. How do the right amino acids attach to tRNA? There are enzymes known as aminoacyl-tRNA synthetases that make sure the correct amino acid is joined to a specific tRNA. There is one enzyme to each of the amino acids, and each enzyme is able to recognize the tRNA as corresponding to that amino acid. The enzyme recognizes structural features of the tRNA—which is not necessarily the anticodon structure—and the relationship between these structural properties of tRNA and the amino acid specificity has been referred to as the **second genetic code**.

FLOW OF INFORMATION FROM DNA TO PROTEIN

We now summarize the important characteristics of the flow of genetic information in the cell (**Figure 3.15**). First, the genetic information is stored in the DNA molecule. It is a double-stranded helical structure with two

Figure 3.15 The flow of genetic information from DNA to protein. On top is a double-stranded DNA molecule. During transcription, one of the DNA strands (the lower strand in the figure) acts as the template for the production of an RNA molecule (mRNA). The mRNA is single stranded. During translation, the sequence of bases in the mRNA specifies what amino acids are to be incorporated into protein. The mRNA is read in the 5′ to 3′ direction, and protein synthesis occurs from the N-terminal to the C-terminal end of the protein. Translation is initiated at an AUG codon. The sequences upstream and downstream of the protein-coding region are referred to as the 5′ UTR (untranslated region) and 3′ UTR, respectively.

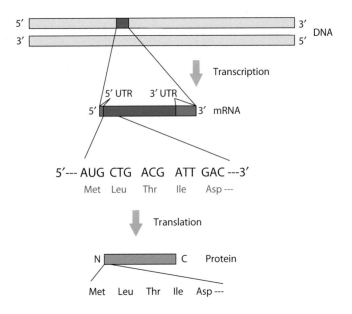

polynucleotide strands running in opposite directions. The DNA does not directly act as a template for the production of protein. Rather this function is carried by mRNA molecules. The mRNA is produced by an enzyme RNA polymerase using one of the strands in DNA as a template. This process is known as **transcription**. The mRNA formed by transcription then acts in protein synthesis to guide the synthesis of proteins. This step of information transfer is also known as **translation**.

Essential for the translation process are tRNA molecules as well as ribosomes. An initiation codon AUG specifies where translation starts on the mRNA and any of the codons UAA, UAG, and UGA marks a stop for translation. In eukaryotes, each of the mRNAs produced by transcription encodes one specific protein.

The whole mRNA molecule does not specify an amino acid sequence. Rather, there are regions upstream and downstream of the actual coding sequence, referred to as the **5′ untranslated region** (**5′ UTR**) and **3′ UTR**, respectively.

SICKLE CELL ANEMIA IS THE RESULT OF A SINGLE NUCLEOTIDE CHANGE IN DNA

As sickle cell anemia is an inherited disorder, there must be changes in DNA responsible for the disease. With the knowledge that we now have about the flow of genetic information and about the genetic code, we may further discuss the molecular background to sickle cell anemia. We know from Ingram's results in the 1950s that the disease resulted from a change from glutamic acid to valine. From the genetic code, we note that glutamic acid is encoded by the codons GAA and GAG. Furthermore, the valine codons are GTT, GTC, GTA, and GTG.

What are the nucleotide sequences of the globin mRNA and its sickle cell variant? How could an mRNA be sequenced? In the 1960s, a laborious protocol was used. DNA complementary to globin mRNA from normal individuals and from sickle cell patients was first synthesized with the enzyme **reverse transcriptase** that uses an mRNA as a template to produce a complementary DNA. Then, this DNA was used as a template to make radioactive RNA by transcription. The resulting RNA was then digested with enzymes to produce small fragments that could be sequenced individually.

Normal

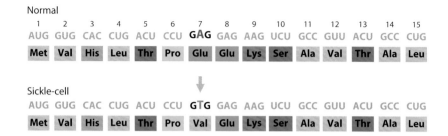

Figure 3.16 A change in one base causes sickle cell anemia. A mutation A to T in DNA causes the seventh codon GAG (glutamic acid) in the mRNA to be changed into GTG (valine).

These RNA sequences were compared to the previously known amino acid sequence of hemoglobin. The results showed that the normal codon specifying glutamic acid was GAG (codon number seven), and in sickle cell anemia it was changed to GTG (**Figure 3.16**). We therefore reach the conclusion that the disorder is caused by a change from A to T in the codon specifying the seventh codon of the β-globin mRNA. Looking at this change from the perspective of the double-stranded DNA, the change A to T in one strand is accompanied by a corresponding change T to A in the other strand. In essence, a base pair A•T in DNA is changed to T•A.

The observant reader may note that in Figure 3.16 the first codon in the globin mRNA encodes methionine, but in Figures 2.9 and 2.13 of Chapter 2, this amino acid is lacking from the protein sequence. The reason for this discrepancy is that the N-terminal methionine encoded by the mRNA is enzymatically removed shortly after protein synthesis.

To better understand how DNA sequences are translated into amino acid sequences in proteins, we need to discuss the structure of genes as well as the organization of the human genome. These issues will be covered in Chapter 4.

SUMMARY

- The nucleic acids DNA and RNA are polymers built from nucleotides, where each nucleotide is composed of a sugar moiety, a phosphate group, and a nitrogenous base.
- Nucleic acids have a distinct polarity with 5′ and 3′ ends.
- The DNA molecule is typically built from two strands forming a helical structure. The two chains are held together by pairing between the bases. The base pairs formed are A•T and G•C.
- The genetic code has 64 three-letter codons, where 61 specify amino acids and 3 represent stop.
- Codons in mRNA are read by tRNAs. The tRNAs have an anticodon able to read one or more codons. At the same time, all tRNAs carry an amino acid that matches the anticodon and the codon read.
- Protein synthesis is carried out by ribosomes, large complexes of RNA, and protein.
- The genetic code was elucidated in the 1960s using a bacterial cell-free translation system.
- The genetic code is the same in virtually all species. Deviations from the universal code occur, for instance, in mycoplasmas and mitochondria.
- Selenocysteine is encoded by UGA codons in a specific environment.
- Sickle cell disease is caused by a mutation A > T, changing a codon GAG (glutamate) to GTG (valine).

QUESTIONS

1. Draw the chemical structure of a nucleotide, indicating the location of the 5′ phosphate, the ribose, and the nitrogenous base.

2. What is meant by the 5′ and 3′ ends of a nucleic acid?

3. Draw a structure of the double-stranded DNA, illustrating the *antiparallel* nature of the structure.

4. What are the chemical differences between DNA and RNA?

5. What was the first nucleic acid to be sequenced?

6. Write the complementary strand of the sequence 5′ GCAAAGGT 3′, and indicate the 5′ and 3′ ends.

7. What are the processes referred to as *transcription* and *translation*, respectively?

8. Consider the DNA sequence where one of the strands is 5′ AACGGCGAGGTACTGGCCG<u>T</u>TGCG 3′. If this strand is the template strand for transcription and the transcription starts at the underlined base, what is the RNA product? Write the RNA sequence in the 5′ to 3′ direction.

9. What is meant by the *one gene–one enzyme hypothesis*?

10. What were the types of experiments carried out in order to elucidate the *genetic code*?

11. What is meant when we say that the genetic code is *degenerate*?

12. How many codons in the genetic code specify an amino acid?

13. How many amino acids are specified by four different codons and six different codons in the genetic code, respectively?

14. How is the reading in the third position of a codon different from the two other positions?

15. How is the amino acid selenocysteine encoded?

16. Give some examples of genetic systems deviating from the standard genetic code.

17. What relationship has been termed the *second genetic code*?

18. We assume that the following sequence represents the start of a mRNA protein coding sequence, with the 5′ AUG as the start codon. Use the genetic code to derive the corresponding amino acid sequence. Some mutations that are single nucleotide substitutions (one nucleotide replaced with another) give rise to changes in the amino acid sequence. Show three examples of such mutations in the following sequence. Also show three nucleotide substitutions that do *not* change the amino acid sequence.

5′ AUG-CAA-GCA-GUU-GGG-AGG-ACA-CGA 3′

19. Sometimes we do not know the correct reading frame of a region of mRNA. In such cases, we may need to consider every possible reading frame. Consider the nucleotide sequence that follows that is part of a protein-coding region in an mRNA. First write a translation product where the first position of the nucleotide sequence is the first position of the first codon. Then consider a different reading frame where translation starts at the second position of the nucleotide sequence. Then also consider reading the sequence with a start at the third position. Finally, produce the complementary sequence to the original nucleotide sequence and produce the three different protein products from that sequence. You will end up with six different protein products.

5′ TGCAAGCAGTTGGGAGGACACGACCCA 3′

FURTHER READING

(Scientific papers listed here are all of an historical nature.)

One gene–one enzyme, mutations in *Neurospora*

Beadle GW, Tatum EL. 1941. Genetic control of biochemical reactions in *Neurospora*. *Proc Natl Acad Sci USA* 27(11): 499–506.

DNA as the genetic material

Avery OT, Macleod CM, McCarty M. 1944. Studies on the chemical nature of the substance inducing transformation of pneumococcal types: Induction of transformation by a desoxyribonucleic acid fraction isolated from pneumococcus type III. *J Exp Med* 79(2):137–158.

Structure of tRNA

Holley RW, Apgar J, Everett GA et al. 1965. Structure of a ribonucleic acid. *Science* 147:1462–1465.

Structural model of DNA

Watson JD, Berry A. 2003. *DNA: The secret of life.* Alfred A. Knopf, New York.

Watson JD, Crick FH. 1953. Molecular structure of nucleic acids—A structure for deoxyribose nucleic acid. *Nature* 171(4356):737–738.

Genetic code

Nirenberg M, Leder P, Bernfield M et al. 1965. RNA codewords and protein synthesis, VII. On the general nature of the RNA code. *Proc Natl Acad Sci USA* 53(5):1161–1168.

Three-dimensional structure of tRNA

Sussman JL, Kim S. 1976. Three-dimensional structure of a transfer RNA in two crystal forms. *Science* 192(4242):853–858.

Nucleotide sequence of β-globin mRNA

Marotta CA, Forget BG, Cohne-Solal M et al. 1977. Human beta-globin messenger RNA. I. Nucleotide sequences derived from complementary RNA. *J Biol Chem* 252(14):5019–5031.

Marotta CA, Wilson JT, Forget BG, Weissman SM. 1977. Human beta-globin messenger RNA. III. Nucleotide sequences derived from complementary DNA. *J Biol Chem* 252(14):5040–5053.

Selenocysteine as twenty-first amino acid of the genetic code

Zinoni F, Birkmann A, Leinfelder W, Bock A. 1987. Cotranslational insertion of selenocysteine into formate dehydrogenase from *Escherichia coli* directed by a UGA codon. *Proc Natl Acad Sci USA* 84(10):3156–3160.

Zinoni F, Birkmann A, Stadtman TC, Bock A. 1986. Nucleotide sequence and expression of the selenocysteine-containing polypeptide of formate dehydrogenase (formate-hydrogen-lyase-linked) from *Escherichia coli*. *Proc Natl Acad Sci USA* 83(13):4650–4654.

The Genome

4

In Chapter 3, we saw how genetic information in DNA guides the production of proteins. This principle is illustrated with DNA sequences that are the blueprint for the β-globin chain, and with sickle cell anemia where a change in the DNA sequence gives rise to an erroneous amino acid sequence. From Chapter 6 onward, we account for more details on the relationship between DNA sequence and proteins, but we first need to discuss some general aspects of human DNA to better understand the flow of genetic information.

Remember that we refer to the complete genetic material of an organism as its **genome**. In this chapter, a variety of aspects of the human genome are covered. We start out with a microscopic view of DNA and progress to a deeper molecular understanding. We examine some of the most important experimental methods that made possible the detailed molecular analysis of the human genome and elucidation of the nucleotide sequence of human DNA. Having access to this sequence, other questions arise. What is the general organization of the human genome, and what elements are quantitatively or functionally important? What is the structure of genes, and what is their environment? For instance, we learn that the human genome contains not only protein-coding genes but a lot of noncoding material, including a huge number of repetitive elements.

DNA IS PRESENT IN THE NUCLEUS OF CELLS

We first turn to the question of where DNA is localized. It is present within a cell, the basic functional unit of all living organisms. All cells have certain characteristics in common. For example, a cell is surrounded by a membrane. The transportation of molecules across the membrane is strictly controlled, and in this manner the composition within a cell is different from the surrounding environment. The total number of cells in a human body is astounding, in the order of 10^{13}. Adding to the complexity of the human body is that even though all cells have common features, cells within an individual are different. For instance, nerve and muscle cells have entirely different properties.

Animals, plants, and fungi are all members of the group of living organisms referred to as **eukaryotes**. These organisms share the same basic cell architecture. Thus, the cells of eukaryotes contain a membrane-surrounded nucleus—a compartment that has the major portion of DNA. One of the most important functions of the nucleus is to protect and "keep track" of the DNA inside. In the cell nucleus, there are important processes going on that deal with DNA. One example is where DNA is copied in the context of cell division. Furthermore, there is RNA transcription—that is, when RNA molecules are formed with DNA as a template. Transcription is a strictly controlled process where in every moment it is decided what parts of the DNA are to be transcribed. Therefore, the cell nucleus is an important genetic control unit within the cell.

In eukaryotic cells, translation occurs outside the nucleus, in a compartment known as the cytosol, defined as the liquid in the cell excluding the nucleus. Eukaryotic cells have characteristic cellular substructures in the cytosol, such as mitochondria, organelles that are specialized energy factories. Like the nucleus, mitochondria have DNA. This DNA is quite different from the nuclear DNA. In a human, it is a molecule about 16,000 nucleotides in length, much smaller than the DNA molecules found in the nucleus, as elaborated on later in this chapter.

DNA AS SEEN UNDER THE MICROSCOPE

What can we learn about DNA by examining human cells under the microscope? By the end of the nineteenth century, **chromosomes**—located in the nucleus—were identified using light microscopy. The name *chromosome* (from the Greek words *chroma* ["color"] and *soma* ["body"]) was derived from the fact that these structures were strongly stained with specific dyes.

By the early 1900s, Walter Sutton and Theodor Boveri suggested that chromosomes carry genetic information. The molecular nature of chromosomes was not known at this time. We now know that chromosomes are composed of DNA and protein and that DNA is the actual carrier of genetic information. Miescher used white blood cells for isolation of DNA (Chapter 3). However, all human cells contain DNA. The only exception is red blood cells that lack both nuclei and mitochondria.

In all cells, the genetic information is distributed among different chromosomes, each containing a distinct DNA molecule. In humans, there are 22 different chromosomes (non-sex chromosomes or **autosomes**) and, in addition, two different sex-specific chromosomes, X and Y. As the human is a **diploid** organism,[1] all individuals have two copies of each of the 22 chromosomes, where one copy originates from the father and the other from the mother. In each of these chromosome pairs, the two chromosomes are similar and are said to be **homologous** to each other. Furthermore, females have two X chromosomes, whereas males have one X and one Y chromosome (**Figure 4.1**). In contrast to the autosome pairs, the X and Y chromosomes are not similar and are of different lengths.

The numbering of the chromosomes is based on the lengths of DNA where the longest chromosome is number 1. At first it was believed that chromosome 22 was smaller than chromosome 21 (most widely known for its role in Down syndrome), but more detailed studies revealed that it is actually 5% longer than chromosome 21. However, the numbering of the chromosomes was not altered by this revision of the length relationship.

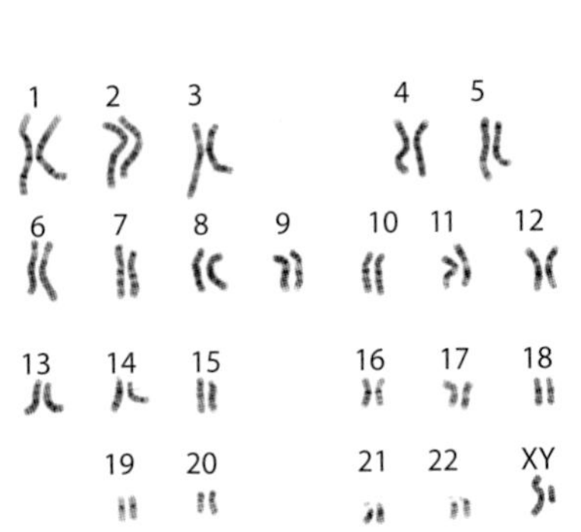

Figure 4.1 Human karyotype. Human chromosomes were stained with Giemsa and sorted in order. Reflecting the diploid nature of the genome, there are two copies of each of the chromosomes 1–22. A male is depicted here, and hence, there is one copy each of the X and Y chromosomes. (Courtesy of National Human Genome Research Institute.)

---

[1]Diploid cells have two sets of the autosomal chromosomes, whereas haploid cells (like human germline cells) have only one set.

In preparation for cell division, all pairs of homologous chromosomes become aligned and are highly condensed as compared to their appearance in other phases of the cell cycle. Chromosomes in a condensed state may be observed using a light microscope. For this purpose, they are often stained such that the different chromosomes may be distinguished from one another. A classic type of staining is G-banding. Chromosomes are treated with the enzyme trypsin (to degrade protein material) and a stain known as Giemsa. This stain is specific for the phosphate groups of DNA and binds to regions in DNA where there is a high proportion of A•T base pairs. The final result of the staining procedure is a series of alternating light and dark bands (see Figure 4.1). A schematic view of the banding for chromosome 16 is shown in **Figure 4.2**.

Every chromosome has two arms, one short (*p*) and one long (*q*), where "p" stands for *petit*, "small" in French, and "q" was chosen as it is the character following "p" in the alphabet. The arms are separated by a centromere, which is a point that plays a critical role when chromosomes are to be separated during cell division. Individual bands in a chromosome may be referred to an expression such as 3p26.3, where 3 is the number of the chromosome, p is the short arm, and 26.3 is the number of the band. At the very ends of each chromosome, there are regions with repetitive sequences, known as **telomeres**.

The view of all chromosomes as obtained with staining is referred to as a **karyotype**. Although the bands do not tell us about the molecular detail of DNA, there are many applications to this type of microscopic analysis. For instance, it is used to determine whether there are extensive chromosomal aberrations in an individual or in a sample of cells. Examples of such chromosomal aberrations are Down syndrome (a common cause is an extra copy of chromosome 21) and Klinefelter syndrome, also known as XXY, where there is more than one copy of the X chromosome in males. Karyotype analysis may also reveal certain cancers that are the result of chromosomal changes. One example is the translocation known as the Philadelphia chromosome, involving an interchange of material between chromosomes 9 and 22. This abnormality is associated with chronic myelogenous leukemia (CML) and acute lymphoblastic leukemia (ALL).

Chromosomes may also be visualized with higher resolution using electron microscopy (**Figure 4.3**). Mitochondrial DNA molecules are not included in karyotyping as they are too small for light microscopy, but these molecules may be observed with electron microscopy (**Figure 4.4**).

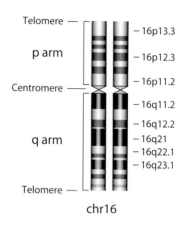

chr16

Figure 4.2 Schematic view of chromosomes with banding pattern. A pair of chromosomes 16 is depicted with p and q arms together with a schematic view of the Giemsa banding pattern. The names of selected bands are indicated.

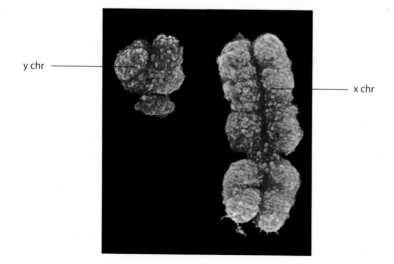

Figure 4.3 Electron micrograph of sex chromosomes. A colored scanning electron micrograph of an X chromosome (pink) and a Y chromosome (blue). (Courtesy of Science Photo Library.)

Figure 4.4 Electron micrograph of mitochondrial DNA. (Courtesy of Jack Griffith at the University of North Carolina at Chapel Hill, and Thomas Nicholls at the University of Gothenburg, Sweden.)

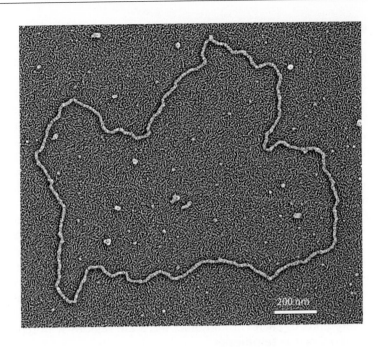

DNA IS COMPACTED WITHIN THE NUCLEUS OF A HUMAN CELL

Human DNA molecules are extremely long polymers. If the DNA contained within the nucleus of a human cell would be completely extended, it covers a length of several centimeters. Assuming that a human body contains 10^{13} cells, the entire length of all DNA would be about 300 million kilometers. This is about the distance between Earth and the planet Mars.

As a cell nucleus is only approximately 6×10^{-6} m in diameter, DNA needs to be packed in order to fit there. With the help of proteins known as **histones**, DNA is packed into **nucleosomes**, where the DNA is wrapped around a core of histone proteins named H2A, H2B, H3, and H4 (**Figure 4.5**). The complex formed with chromosomal DNA, proteins and other molecules is known as **chromatin**.

The nucleosomes are, in turn, packed into more condense structures that include chromatin loops (**Figure 4.6**, see also Chapter 10 and Figure 10.18).

There is a limit as to what we can learn about DNA with light microscopy. Although electron microscopy is getting more and more powerful, it cannot reveal details such as the sequence of nucleotides. For such information, techniques of molecular biology must be applied, like cloning and DNA sequencing methods.

TOOLBOX OF THE MOLECULAR BIOLOGIST

How is the sequence of nucleotides in DNA determined? It relies on a series of different methods used by molecular biologists. Basic DNA technology started in the 1970s. The DNA that we are able to purify from cells is of high molecular weight and is difficult to analyze as it is. Typically we want to isolate a smaller piece of DNA, and we need to obtain that DNA in pure form, isolated from the rest of the DNA. We also need to have it in a large quantity, because single molecules are difficult to work with from a practical point of view. There were important techniques developed in the 1970s

(a)

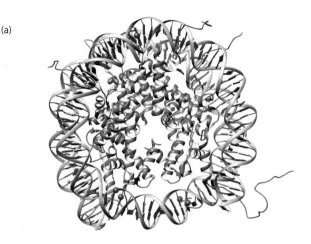

Figure 4.5 Nucleosome structure. DNA is wrapped around a core of histone proteins, two subunits each of histones H2A, H2B, H3, and H4. (a) Histones are shown in green. (b) A cartoon version of the nucleosome structure. The "linker" DNA is the DNA in between nucleosome particles. The histone H1 is at the site where DNA enters and exits the nucleosome. ([a] Adapted from Harp JM et al. 2000. *Acta Crystallogr* (Section D) 56:1513–1534. PBD ID: 1EQZ. With permission.)

(b)

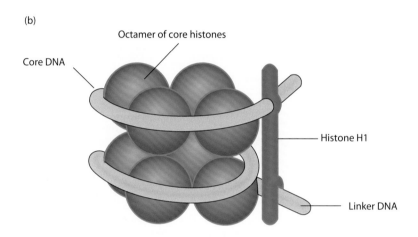

Octamer of core histones

Core DNA

Histone H1

Linker DNA

and 1980s that aim at producing multiple copies of a specific region of DNA sequence. For instance, there is the procedure of **molecular cloning**. Thus, a genomic DNA may be fragmented physically or cut up with enzymes to generate shorter fragments. Then, a fragment may be produced in multiple copies by having it replicated many times within a bacterial cell (**Figure 4.7**).

Another widely used technique to amplify a piece of DNA is the **polymerase chain reaction** (**PCR**), invented in the 1980s and used to amplify a particular small region of interest. For instance, a small portion of the genome suspected to contain a mutation may be amplified. At an appropriate distance from the mutation, PCR **primers** are used to specify exactly what DNA region to amplify. The resulting PCR product will then be sequenced. PCR is a technique extensively used in DNA research, and its inventor Kary Mullis was awarded the Nobel Prize in Chemistry in 1993 (**Figure 4.8**).

DEVELOPMENT OF DNA SEQUENCING

Important technology for the determination of DNA nucleotide sequences was developed in the late 1970s by Frederick Sanger. He was awarded the Nobel Prize in Chemistry in 1980 for this work.[2] Sanger invented methods where DNA replication is allowed to occur *in vitro*, using specific primers

[2]This was actually the second Nobel Prize in Chemistry for Frederick Sanger; in 1958 he received the prize for his work on protein sequencing, see also Chapter 12.

Figure 4.6 DNA packing. The first level of DNA compaction is the formation of nucleosomes. At a higher level of packing a looped structure is observed where regions that are distant at the level of DNA sequence come together in three-dimensional space. Additional packing gives rise to the structure as observed using electron microscopy.

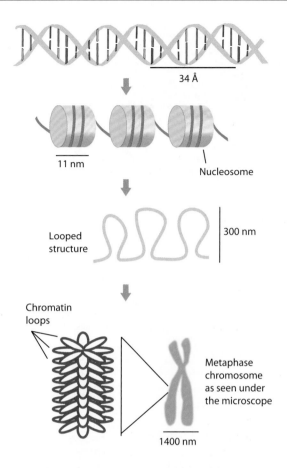

34 Å

11 nm

Nucleosome

Looped structure

300 nm

Chromatin loops

Metaphase chromosome as seen under the microscope

1400 nm

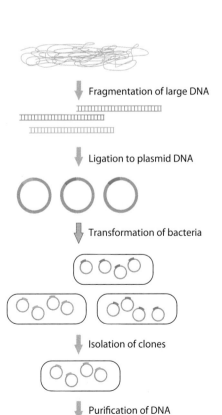

Fragmentation of large DNA

Ligation to plasmid DNA

Transformation of bacteria

Isolation of clones

Purification of DNA

specifying where replication is to start. These methods were applied in the sequencing of the bacteriophage phiX174 genome, with a total of 5,368 nucleotide pairs. This was the first genome to be sequenced. The methodology was soon further improved by using nucleotide derivatives to terminate replication (**Figure 4.9**). The original Sanger sequencing method using such terminators of replication is outlined in **Figure 4.10**.

Characteristic of the commonly used DNA sequencing methods is that in a single experiment only a short piece of DNA may be sequenced. In the classic Sanger sequencing method, a maximum of about 1,000 nucleotides may be read. Therefore, the task of sequencing a longer DNA, like that of phiX174, has to be divided up into several different sequencing experiments. In the case of phiX174, a **restriction enzyme map** was first created as explained in **Figure 4.11**.

In the early days of DNA sequencing, scientists sequenced fairly small regions of DNA, like those encompassing a single gene. However, in the 1990s as sequencing technology improved, much longer pieces of DNA

Figure 4.7 Molecular cloning. There are two important elements of molecular cloning. First, a specific sequence is isolated from a background of other sequences. Second, the specific sequence is produced in quantities that allow further analysis using biochemical or molecular biology technology. The scheme shown here is to illustrate steps during cloning of a genomic DNA fragment. First, genomic DNA is fragmented and each of the fragments is introduced into circular plasmid DNAs, small DNA molecules that are able to replicate independently within bacteria. The plasmid DNAs are then mixed with bacteria that have been made competent to take up foreign DNA. A suspension of bacteria containing plasmid DNA is then transferred to an agar plate. The suspension is appropriately diluted such that each of the original cells (that each contains only one plasmid DNA molecule) grows overnight to colonies of cells that are well separated from each other. In each of the colonies all cells are genetically identical and contain the same plasmid DNA. Such a colony may now be transferred to bacterial liquid medium and grown in larger quantity. Finally, plasmid DNA is prepared from the resulting bacterial cells.

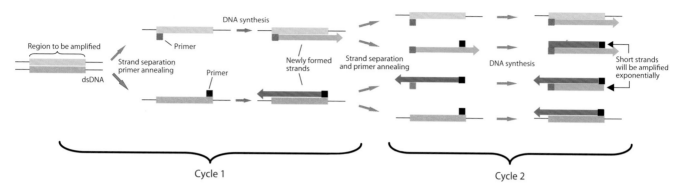

Figure 4.8 Principle of polymerase chain reaction (PCR). In PCR a region of a DNA is amplified. The double-stranded DNA is first heated to about 95°C to allow the strands to separate. The temperature is then lowered to 50°C–65°C to allow pairing of two different oligonucleotides (shorter pieces of single-stranded DNA) to the DNA strands. These oligonucleotides, known as primers, specify where replication is started. Primers are necessary to initiate any DNA replication reaction. One primer pairs with the forward strand and the other with the reverse strand. Then DNA polymerase produces new DNA by extending the primers in the 5′ to 3′ direction. For this step, the temperature is raised to a level suitable for the DNA polymerase being used. The strand separation, primer pairing, and chain elongation steps form one cycle of PCR. This cycle may now be repeated any number of times by adequately adjusting the temperature. The first two cycles are shown here. After many more cycles, the short fragments as indicated have by far outnumbered all other DNA molecules in the reaction mixture. The whole PCR is carried out in a test tube. A thermostable enzyme is typically used such that it will survive the high temperatures of the strand separation step. In this way, there is no need to add new enzyme in each of the replication steps.

were examined with respect to their nucleotide sequence. Thus, the first complete bacterial genomes were reported in 1995, those of *Mycoplasma genitalium* of about 580,000 base pairs and *Haemophilus influenzae*, with about 1,830,000 base pairs. The sequencing of the two bacteria was carried out at the Institute for Genomic Research (TIGR), founded by Craig Venter. Venter used a sequencing strategy where the original genome was randomly fragmented into smaller pieces, and sequencing was performed with these smaller pieces (an approach referred to as **whole genome shotgun sequencing**). The reconstruction of the complete genome, **sequence assembly**, was carried out on the basis of overlaps between the shorter sequences (**Figure 4.12**). This protocol saved some work as the first step involving a restriction map of the original DNA was avoided.

Shortly after the sequencing of the bacteria *M. genitalium* and *H. influenzae*, the first complete sequence of a eukaryotic species was presented. *Saccharomyces cerevisiae*, also known as baker's or brewer's yeast, has a genome of about 12.1 million base pairs (MB). The sequencing was made possible through an international consortium of scientists. In 1998, the first animal was sequenced, the worm *Caenorhabditis elegans* (100 MB). Genomes of the first insect, the fruit fly *Drosophila melanogaster* (165 MB), and of the first plant, *Arabidopsis thaliana* (119 MB), were reported in 2000. More details of the history of genome sequencing are shown in **Table 4.1**. However, that table shows only a small subset of the species that have been sequenced today (**Table 4.2**).

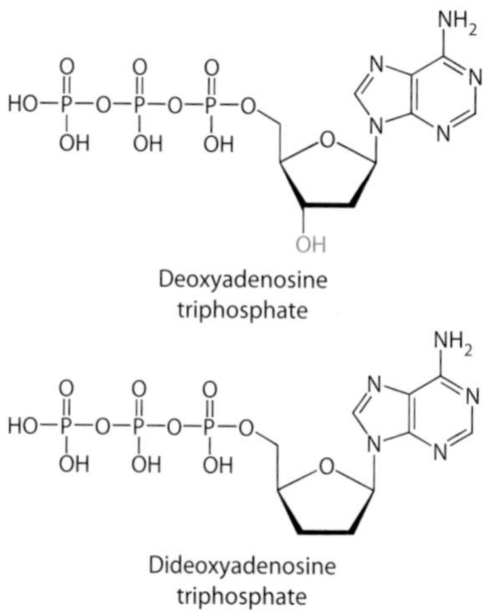

HUMAN GENOME SEQUENCING PROJECT

The size of the human genome is some three billion base pairs, about 300 times larger than the yeast genome. When scientists started to think about the sequencing of the genome, it was therefore a task of considerable magnitude. A project to sequence the entire human genome was nevertheless initiated in 1990. The project was formally founded by the US Department of Energy (DOE) and the National Institutes of Health (NIH) but involved a collaboration of many different laboratories around the world. James Watson was first heading the NIH effort, and in 1993 he was succeeded by Francis

Deoxyadenosine triphosphate

Dideoxyadenosine triphosphate

Figure 4.9 Chemical structures of deoxynucleotides and dideoxynucleotides used in Sanger sequencing.

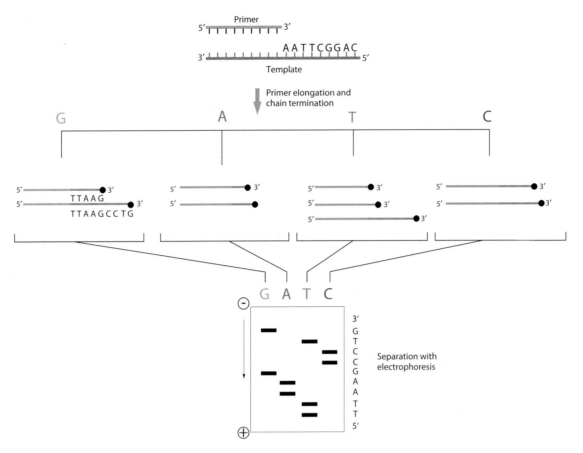

Figure 4.10 Sanger protocol of DNA sequencing. Sanger sequencing relies on a method where DNA is replicated *in vitro*. On top is shown a primer that has annealed to a DNA strand acting as template for DNA replication. The primer specifies where synthesis is to start, and the sequence of the newly produced DNA is specified by the template DNA strand. Four separate replication reactions are carried out. In each of these reactions, DNA replication is allowed to occur with all four deoxynucleotides (dNTPs) as substrates, but in addition one specific dideoxynucleotide (ddNTP) (see Figure 4.9) is added. The ddNTP causes termination of DNA replication. The concentration of the ddNTP is adjusted so that the DNA replication reaction will give rise to a mixture of different DNAs, each terminating with the specific ddNTP. Consider for instance the reaction with a dideoxy-G (leftmost reaction). Two different DNA fragments are generated, ending in sequences TTAAG and TTAAGCCTG, respectively. The products of the four different reactions are then separated according to size using electrophoresis, where smaller DNA molecules migrate faster toward the cathode (the positively charged electrode). After electrophoresis the DNA fragments may be visualized, for instance based on radioactive labeling of the DNA. The sequence of the DNA is finally deduced by reading bands from bottom to top. In a further development of Sanger sequencing (not shown here), each of the ddNTPs is labeled with fluorescent dyes, each of which emits light at different wavelengths. Thereby, only one sequencing reaction is required, rather than four different reactions as shown here.

Collins who from that year came to head the human genome project as the director of the National Center for Human Genome Research (later known as the National Human Genome Research Institute).

As the human genome is large and contains a lot of repeated sequences, it was difficult to sequence in one go with the shotgun method. It was therefore decided by the human genome project that a physical map of large (up to 200,000 nucleotides) restriction fragments of the genome was first to be constructed. Then, each of these large fragments would be sequenced separately using the shotgun approach (**Figure 4.13**) and using the Sanger sequencing protocol. With the help of the map, the fragment sequences could finally be put together to reconstruct the entire genome.

In 1998 Craig Venter appeared in the context of a commercial enterprise, Celera Genomics, as a competitor to the academic genome sequencing initiative. Venter intended to use shotgun sequencing directly, thereby eliminating the physical map step entirely. This started a competitive period in the history of human genome sequencing.

The first version of the human genome sequence was presented in June 2000, and both the academic project (Human Genome Project [HGP])

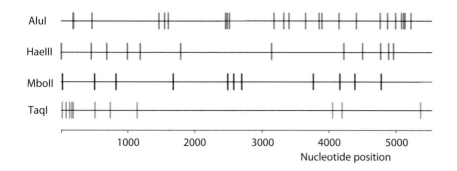

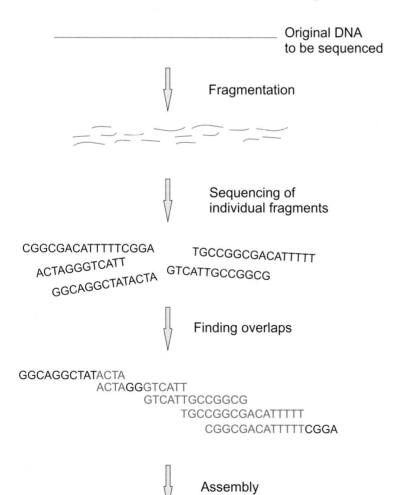

Figure 4.11 Restriction enzyme map of bacteriophage phiX174. Restriction enzymes are able to cut a double-stranded DNA at specific sequences. The restriction enzymes indicated AluI, HaeIII, MboII, and TaqI cleave at the DNA sequences AGCT, GGCC, GATC, and TCGA, respectively. Without access to the actual nucleotide sequence of the phiX174 genome, a map to show the location of restriction enzyme sites (a "restriction map") may nevertheless be obtained by experiments with restriction enzyme cleavages. Consider a simple example with only two restriction enzymes, in which two reactions are carried out, each with a separate enzyme. In a third reaction, both of the enzymes are used in the same reaction. The sizes of the resulting DNA fragments in each experiment are determined by separation with electrophoresis. The sizes may be used to infer where in the original DNA the specific enzymes are cleaving and a restriction map may thereby be obtained. When the original DNA is to be sequenced, the first step is to sequence individual DNA fragments obtained by restriction enzyme cleavage. These sequences may finally be put in the right order with the help of the restriction enzyme map.

and Celera presented their results in publications in June 2001. Venter's project had completed the task in a shorter time than the academic effort, but their crew had the advantage of using the physical map information made available by the academic project.

The total cost of the academic human genome project was about $2.7 billion (financial year 1991). All of this money was not spent on actual sequencing but on other molecular biology procedures. Therefore, the project led to significant development of many methods of life science, including cloning and PCR. Furthermore, the sequencing of the human genome was indeed motivated, as we now see that it has had a significant medical

Figure 4.12 Shotgun sequencing and sequence assembly. In shotgun sequencing, a larger DNA to be sequenced is first divided up into a set of random, overlapping fragments. These fragments are then sequenced individually. During sequence assembly, the different shorter sequences are put together on the basis of overlaps to reconstruct the sequence of the original larger DNA. Please note that in reality the individual sequences shown here, as well as the overlaps, are much longer.

Table 4.1 Selected organisms whose genomes have been sequenced

Species, latin name	Species, common name	Year of sequencing	Number of bases
Bacteriophage MS2[a]		1976	3,569
Bacteriophage phiX174[b]		1977	5,386
Haemophilus influenzae[c]		1995	1.8 MB
Mycoplasma genitalium[d]		1995	0.58 MB
Methanococcus jannaschii[e]		1996	1.7 MB
Saccharomyces cerevisiae[f]	Baker's or brewer's yeast	1996	12.1 MB
Caenorhabditis elegans[g]	(A nematode)	1998	100 MB
Drosophila melanogaster[h]	Fruit fly	2000	165 MB
Arabidopsis thaliana[i]	Thale cress	2000	119 MB
Homo sapiens[j,k]	Man	2001	3,200 MB
Mus musculus[l]	Mouse	2002	3,480 MB
Gallus gallus[m]	Chicken	2004	1,070 MB
Populus trichocarpa[n]	Poplar	2006	510 MB
Xenopus tropicalis[o]	Western clawed frog	2010	1,360 MB
Danio rerio[p]	Zebrafish	2013	1,500 MB
Picea abies[q]	Norway spruce	2013	20,000 MB
Triticum aestivum[r]	Bread wheat	2017	15,000 MB
Ambystoma mexicanum[s]	Axolotl	2018	32,000 MB

Note: In the case of bacteriophage MS2, the genome is an RNA molecule that was sequenced with RNA sequencing technology. The other genomes are DNA and analyzed using DNA sequencing methods. The table shows only a small subset of all organisms sequenced today.

[a] Fiers W et al. 1976. Complete nucleotide sequence of bacteriophage MS2 RNA: primary and secondary structure of the replicase gene. *Nature* 260(5551):500–507.
[b] Sanger F et al. 1977. Nucleotide sequence of bacteriophage phi X174 DNA. *Nature* 265(5596):687–695.
[c] Fleischmann RD et al. 1995. Whole-genome random sequencing and assembly of *Haemophilus influenzae* Rd. *Science* 269(5223):496–512.
[d] Fraser CM et al. 1995. The minimal gene complement of *Mycoplasma genitalium*. *Science* 270(5235):397–403.
[e] Bult CJ et al. 1996. Complete genome sequence of the methanogenic archaeon, *Methanococcus jannaschii*. *Science* 273(5278):1058–1073.
[f] Goffeau A et al. 1996. Life with 6000 genes. *Science* 274(5287):546, 563–567.
[g] The Washington University Genome Sequencing Center (This is a Consortium). Genome sequence of the nematode C. elegans: a platform for investigating biology. *Science.* 1998. 282(5396):2012–2018.
[h] Adams MD et al. 2000. The genome sequence of *Drosophila melanogaster*. *Science* 287(5461):2185–2195.
[i] The Arabidopsis Genome Initiative. Analysis of the genome sequence of the flowering plant *Arabidopsis thaliana*. *Nature*, 2000. 408(6814):796–815.
[j] Lander ES et al. 2001. Initial sequencing and analysis of the human genome. *Nature* 409(6822):860–921.
[k] Venter JC et al. 2001. The sequence of the human genome. *Science* 291(5507):1304–1351.
[l] Waterston RH et al. 2002. Initial sequencing and comparative analysis of the mouse genome. *Nature* 420(6915):520–562.
[m] Hillier LW et al. 2004. Sequence and comparative analysis of the chicken genome provide unique perspectives on vertebrate evolution. *Nature* 432(7018):695–716.
[n] Tuskan GA et al. 2006. The genome of black cottonwood, *Populus trichocarpa* (Torr. & Gray). *Science* 313(5793):1596–1604.
[o] Hellsten U et al. 2010. The genome of the Western clawed frog *Xenopus tropicalis*. *Science* 328(5978):633–636.
[p] Howe K et al. 2013. The zebrafish reference genome sequence and its relationship to the human genome. *Nature* 496(7446): 498–503.
[q] Nystedt B et al. 2013. The Norway spruce genome sequence and conifer genome evolution. *Nature* 497(7451):579–584.
[r] Zimin AV et al. 2017. The first near-complete assembly of the hexaploid bread wheat genome, *Triticum aestivum*. *GigaScience* 6(11):1–7.
[s] Nowoshilow S et al. 2018. The axolotl genome and the evolution of key tissue formation regulators. *Nature* 554:550.

impact—it has, for instance, generated detailed information regarding cancer mutations as we will see in Chapter 5.

The first version of the human genome covered about 94% of the genome. Since then, further sequencing has led to better coverage and accuracy, although we still today are lacking small pieces of the genome. Most of the regions that remain unsequenced are difficult to access because they contain highly repetitive sequences, for instance, sequences where a dinucleotide, like CA, is repeated numerous times. The current status of sequencing with respect to the 24 chromosomes is shown in **Table 4.3**.

WHOSE GENOME?

Who were the individuals whose genomes were sequenced? In the case of HGP, researchers collected blood (female) or sperm (male) samples from a large number of volunteer donors. Only a few of these were processed as DNA resources for the genome project. Donor identities were protected so that neither donors nor scientists could know whose DNA was sequenced. For technical reasons related to the quality of DNA libraries, much of the sequence (>70%) of the HGP genome came from a single anonymous male donor from Buffalo, New York, whose code name was RP11.

In the Celera Genomics project, five individuals out of a total of 21 donors were selected for sequencing. Craig Venter communicated in 2003 that his DNA was one of the 21 samples.

The genome sequences reported in 2001 were therefore mosaics of different individuals. Furthermore, they were "haploid" genomes, meaning that only one sequence was presented, although in reality individuals have two copies of each of the autosomal chromosomes with nonidentical sequences. The sequence of the human genome first presented in 2001 and that has since been improved is referred to as the **reference** version of the human genome. As human individuals are genetically different, they will all be different from the reference sequence. The reference genome is not static but is now and then updated. For instance, gaps of previously unsequenced regions are filled in, and assembly errors are corrected. One recent version is GRCh38, released by the Genome Reference Consortium (GRC) in 2013. This version is derived from 13 anonymous volunteers from Buffalo, New York. For more versions, see **Table 4.4**. Now and then in human sequencing projects, new population-specific regions are discovered that are not part of the reference assembly.

Table 4.2 Number of sequenced genomes

Organism group	Number of species sequenced
Eukaryotes	3,324
Animals	1,097
Plants	384
Fungi	1,528
Protists	312
Archaea	1,950
Bacteria	36,544
Viruses	11,544

Note: Table shows the approximate number of species whose genomes have been sequenced. Information is derived from the NCBI collection of genome assemblies in 2018 (https://www.ncbi.nlm.nih.gov/assembly).

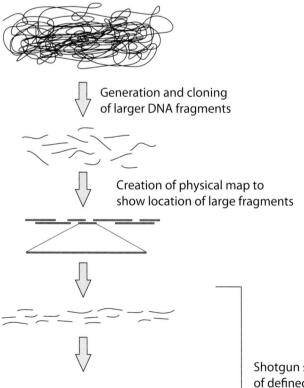

Generation and cloning of larger DNA fragments

Creation of physical map to show location of large fragments

Shotgun sequencing of defined genome fragments

CAGGCGAGCG**TACTATCGGAACTG**
      TACTATCGGAACTGCCACGCGAC

CAGGCGAGCG**TACTATCGGAACTG**CCACGCGAC

Figure 4.13 Sequencing strategy of the human genome project. Genomic DNA was first divided into larger restriction enzyme fragments, and a map to show the physical location of the different fragments was created. Each of the larger fragments was then sequenced according to the shotgun sequencing method as described in Figure 4.12.

Table 4.3 Human chromosome statistics

Chromosome	Total nucleotides	Total non-N nucleotides	Number of protein coding genes
1	248,956,422	230,481,012	2,050
2	242,193,529	240,548,228	1,301
3	198,295,559	198,100,135	1,079
4	190,214,555	189,752,667	753
5	181,538,259	181,265,378	884
6	170,805,979	170,078,522	1,045
7	159,345,973	158,970,131	992
8	145,138,636	144,768,136	665
9	138,394,717	121,790,550	778
10	133,797,422	133,262,962	731
11	135,086,622	134,533,742	1,316
12	133,275,309	133,137,816	1,036
13	114,364,328	97,983,125	321
14	107,043,718	90,568,149	820
15	101,991,189	84,641,325	616
16	90,338,345	81,805,943	862
17	83,257,441	82,920,204	1,188
18	80,373,285	80,089,605	269
19	58,617,616	58,440,758	1,474
20	64,444,167	63,944,257	543
21	46,709,983	40,088,619	232
22	50,818,468	39,159,777	492
X	156,040,895	154,893,029	846
Y	57,227,415	23,636,355	45
Total	3,088,286,401	2,934,876,993	20,358

Note: For each chromosome is shown the number of protein coding genes, the total number of bases covered by the assembly, and the number of bases that are not unknown (not N). Data are from Ensembl, October 2018 and refer to the human genome assembly GRCh38.p12.

Table 4.4 Versions of the human genome assembly

Release name	Date of release	Equivalent UCSC version
NCBI Build 34	July 2003	hg16
NCBI Build 35	May 2004	hg17
NCBI Build 36.1	March 2006	hg18
GRCh37	February 2009	hg19
GRCh38	December 2013	hg38

For instance, the sequencing of two Swedish individuals revealed more than 10 million nucleotides in each individual that were absent from the human GRCh38 assembly.

DNA molecules are double stranded, and scientists have used different nomenclature to refer to the two different strands—forward/reverse, plus (+)/minus (−), upper/lower, top/bottom, and even Watson/Crick. Whenever the human reference sequence is presented in databases and in many other contexts, only one of the strands is shown and the other

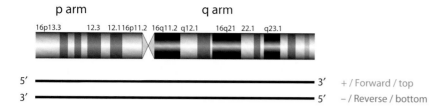

Figure 4.14 Each chromosome corresponds to a linear double-stranded DNA molecule. A karyotype representation of chromosome 16 is shown together with the corresponding double-stranded DNA. The human reference genome represents a double-stranded DNA, but it is the forward strand that is commonly presented. This strand has its 5′ end at the telomeric end of the p arm.

is implicit. This is to save space and because the other strand may be inferred directly on the basis of base complementarity. The strand typically shown, the "forward" strand, is for the human genome the strand that has its 5′ end at the telomeric end of the p arm and its 3′ end at the telomeric end of the q arm (**Figure 4.14**). Conversely, the "reverse" strand is the complementary strand with its 5′ end at the telomeric end of the q arm.

THE MITOCHONDRIAL GENOME

Genetic information is contained not only in nuclear chromosomes, but also in mitochondria that are located in the cytoplasm. The human mitochondrial genome is small as compared to the nuclear genome, only 16,569 bases. It is a circular genome and encodes 13 different proteins that are all essential components of the mitochondrial machinery for producing ATP. In addition to these protein genes, there are ribosomal RNA (rRNA) and **tRNA** genes (**Figure 4.15**).

Mitochondria originally evolved from early bacteria that came to exist within an early eukaryote or archaebacterium. During evolution, portions

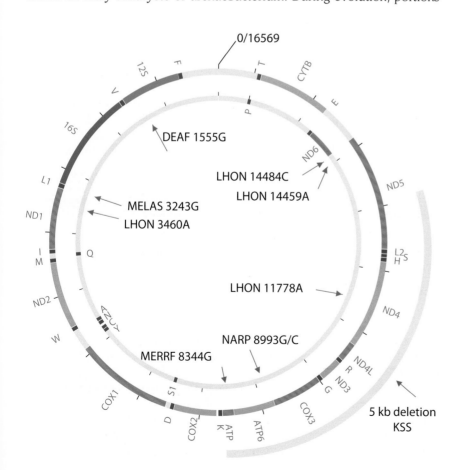

Figure 4.15 The mitochondrial genome and disorders. Protein genes are indicated in red or orange. Single letters represent tRNA genes. Ribosomal RNA genes are marked in green. The outer circle depicts the strand of mitochondrial DNA that encodes the large majority of genes. The disorders indicated along with mutations are (1) Leber's hereditary optic neuropathy (LHON); (2) neuropathy, ataxia, retinitis pigmentosa, and ptosis (NARP); (3) mitochondrial myopathy, encephalomyopathy, lactic acidosis, stroke-like symptoms (MELAS); (4) myoclonic epilepsy with ragged red fibers (MERRF syndrome); (5) deafness (DEAF); and (6) Kearns-Sayre syndrome (KSS). (Based on information at https://mitomap.org.)

of the bacterial genome were either lost or transferred to the nuclear genome of the host organism. Therefore, many gene products that are required inside mitochondria are now specified by nuclear genes. An import machinery makes sure the right proteins are imported into mitochondria.

Mitochondrial DNA mutates at least about 10 times the rate of the nuclear genome. Possible reasons for this difference are that DNA damage occurs more frequently in mitochondria because of free radicals and that mitochondria lack some of the repair mechanisms available to nuclear DNA. It has also been observed that there is an accumulation of mutations in mitochondrial DNA with age.

Each human cell typically contains thousands of copies of the mitochondrial genome. However, all mitochondrial DNA molecules present in a cell may not have identical sequences. This is referred to as mitochondrial **heteroplasmy**.

The high mutation rate of mitochondrial DNA has the effect that its sequence is variable between human individuals. In particular, there is a highly variable noncoding region, known as the D-loop. This is a sequence highly useful for evolutionary studies, for instance to examine relationships between different geographic human populations.

Following fertilization of the egg, paternal mitochondrial DNA is effectively destroyed. This leads to another characteristic of mitochondria—DNA is maternally inherited. Therefore, analysis of mitochondrial DNA sequences may be used to reveal maternal lineages.

Finally, a number of inherited diseases are associated with changes in mitochondrial DNA. Such disorders are relatively frequent—they are known to affect about 1 in 4,300 of the population. Examples are shown in Figure 4.15. (For more details on a specific example of a mitochondrial disease see also Chapter 11.)

ACCESSING THE GENOME—GENOME BROWSERS

After completion of the human genome in 2001, a number of important tools to navigate through the vast information collected in the project were developed. The object of a **genome browser** is to display information pertaining to a selected region of the genome. One such useful tool is the UCSC (University of California Santa Cruz) Genome Browser (**Figure 4.16**).

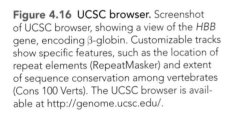

Figure 4.16 UCSC browser. Screenshot of UCSC browser, showing a view of the *HBB* gene, encoding β-globin. Customizable tracks show specific features, such as the location of repeat elements (RepeatMasker) and extent of sequence conservation among vertebrates (Cons 100 Verts). The UCSC browser is available at http://genome.ucsc.edu/.

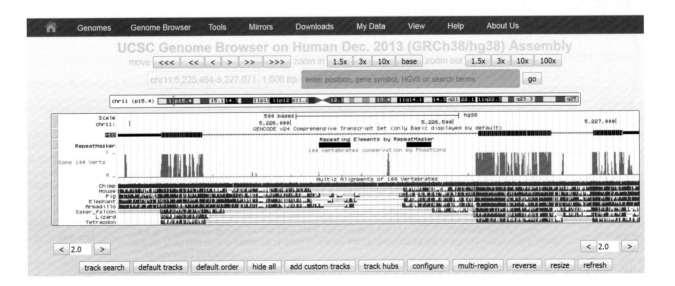

Similar tools are provided by Ensembl at the European Bioinformatics Institute (EBI) and the National Center of Biotechnology Information (NCBI). All information at these sites, including the human genome sequence, is freely available for examination and download to everyone.

THE HUMAN GENOME SPECIFIES PROTEINS AND NONCODING RNAs

The human genome has about 3.2 billion nucleotides. This overwhelming complexity reflects a long process of evolution. The genetic message is a long string of the letters A, T, C, and G. Looking at these long strings, it would seem they are just letters in a random order. It is quite impossible, without further knowledge, to draw any conclusions on the function of different parts of the genome. We now reach an important problem. What is the meaning or biological significance of all three billion letters of the human genome? Imagine that we are able to move along a long DNA molecule from one end to the other, reading off its sequence. What biologically significant elements are encountered? Throughout this book, we address this important question.

Essential functional parts of the genome are **genes**. A gene is a concept used early on in the history of genetics, but we may now define it in molecular terms. A gene may be viewed as a portion of genomic DNA that contains all of the information for production of a protein or RNA. One category of genes specifies messenger RNAs (mRNAs) that are used as templates for the production of proteins. A second category of genes specifies **noncoding[3] RNAs (ncRNAs)**, which is to say RNAs that are not mRNAs but have other functions. Human protein and noncoding RNA genes all have acronym names, for instance, the protein genes ABL1, BRCA1, and TAS2R38, and the ncRNA gene RMRP, following a nomenclature as decided by the HUGO Gene Nomenclature Committee. We focus on the protein genes here and in the next few chapters and return to the ncRNAs in Chapter 11.

About 20,000 different protein-coding genes have been identified in the human genome (**Table 4.5**). An important element of a protein-coding gene is a region that specifies the sequence of amino acids in the protein. All eukaryotic genes also have a lot of regulatory signals to say when and where the protein is to be made. These are signals that regulate transcription and translation.

Table 4.5 Number of human genes	
Category of gene	**Number of genes (Ensembl 2018)**
Coding genes	20,418
Noncoding genes (ncRNA)	22,107
Pseudogenes	15,195

Note: Numbers are based on the human genome assembly GRCh38.p12. Pseudogenes are genes that resemble true genes but have lost their function through mutations. It should be noted that there are studies that suggest an even larger number of noncoding RNA genes than what is reported in this table.

MOST GENES ARE MOSAICS OF EXONS AND INTRONS

Most eukaryotic genes are divided between **exons** and **introns**, where the exons are the parts that form the mature mRNA. Shown in **Figure 4.17** is the gene considered in previous chapters in the context of sickle cell

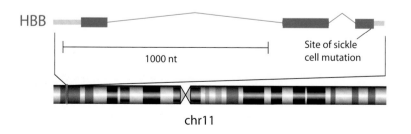

Figure 4.17 Exon-intron structure of *HBB* gene. Green bars represent protein coding parts of exons, and cyan bars are untranslated regions (UTRs) of exons. Red lines indicate introns. The *HBB* gene is located on the reverse strand of the human genome reference sequence. For this reason, the 5′ end of the β-globin mRNA is to the right in this view.

---

[3]With the established expression "noncoding," we actually mean "non-protein-coding."

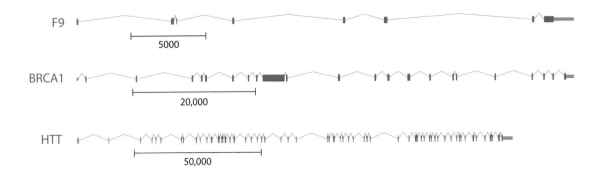

Figure 4.18 Selected human protein coding genes and their exon-intron structure. The genes F9, BRCA1, and HTT are shown. Numbers with scale bars represent numbers of nucleotides. Exons and introns are shown as in Figure 4.17.

disease, the β-globin gene. This gene is located on chromosome 11 and has been named *HBB* (HB for hemoglobin and B for beta). It has a relatively simple structure as it contains only two introns, and these introns are both relatively small. However, most human genes have many more introns, and they are typically much longer than in the *HBB* gene. The reader is referred to **Figure 4.18** for examples of such larger genes and their exon/intron structures.

The genomic structure with exons and introns is kept during transcription, but in a process called **splicing** the exons are combined while the introns are removed. Both transcription and splicing occur in the cell nucleus. Introns are often waste materials that are degraded to nucleotides that may be reused as building blocks for new transcripts. The final result of the splicing reaction is the mature mRNA that is eventually being used as a template for the production of proteins. It is through splicing that the actual coding sequence will be formed, as shown in **Figure 4.19**. In light of splicing, we need to revise our model of human gene expression somewhat, as shown in **Figure 4.20**.

ONLY ABOUT 1.6% OF THE GENOME CODES FOR PROTEIN

The total length of a human gene varies quite a lot. It may be as short as a few thousand bases but is typically much longer, and the longest genes are more than a million nucleotides. The mean length for a protein

Figure 4.19 The mRNA coding sequence is formed by splicing. (a) An example of how two exons are combined in the process of splicing to generate the coding sequence of the mature mRNA. In the case shown here, the first two positions of the alanine codon GCA are from the first exon, and the third position is from the second exon. (b) In general, a codon may be related to the intron in three different ways—the intron is said to have the phases 0, 1, or 2. Thus, the intron in (a) is an example of the intron of phase 2.

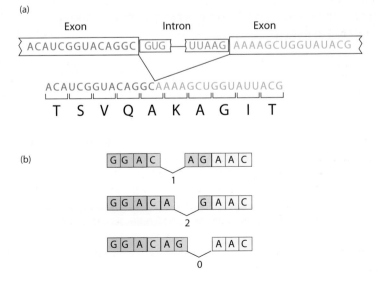

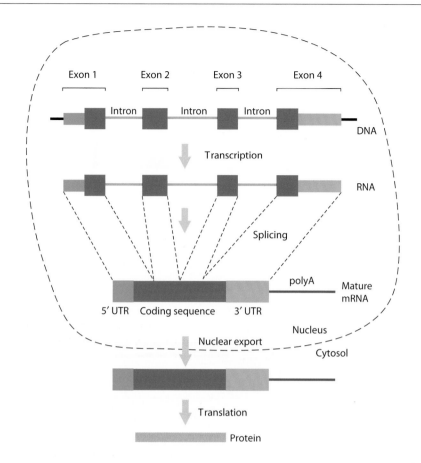

Figure 4.20 Flow of genetic information, including splicing. A protein gene with four exons is depicted on top with protein coding sequences in green and untranslated regions in blue. The three introns are shown in orange. After transcription, the four exons are joined in the process of splicing. Thereby the mature mRNA is formed, and this mRNA is finally transported from the nucleus to the cytosol to be used in protein synthesis (translation).

coding gene is about 62,000 nucleotides; on average, a gene has about nine exons.

In a typical human gene, the introns are much longer than the exons—in protein coding genes, the mean sizes of introns and exons are about 6,000 and 250 nucleotides, respectively (see Figure 4.18).

The marked size difference between exons and introns contributes to the effect that only a minor portion of the human genome is made up of sequences that encode protein. In addition, about half of the genome is intergenic regions—that is, noncoding regions that are in between genes. Protein and ncRNA exons together make up about 3.8% of the whole genome. Out of this, only 1.6% are protein coding sequences. An overview of compositional statistics for human genome exon and intron elements is shown in **Figure 4.21**.

Protein genes are by and large randomly distributed throughout the different chromosomes. For instance, two genes with closely related sequences and biological functions need not be on the same chromosome.

There is also a polarity or strand specificity to each gene. Remember that DNA is a double-stranded molecule. For each of the genes, only one of the two strands is read during transcription to an mRNA and gives rise to a protein. Some genes are encoded by one strand and others by the other strand of the double-stranded DNA molecule. This is illustrated in **Figure 4.22**, where the exon/intron structure for clarity has been left out—instead the whole transcribed regions for each of the genes are shown. There is no particular rule as to what genes are on what strands. Rather they seem to be more or less randomly distributed with respect to their strand location.

Figure 4.21 Only a small portion of the human genome is exonic material. The pie chart shows composition of nucleotide sequences of the human genome being part of exonic, intronic, and intergenic regions. Of the whole genome, protein coding sequences and ncRNA exons make up only about 1.6% and 0.55%, respectively.

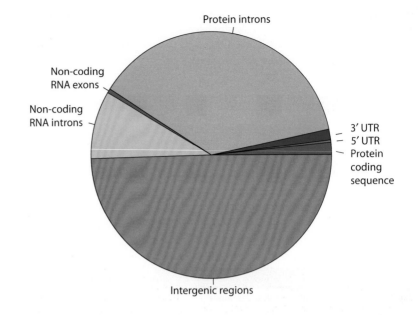

Figure 4.22 Protein coding genes are distributed on both strands of DNA. A portion of chromosome 16 is shown containing seven different protein genes. The green rectangles represent the gene transcripts (including both exons and introns), and the directions are indicated with the arrows. Thus, ATXN2L, SH2B1, ATP2A1, CD19, and NFATC2IP are proteins encoded by one (forward) strand and the other two proteins by the other (reverse) strand. Numbers refer to nucleotide positions in chr16 in the hg19 version of the human genome.

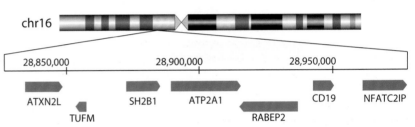

REPETITIVE SEQUENCES MAKE UP MORE THAN HALF OF THE HUMAN GENOME

If only a small portion of the human genome codes for protein and ncRNA, what is the significance of the rest? As we see in Chapter 10, there is evidence that a large fraction of the genome is involved in the regulation of transcription. With regard to the large size of the human genome, it is also important to note that more than half of the genome consists of repetitive sequences, most of which are noncoding and not involved in regulation (**Figure 4.23**, **Table 4.6**).

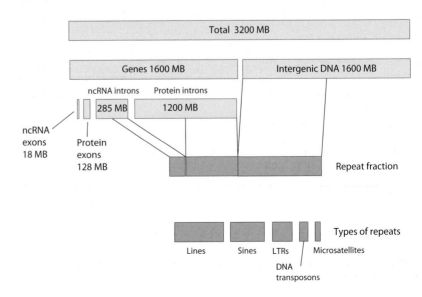

Figure 4.23 Distribution of different elements in the human genome, including repeats. About half of the genome is intergenic DNA—that is, DNA in between protein or ncRNA genes. Repeats are mainly found in introns and in intergenic regions. At bottom is shown the proportion of different repeats, where the lengths of rectangles indicate relative total length of each type of repeat. See also Table 4.6 and Figure 4.25.

Table 4.6 Repeats in the human genome		
Repeat element	Total nucleotides	Fraction of genome
Nonrepeat fraction	1,534,775,888	48.8
LINE1	536,111,998	17.1
LINE2	115,120,085	3.7
SINE/Alu	321,968,603	10.2
SINE/MIR	87,678,939	2.8
LTR/ERV1	89,021,720	2.8
LTR/ERVL	59,920,359	1.9
DNA	113,446,795	3.6
Satellite	78,362,451	2.5
Simple repeat	38,687,958	1.2

Note: LINEs (LINE1 and LINE2) and SINEs are long and short interspersed elements, respectively. SINE/ALU and SINE/MIR (mammalian-wide interspersed repeat) are subgroups of SINEs. LTR/ERVL and LTR/ERV1 are endogenous retroviruses with long terminal repeats. "DNA" represents DNA transposons. Simple repeats and satellites are explained in the text. Data are based on analysis with RepeatMasker (http://www.repeatmasker.org/genomicDatasets/RMGenomicDatasets.html).

The most abundant repeats are **transposons**, DNA elements that during evolution are able to move around in the genome. **Retrotransposons** are transposons that use reverse transcription in their mode of propagation. Thus, a genomic retrotransposon DNA is first transcribed into RNA, the RNA is reverse transcribed into DNA, and this DNA is inserted into the genome at a site different from its original location. The two important classes of retrotransposons are LINEs and SINEs, long and short interspersed elements, respectively. A LINE encodes two proteins— one of them is the reverse transcriptase enzyme. LINEs are up to 7000 nt in size and are present in about one million copies in the genome. SINEs do not encode any protein. They are in a size range about 100–700 nt and are present in nearly two million copies. Among different subclasses of SINEs, one abundant category is the Alu element, present in about one million copies and constituting about 10% of the genome (**Figure 4.24**). Another class of retrotransposons includes LTR repeats. They are similar to LINEs and SINEs but contain long terminal repeats of retroviral origin.[4] Retrotransposons are distributed throughout the genome, but they are rare in exonic regions.

Another type of repeat in the human genome is one where a short motif, typically 1–13 nt in length, is repeated multiple times. Examples of such repeats are **simple sequence repeats** (SSRs), also referred to as **microsatellites**, found throughout the genome. **Satellites** are repeats whose average length is much larger than the SSRs, and they typically occupy centromeric and telomeric regions. With the exception of satellites, all repeat elements are more or less randomly distributed throughout the genome (**Figure 4.25**).

What are the biological functions of repeat elements? The transposable elements presumably played a role during evolution by reshaping the genome and creating new functions. Furthermore, SINEs have been implicated in transcriptional regulation, as they, for instance, are able to modify chromatin structure (Chapter 10). In addition, a number

---

[4]A retrovirus is a virus whose genome is an RNA molecule. The virus uses a reverse transcriptase enzyme to produce DNA from its RNA genome. This is a mechanism opposite that of normal transcription, hence the term *retro*. All retrotransposons are of retroviral origin.

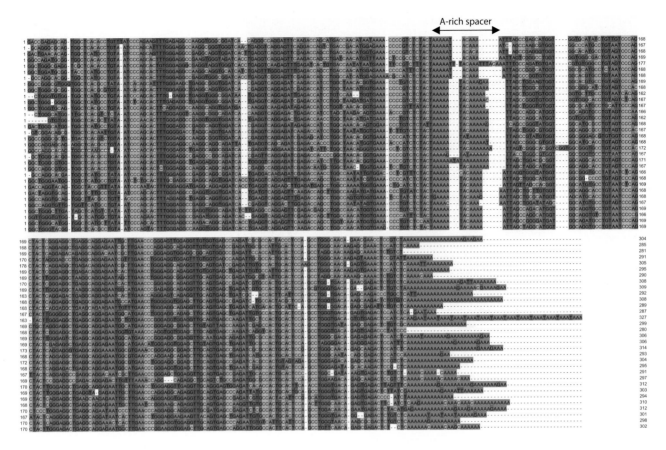

Figure 4.24 Alu sequence is an abundant class of repeat elements in the human genome. An alignment of a small selection of Alu element sequences is shown. The Alu sequence is from an evolutionary point of view derived by reverse transcription of a noncoding RNA, the RNA component of the signal recognition particle, a RNA-protein complex directing proteins to membranes. Two regions of this RNA are represented in the Alu sequence—the regions are separated by an A-rich sequence (marked with arrow). Note that the Alu sequences, just as the other transposable elements in the human genome, show a certain extent of sequence conservation but are nonidentical as a result of mutations. The color code of the nucleotides are A = green, T = blue, C = yellow, and G = red.

Figure 4.25 The human genome is dense with repeat elements. A region of chromosome 11 that includes a cluster of β-globin genes is shown. The repeats as shown were identified with the computational tool RepeatMasker (http://www.repeatmasker.org).

of disorders including cancer and neurological disorders are associated with retrotransposon insertions. One of the first disorders in this category to be identified is a form of hemophilia, caused by LINE insertion. The insertion disrupts the gene *F8* that encodes an essential blood coagulation factor.

A COMPLEX PROTEIN UNIVERSE

We have seen that the human genome encodes some 20,000 different proteins. What is the biological function of all of these proteins? Quite a few of them have been examined through the years, and their biological functions have been elucidated. The proteins may be classified in different ways, for instance with respect to molecular function (**Figure 4.26**). Examples of predominant classes of proteins are enzymes and transcription factors (proteins that regulate transcription, see also Chapter 10). However, while many human proteins have been characterized experimentally, we are still lacking functional information for as much as about 25% of all human proteins.

Whereas experimental studies are important to examine the biological function of proteins, functions may also be predicted from their amino acid sequences using computational tools. There are, for instance, tools that are

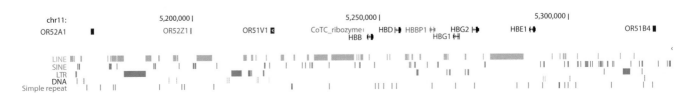

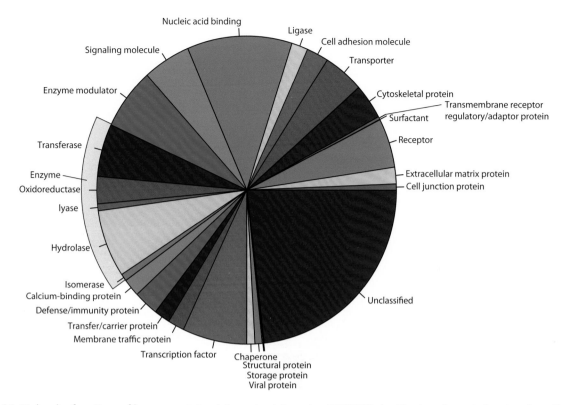

Figure 4.26 Molecular functions of human proteins. Information is based on PANTHER classification of protein functions (http://www. pantherdb.org/).

able to identify distinct structural and functional units of proteins. There are also specialized tools, such as those that allow prediction of what proteins are part of cellular membranes. Such methods inform us that about 26% of all human proteins are integrated into membranes. (For more on computational tools, see Chapter 12.)

Knowing that 20,000 different proteins may be formed in a human cell and that each of these proteins has a specific task, this would seem complex enough. But things are more involved. Each of the 20,000 proteins is not independent of each other. Different types of complex protein networks show how proteins are connected, either by function or by physical interaction. As one example, many proteins are enzymes, and each of these is responsible for a specific chemical reaction in the cell. Complex graphs show these reactions. (For metabolic pathways see **Figure 4.27**.) Regulatory mechanisms that influence the rate of individual reactions add a level of complexity. As an example of other complex protein networks, there are signaling pathways initiated by hormones, whereby the binding of an hormone to a receptor located on the surface of the cell may give rise to a complex response in the cell, involving a series of regulated steps.

One effect of complex regulatory networks of the cell is that a small disturbance in one step may have results that are difficult to predict. This is one reason why the functional consequence of a genetic defect in a protein gene is often difficult to predict at a molecular level.

SUMMARY

- The major part of the human genome is encoded by DNA located in the cell nucleus. In addition, mitochondria contain DNA.

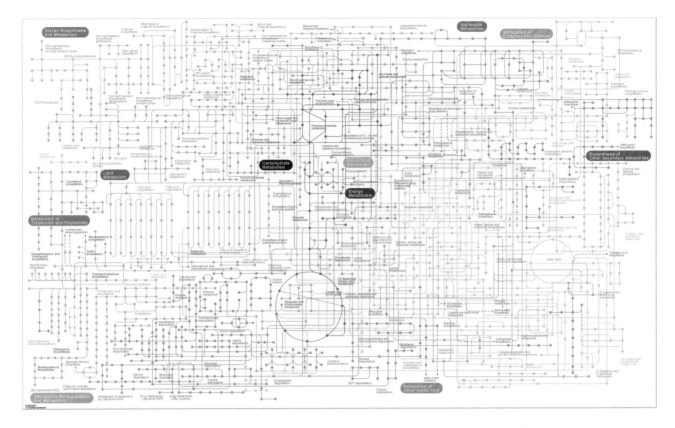

Figure 4.27 Human metabolic pathways. There are about 5,000 different enzymes in a human cell, each catalyzing a specific chemical reaction. A fraction of the reactions involved in metabolism is shown schematically here, where each node represents a substrate or product in an enzymatic reaction. (From Kanehisa M et al. 2017. *Nucleic Acids Res* 45, D353–D361. With permission from KEGG.)

- DNA is highly compacted within the nucleus. One level of packing is through nucleosomes, where DNA is wrapped around histone proteins.
- Human chromosomes may be visualized with staining and light microscopy, thereby revealing any major chromosomal aberrations.
- Cloning and PCR are widely used methods when working in the laboratory with DNA. The object of these methods is to produce a specific DNA fragment in large quantities, allowing subsequent analysis like DNA sequencing.
- A commonly used protocol for DNA sequencing was developed by Frederick Sanger, a protocol where DNA is replicated in the test tube and where specific nucleotide derivatives are used to terminate replication.
- The first draft of the human genome was presented in the early 2000s.
- The mitochondrial DNA genome is small and codes for only a few proteins and RNA. Mitochondrial DNA is maternally inherited.
- The human genome contains genes encoding proteins and noncoding RNAs. These genes are typically mosaics of exons and introns.
- Only about 1.6% of the genome codes for protein.
- Nearly half of the human genome is made up of repeat elements, mainly retrotransposons.
- The human genome encodes some 20,000 different proteins. These proteins form complex networks, where proteins are associated either by function or physical interaction.

QUESTIONS

1. What is a *genome*?
2. Where in the cell is DNA found?
3. What is an *autosome*?

4. What is the relationship between two *homologous chromosomes*?

5. How can two different types of chromosomes (like chromosomes 1 and 2) be distinguished using the microscope? What types of changes in chromosome structure may be identified using the microscope?

6. Describe the structure of a *nucleosome*.

7. What is meant by *molecular cloning*?

8. Describe the procedure of a *PCR*. What are the different molecules required in a PCR? What could be the practical advantages of PCR over molecular cloning?

9. What is characteristic of the Sanger sequencing protocol?

10. Discuss the advantages of whole genome *shotgun sequencing* as compared to older methods where a physical map is first created. Are there disadvantages?

11. What is the meant by the *reference* version of the human genome?

12. In what respects is *mitochondrial* DNA different from the DNA in nuclear chromosomes?

13. What is meant by a *noncoding RNA* (ncRNA)?

14. Approximately how many proteins and ncRNAs are encoded in the human genome? What fraction of the human genome do the protein and ncRNA exons make up?

15. Draw a schematic cartoon of the *HBB* gene, showing exons and introns. Also show how the exons are combined during splicing to form the mature mRNA.

16. What is the mutation in DNA responsible for sickle cell anemia? What is the normal codon, and what is the codon caused by the mutation?

17. If we say that a gene has a specific *direction* or *polarity* in the genome, what is meant by that?

18. What is characteristic of a *pseudogene*?

19. What are the quantitatively most important *repeat elements* of the human genome? What are their possible functions?

20. Give examples of important functional classes of proteins. What are the most predominant groups in terms of function?

URLs

UCSC Genome Browser (http://genome.ucsc.edu/)

Ensembl (www.ensembl.org, www.ebi.ac.uk)

NCBI (ncbi.nlm.nih.gov)

HUGO Gene Nomenclature Committee (www.genenames.org)

RepeatMasker (www.repeatmasker.org)

KEGG (Kyoto Encyclopedia of Genes and Genomes) database (www.genome.jp/kegg)

FURTHER READING

Compaction of DNA

Hansen JC. 2012. Human mitotic chromosome structure: What happened to the 30-nm fibre? *EMBO J* 31(7):1621–1623.

Nishino Y, Eltsov M, Joti Y et al. 2012. Human mitotic chromosomes consist predominantly of irregularly folded nucleosome fibres without a 30-nm chromatin structure. *EMBO J* 31(7):1644–1653.

Ou HD, Phan S, Deerinck TJ et al. 2017. ChromEMT: Visualizing 3D chromatin structure and compaction in interphase and mitotic cells. *Science* 357(6349):eaag0025.

Invention of PCR

Mullis K, Faloona F, Scharf S et al. 1986. Specific enzymatic amplification of DNA in vitro: The polymerase chain reaction. *Cold Spring Harb Symp Quant Biol* 51(Pt 1):263–273.

Mullis KB. 1998. *Dancing naked in the mind field*. Pantheon Books, New York.

Saiki RK, Gelfand DH, Stoffel S et al. 1988. Primer-directed enzymatic amplification of DNA with a thermostable DNA polymerase. *Science* 239(4839):487–491.

Sanger sequencing

Sanger F, Nicklen S, Coulson AR. 1977. DNA sequencing with chain-terminating inhibitors. *Proc Natl Acad Sci USA* 74(12):5463–5467.

Sequencing of phiX genome

Sanger F, Air GM, Barrell BG et al. 1977. Nucleotide sequence of bacteriophage phi X174 DNA. *Nature* 265(5596):687–695.

Sequencing of *Mycoplasma genitalium* genome

Fraser CM, Gocayne JD, White O et al. 1995. The minimal gene complement of *Mycoplasma genitalium*. *Science* 270(5235):397–403.

Sequencing of *Haemophilus influenzae* genome

Fleischmann RD, Adams MD, White O et al. 1995. Whole-genome random sequencing and assembly of *Haemophilus influenzae* Rd. *Science* 269(5223):496–512.

First versions of the human genome

Lander ES, Linton LM, Birren B et al. 2001. Initial sequencing and analysis of the human genome. *Nature* 409(6822)0:860–921.

Venter JC, Adams MD, Myers EW et al. 2001. The sequence of the human genome. *Science* 291(5507):1304–1351.

Population-specific regions of the human genome

Ameur A, Che H, Martin M et al. 2018. *De novo* assembly of two Swedish genomes reveals missing segments from the human GRCh38 reference and improves variant calling of population-scale sequencing data. *bioRxiv* 267062. doi: https://doi.org/10.1101/267062.

UCSC genome browser

Kent WJ, Sugnet CW, Furey TS et al. 2002. The human genome browser at UCSC. *Genome Res* 12(6):996–1006.

Exon and intron statistics

Sakharkar MK, Chow VT, Kangueane P. 2004. Distributions of exons and introns in the human genome. *In Silico Biol* 4(4):387–393.

Prediction of the human membrane proteins

Fagerberg L, Jonasson K, von Heijne G et al. 2010. Prediction of the human membrane proteome. *Proteomics* 10(6):1141–1149.

Hemophilia A and insertion of LINE1 elements

Kazazian HH, Wong C, Youssoufian H et al. 1988. Haemophilia A resulting from de novo insertion of L1 sequences represents a novel mechanism for mutation in man. *Nature* 332(6160):164–166.

Mobile elements and disease

Solyom S, Kazazian HH, Jr. 2012. Mobile elements in the human genome: Implications for disease. *Genome Medicine* 4(2):12.

Panther classification of proteins

Mi H, Huang X, Muruganujan A et al. 2017. PANTHER version 11: Expanded annotation data from Gene Ontology and Reactome pathways, and data analysis tool enhancements. *Nucleic Acids Res* 45(D1):D183–D189.

Variants in the Human Genome Sequence and Their Biological Significance

5

This chapter covers some more background as to the human genome, before we go on to discuss the role of the protein coding sequences in Chapter 6.

The genomes of humans and all other living organisms are not static over time but are subject to changes. We refer to such alterations of the nucleotide sequence as **mutations** or **variants**. Mutations are necessary for evolution to proceed, but sometimes they are harmful. One example of such a mutation is the A to T nucleotide substitution associated with sickle cell anemia. Throughout this book, we examine the effect of mutations in the human genome, because they give important clues as to the normal function of different sequence elements.

In this chapter, we look into the consequences of mutations in the human genome. Human individuals are genetically different, and we need to understand the nature of these differences. Next, we see how the human genome sequence is related to disorders and traits. We also discuss cancer, one important category of serious diseases with a genetic background. But first we examine some general aspects of mutations and their molecular causes.

TYPES OF MUTATIONS

Mutations involve one or a few nucleotides only or they may involve larger regions (say 1,000 nucleotides or more) of DNA sequence. The mechanisms by which these two categories of mutations arise are different. In this chapter, we focus on the smaller changes that may be classified into **substitution** mutations (or **point** mutations), where one nucleotide is replaced with another, and **insertions/deletions**, also known as **indels** where a small number of nucleotides (mean length about 10, median 6) are inserted or deleted as compared to the original sequence. Throughout this book a mutation is represented with an expression like "A > G," meaning that an A has changed into G. The reader should be aware that a mutation often first appears on one of the DNA strands, but after replication of DNA the mutation is observed on both strands. For instance, for the "A > G" mutation, an A•T base pair has been replaced by a G•C base pair.

GERMLINE AND SOMATIC MUTATIONS

Mutations have different effects depending on when and where they come about. When mutations occur in sperm or egg cells, they may be carried over from one generation to the next. Also mutations that occur in the fertilized egg at a very early stage of embryonic development may be carried over to the next generation, although they are not inherited from any of the parents. We refer to all these events as **germline** mutations. Some of them may give rise to inherited disorders. A mutation is **somatic** when it occurs in any other cell, such as a liver, muscle, or nerve cell. Thereby, it is not passed on to the next generation, it is typically restricted to a single type of cells or tissue, and it will not be propagated to all cells of the adult body. However, somatic mutations are nevertheless a considerable threat to human health as they may cause diseases such as cancer.

MUTATION RATE IN THE HUMAN GENOME

It is estimated that in the human germline about 0.3 mutations occur per genome replication (or per cell division). However, as compared to the egg, sperm cells divide many times in the development from zygote to sperm cell. This has the effect that there are an estimated 30–200 mutations in every new generation and that most mutations in the germline that arise from errors during DNA replication are inherited from the father. In addition, the number of mutations increases with paternal age. Children to 40-year-old fathers carry twice as many substitution and indel mutations compared with children of 20-year-old fathers. Maternal age effects are less pronounced for substitution mutations, but in the egg there are instead errors during chromosome separation during cell division in meiosis that give rise to abnormal chromosome numbers. For instance, there is an association of maternal age and trisomy 21 (Down syndrome).

There are studies of the mutation rate in somatic cells that suggest it is one order of magnitude higher than in germline cells, although it is not clear what causes this difference. A significant mutation rate in somatic cells implies that the cells in our body are genetically diverse. Somatic mutations are intimately linked to cancer, as discussed later in this chapter. However, there are also other examples of diseases caused by somatic mutations. One example is Proteus syndrome, a rare disorder. It is characterized by overgrowth of bones, skin, and other tissues. The organs affected grow out of proportion to the rest of the body. Proteus syndrome is caused by a mutation that changes a lysine to glutamine in AKT1 kinase, an enzyme mediating processes such as cell proliferation and **apoptosis** (programmed cell death). It is an activating mutation, stimulating a signalling pathway important for the cell cycle. The play and film *The Elephant Man* tells the story of Joseph Merrick (**Figure 5.1**), an Englishman who lived in the late nineteenth century. It has been suggested that Merrick had Proteus syndrome, although it has also been put forward that he

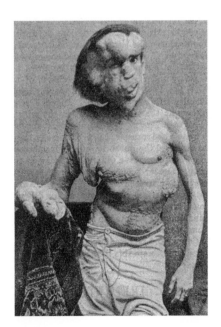

Figure 5.1 *Joseph Merrick.* Photograph of 1889.

suffered from Neurofibromatosis type I (NF-1), yet another disease caused by somatic mutation.

HOW DO MUTATIONS COME ABOUT?

Mutations are critical for evolution to proceed, but a disadvantage of this type of unstable genetic system is that some mutations have a bad effect on the organism. For instance, a long range of inherited disorders as discussed throughout this book is a result of mutations.

What are the mechanisms that are responsible for substitution mutations and indels? There are a number of them to consider. First, mutations occur naturally as a result of normal cellular biochemical activity. Examples are errors during DNA replication and spontaneous chemical changes in DNA. Second, errors may arise from external influence, such as irradiation and chemicals. Damage to DNA is common—in a human cell tens of thousands of DNA damages occur in one day. These would be highly vulnerable had it not been for mechanisms of DNA repair.

DNA REPLICATION ERRORS

The first category of errors we consider are those that arise during DNA replication. The fidelity during DNA replication is high—it has been estimated that the error frequency is as low as 1×10^{-9}. What is the source of the errors? One spontaneous mechanism during replication is **tautomerism** where a base may exist in two different isomeric forms with different base-pairing properties. For instance, a rare form of adenine can pair with cytosine. In this manner cytosine may be misread during replication, eventually resulting in a G•C pair replaced with an A•T pair, a substitution mutation (**Figure 5.2**).

Another error during DNA replication is the incorporation of ribonucleotides instead of deoxyribonucleotides. This misincorporation will negatively influence replication and transcription of DNA, and it will reduce the chemical stability of DNA.

There is also an error mechanism during replication whereby the length of short repeats in DNA is altered. This mechanism will be further discussed in the context of disorders resulting from this type of error (Chapter 7).

SPONTANEOUS DNA DAMAGE

DNA may be damaged directly by reactions occurring spontaneously in the cell. These reactions are caused by chemical instability of bases or bonds in DNA. One such event is **depurination**, where spontaneous cleavage of the glycosidic bond between guanine or adenine and deoxyribose leads to the loss of purine bases from DNA. The result is an **apurinic site** (**Figure 5.3**).

Another spontaneous modification of DNA is the deamination of cytosine to produce uracil. During DNA replication uracil is recognized by A, so that ultimately a G•C pair will be replaced with an A•T pair. There is a DNA repair pathway to correct the deamination error, but in case it fails, the deamination will result in a mutation (**Figure 5.4**).

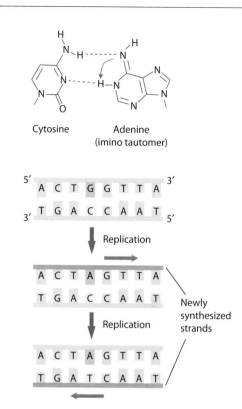

Cytosine   Adenine
                  (imino tautomer)

Figure 5.2 *C•A mispairing error.* A tautomer of adenine may be formed that can pair with cytosine. Once the adenine is incorporated into the DNA chain, it pairs with thymine in the next round of replication. The net result is that a G•C pair is replaced with an A•T pair. Horizontal arrows show direction of DNA replication.

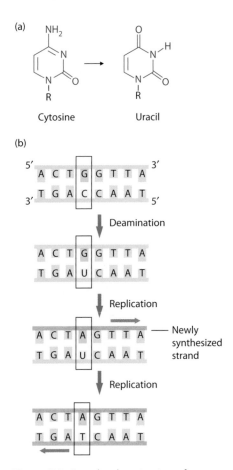

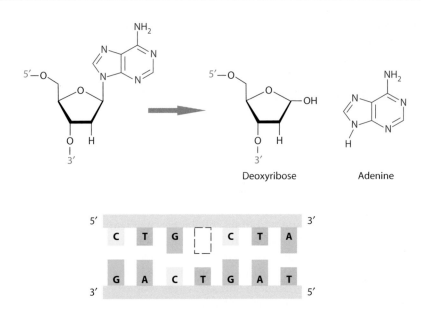

Figure 5.3 *A depurination reaction.* In this example adenine is removed by cleavage of the glycosidic bond to deoxyribose. The result is an apurinic site.

Figure 5.4 *Error by deamination of cytosine.* (a) Deamination of cytosine to uracil. A site of deoxyribose attachment is shown with "R." (b) After a cytosine has been deaminated to a uracil, the uracil will pair with an adenine in the following round of replication. The final outcome is that a G•C pair has been replaced with an A•T pair. Newly synthesized DNA strands are shown in orange, and horizontal arrows indicate direction of DNA replication.

DNA DAMAGE BY REACTIVE OXYGEN SPECIES AND BY IRRADIATION

A number of changes in DNA are caused by chemicals. Many reactions that damage DNA are a result of reactive oxygen species (ROS). These are derived from molecular oxygen (O_2) and include hydrogen peroxide and OH radicals. Many ROS are generated endogenously, but they may also be supplied exogenously—examples include tobacco and irradiation. ROS give rise to single-strand breaks and changes to bases, such as the modification of guanine to 8-oxo-guanine (**Figure 5.5**). The guanine modification can result in mispairing with adenine, giving rise to a replacement of G•C with A•T base pairs.

Ultraviolet (UV) light or ionizing radiation is another cause of DNA damage. A common aberration of this sort is when two adjacent pyrimidines in one strand of DNA are covalently joined to form pyrimidine dimers (**Figure 5.6**). The most common dimers of this type are thymine dimers. Many of the eukaryotic DNA polymerases have difficulties replicating DNA with pyrimidine dimers.

Figure 5.5 *Modification of guanine to 8-oxo derivative.* Guanine may be modified by reactive oxygen species (ROS) to 8-oxo-guanine (also known as 8-hydroxyguanine).

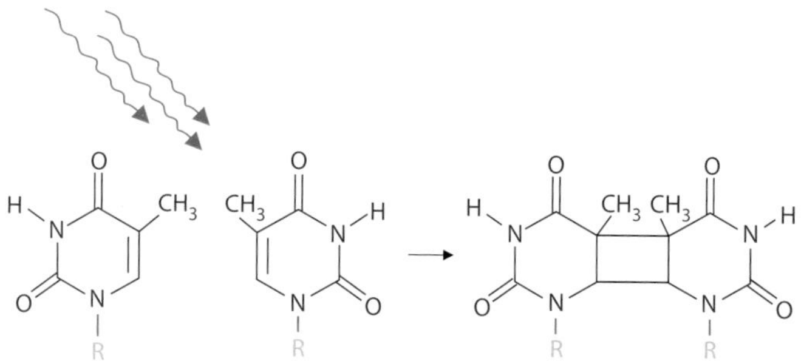

Figure 5.6 *Formation of a pyrimidine dimer.* UV irradiation may cause two adjacent pyrimidines in one strand of DNA to be covalently joined to form a pyrimidine dimer. Shown here is an example where two adjacent thymines are joined to form a cyclobutane dimer. A site of deoxyribose attachment is shown with "R."

HOW ARE DNA ERRORS REPAIRED?

DNA damage would pose a serious threat to the cell, had it not been for DNA repair mechanisms. There are three different mechanisms that are all able to reconstruct the correct sequence by replication of the (correct) complementary template strand (**Figure 5.7**). For instance, in cases where a base is damaged (such as when uracil is formed from cytosine), it may be removed by specific repair enzymes—DNA glycosylases—in a process called **base excision repair**. In **nucleotide excision repair**, bulky groups like pyrimidine dimers are removed in 12–24 nucleotide long strands. In **mismatch repair**, incorrectly paired bases are removed.

While DNA repair mechanisms typically provide efficient repair of DNA, they are also a source of mutations in the instances where they do not perfectly restore the correct base sequence. We consider here two different repair mechanisms that are prone to errors. First, **translesion synthesis** is a mechanism of DNA repair that allows synthesis of past sites of DNA damage such as pyrimidine dimers or sites where a purine or pyrimidine has been lost (apurinic or apyrimidinic sites, respectively). Specialized DNA polymerases are used to incorporate nucleotides in a region containing the damage. Some of these enzymes are able to correct the error, but others incorporate an incorrect base (**Figure 5.8**).

A second error-prone repair mechanism is one that is responsible for repair of double-strand DNA breaks. Such breaks can arise when single-strand breaks occur on both strands in close proximity, for instance in the context of nucleotide excision repair or as a result of irradiation.

Double-strand breaks may be repaired using three different pathways. There are two pathways that use the correct information on the complementary strand for repair—these methods are known as microhomology-mediated end joining (MMEJ) and **homologous recombination.**[1] And, there is nonhomologous end joining (NHEJ)—this mechanism is particularly error prone and will typically generate indel mutations (**Figure 5.9**).

---

[1]Homologous recombination is an event where sequences are exchanged between two similar DNA molecules. It is used not only to repair double-strand breaks but also during meiosis to create genetic variation.

Figure 5.7 *Three mechanisms of DNA repair.* (a) In base excision repair, a damaged base is removed followed by endonuclease cleavage and insertion of the correct base with DNA polymerase. (b) Nucleotide excision repair is important for removal of bulky groups like pyrimidine dimers. These are removed in 12–24 nucleotide long strands, and the gap is filled in with DNA polymerase. (c) Mismatch repair mechanisms are responsible for correcting errors that arise by mispairing during DNA replication. Mismatch repair is strand specific. It is believed that in eukaryotes the newly synthesized strand is distinguished from the other strand as it has unsealed nicks. After removal of a piece of the strand with the error, DNA polymerase synthesizes the correct sequence.

(a)

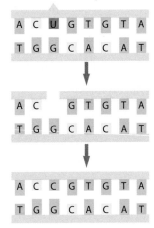

(b)

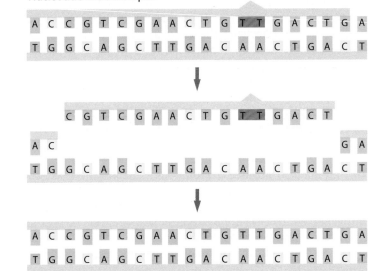

(c)

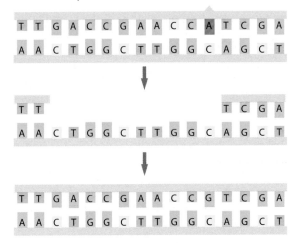

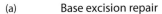

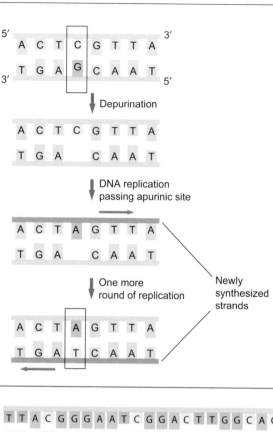

Figure 5.8 *Repair of apurinic site.* Translesion synthesis is a repair mechanism that will allow DNA synthesis past lesions such as a site where a purine has been lost. Enzymes active in translesion synthesis will sometimes incorporate an incorrect base. The overall result in this example is that a C•G base pair will be replaced with an A•T base pair. Newly synthesized DNA strands are shown in orange, and horizontal arrows indicate direction of DNA replication.

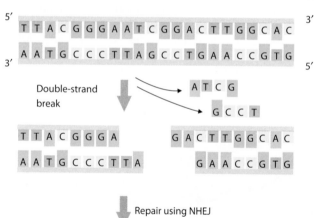

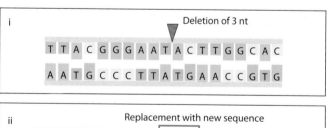

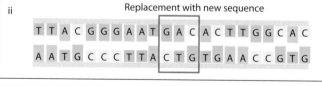

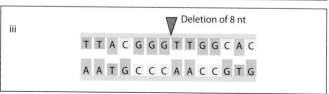

Figure 5.9 *Repair using nonhomologous end joining generates multiple erroneous products.* A double-strand break in DNA like shown on top may be repaired with different mechanisms. Depicted here is NHEJ. A number of different proteins, such as Ku70, DNA-PKcs, nucleases, DNA polymerases, and DNA ligase IV, take part in the repair pathway. There is a large number of possible erroneous products. Shown in boxes labeled i, ii, and iii are three examples—two deletions and one change where one triplet is replaced with another.

HUMAN GENETIC VARIATION

One effect of mutations is that human individuals are genetically different. We now look more closely into the nature of this particular variation. We know a great deal about it from analyzes of many different human genomes.

During the years that have passed since the publication of the first draft of the human genome, DNA sequencing technology has developed dramatically (**Figure 5.10**) as discussed in more detail in Chapter 13. A human genome can be sequenced within days. Furthermore, whereas the first human genome cost about US$3 billion to sequence, the price has now been reduced several orders of magnitude (**Figure 5.11**). This has made possible large-scale sequencing such that more than a million genomes now have been analyzed. And there is more to come. As one example, there are plans for the National Health Service in England to sequence 5 million genomes over a 5 year period. There are a number of important applications to all of this sequencing. It can be used to address a lot of basic scientific questions, such as those related to human evolution. Genome sequencing may also identify individual differences of medical and pharmacological importance. For instance, we see that there are human genetic variants that influence how we react to a certain drug. Then again, there are variants that cause disease or give rise to an altered risk for specific diseases.

Comparing one human individual to another, what are the differences in the genetic makeup? First, we are all similar. Taking into account substitution mutations for instance, we are about 99.9% identical. As to the differences, we consider two major categories. First, there are small changes affecting only one or a few nucleotides, substitutions, and small indels. Second, there are differences involving larger pieces of DNA, referred to as **structural variation.** We begin by considering the smaller changes.

Figure 5.10 *A handheld DNA sequencing device.* An example of remarkable technical development within the area of DNA sequencing is this portable device named MinION, developed by Oxford Nanopore Technologies. (Image courtesy of Oxford Nanopore Technologies.)

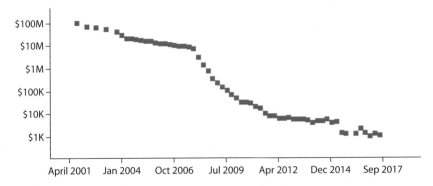

Figure 5.11 *A dramatic drop in the cost of DNA sequencing.* The cost of sequencing a human genome in US dollars. (Data are from the National Human Genome Research Institute.)

INDIVIDUAL DIFFERENCES: SINGLE NUCLEOTIDE POLYMORPHISMS AND SINGLE NUCLEOTIDE VARIANTS

The sites where human individuals are different are known as sites of **poly-morphism**. A **single nucleotide polymorphism** (**SNP**, pronounced "snip") is a variation in a single nucleotide position in the genome. For instance, one individual may have an A in a specific position in the genome, whereas another individual has a G in that position. Most variations are simple nucleotide substitutions but in about 10% of the cases insertion or deletion of a small number of nucleotides (indels) (**Figure 5.12**). Mutations responsible for these types of individual variations arise by mechanisms that were previously discussed.

For a variation to qualify as a SNP, it should be present in a significant fraction in a human population. Thus, for a variation in a genomic position to be classified as a SNP, more than 1% of a population does not carry the same nucleotide in that position.

When we consider just any variant, regardless of frequency, we refer to it as a **single nucleotide variant** (**SNV**). The majority of variants are in fact rare. SNVs are more or less randomly distributed over the human

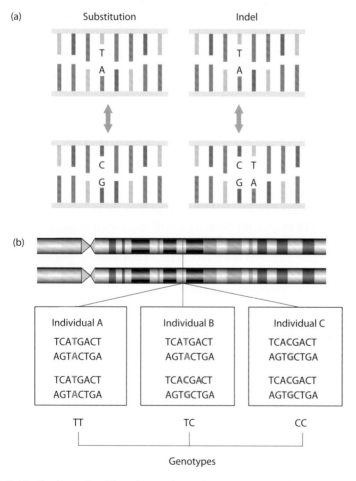

Figure 5.12 *Single nucleotide polymorphism (SNP).* (a) Two examples are shown of DNA variants—one substitution (base pair T•A as compared to C•G) and one indel (T•A as compared to C•G/T•A). (b) The relationships of SNPs to the diploid nature of the human genome are illustrated. Two homologous chromosomes are shown. In the boxes are shown examples of genotypes for a SNP site in three different individuals. In each box are shown the two different alleles. With respect to the SNP shown, three genotypes are possible, TT, TC, and CC.

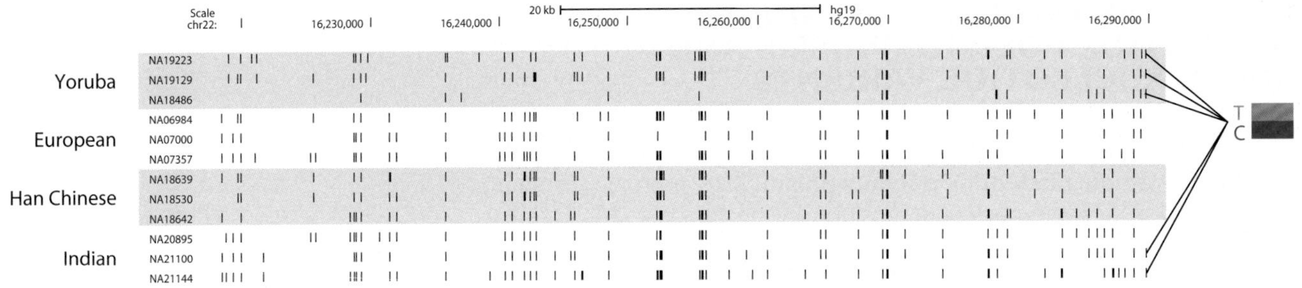

Figure 5.13 *Genomic landscape of SNVs.* SNVs are randomly distributed as illustrated with a region of chromosome 22. Twelve different anonymous individuals are compared with respect to SNVs. Bars represent situations where an individual has at least one allele not identical to the allele of the reference genome. To the right is shown for one of the SNVs more details of the genotype. The reference genome has a T in this genomic position. For the individuals indicated (Yoruba and Indian), the genotype is TC. For the remaining individuals the genotype is TT. "European" are Utah residents with Northern and Western European Ancestry, and "Indian" represents Gujarati Indian from Houston, Texas. (Based on information from the "1000 Genomes Project," a large-scale project to examine human genetic variation, run between 2008 and 2015.)

genome (a portion of chromosome 22 with SNVs is shown in **Figure 5.13**). Exceptions are, for instance, at the ends of chromosomes and at the HLA locus (a locus holding some of the genetic variation of the immune system), where variants are more common.

The database dbSNP of the National Center of Biotechnology Information (NCBI) collects SNP and SNV information. As more and more human individuals are analyzed, the number of SNVs increases. In 2018, the dbSNP had nearly 700 million human SNVs. About 37 million of these correspond to common variants, present in more than 1% of a population, and are therefore classified as SNPs. We should therefore expect a SNP on average about every 100 positions in the human genome, whereas SNVs are encountered about 20 times more frequently.[2] The number of SNV differences between two unrelated individuals is typically about 5 million.

As referred to previously, the majority of SNVs are rare—in fact most SNVs occur only in one individual or within a single family. For instance, in one study 202 genes in 14,002 people were sequenced. More than 74% of the variants identified were carried by only one or two of the individuals.

SNVs may be identified with genome sequencing, but there are also other less expensive methods, like **SNP arrays**, that directly monitor a certain set of SNPs in an individual. The SNP array technology makes use of a chip with immobilized short DNA sequences, each corresponding to a certain SNP. When this chip is subjected to a DNA sample from an individual, all perfectly matching sample sequences will give rise to a signal based on base-pairing to the chip sequences.

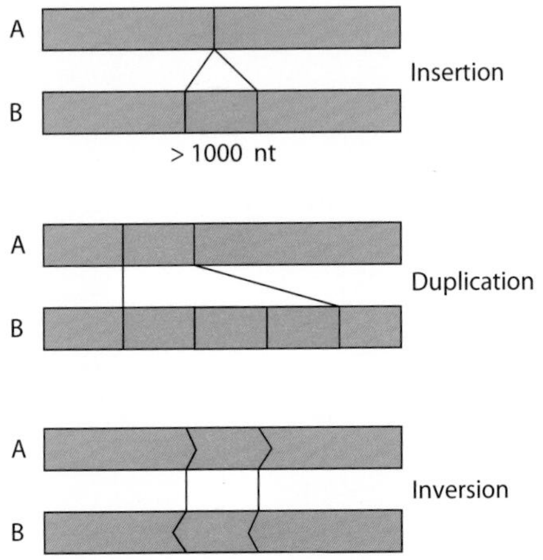

INDIVIDUAL DIFFERENCES: STRUCTURAL VARIATION

Another category of individual genetic variation is structural variation, involving a much longer piece of nucleotide sequence. There is no well-defined lower size limit for a variation to qualify as structural variation, but typically this is the limit set by the methods used to identify SNVs, in which case a too large indel (larger than about 50 nucleotides) cannot be identified. Many structural variations involve segments of DNA larger than 1,000 nucleotides. Structural variation may be the result of deletions, duplications, insertions, or inversions (**Figure 5.14**). Inversions arise when a piece of DNA appears

Figure 5.14 *Examples of structural variation.* Two individuals A and B are compared. Insertion, duplication, and inversion events are illustrated. An inversion arises when a piece of DNA changes its orientation relative to its genomic context.

[2]It should be noted that we expect to see the number of SNVs to increase as more and more people are sequenced. In the end, virtually every position of the genome may be covered by an SNV.

in a reverse orientation. In the case of deletions, insertions, and duplications, these events give rise to **copy number variation** (**CNV**). A common case of CNV is when portions of the genome are repeated because of duplication events, and the number of repeats in the genome varies between individuals. The molecular background to a CNV is often homologous recombination, an exchange of genetic material that occurs, for instance, during the formation of germline cells. A CNV difference between two individuals may be revealed as a genomic region where the number of DNA sequencing reads that map to this region is markedly different in the two individuals.

Other types of structural variation are changes in chromosome number and **chromosomal translocations**. Translocations are rearrangements where a portion of one chromosome is moved to another nonhomologous chromosome. Chromosomal translocations are quite frequent during meiosis—they occur in about 1 of 600 human newborns. Sometimes translocations give rise to diseases such as several forms of cancer (see section Cancer—Translocations and More Dramatic Genomic Aberrations) and Down syndrome. However, most translocations are associated with a normal phenotype because they do not disturb the organization and expression of individual genes.

How common is structural variation? It has been estimated that the total amount of sequences of the human genome involved in structural variation corresponds to 5%–10% of the genome. Thus, the nucleotide content of structural variation is much larger than that of SNVs. When comparing two unrelated individuals, the fraction of the genome that is different in terms of structural variation is in the order of 1%. Structural variation is not as well studied as SNVs. For elucidation of such variation, we require a careful assembly of the human genome from sequencing reads. However, as this is technically demanding, the most common method of reconstructing a genome is by **alignment** of sequencing reads to the reference genome. The alignment method has the disadvantage that it is far from ideal for identifying structural variation.

Figure 5.15 illustrates the differences observed in human individuals with respect to structural variation. As for SNVs, structural variation is distributed throughout the genome. We also know that there is a substantial structural variation, such as CNVs, between different tissues in a human individual. Even between monozygotic twins there are CNV differences.

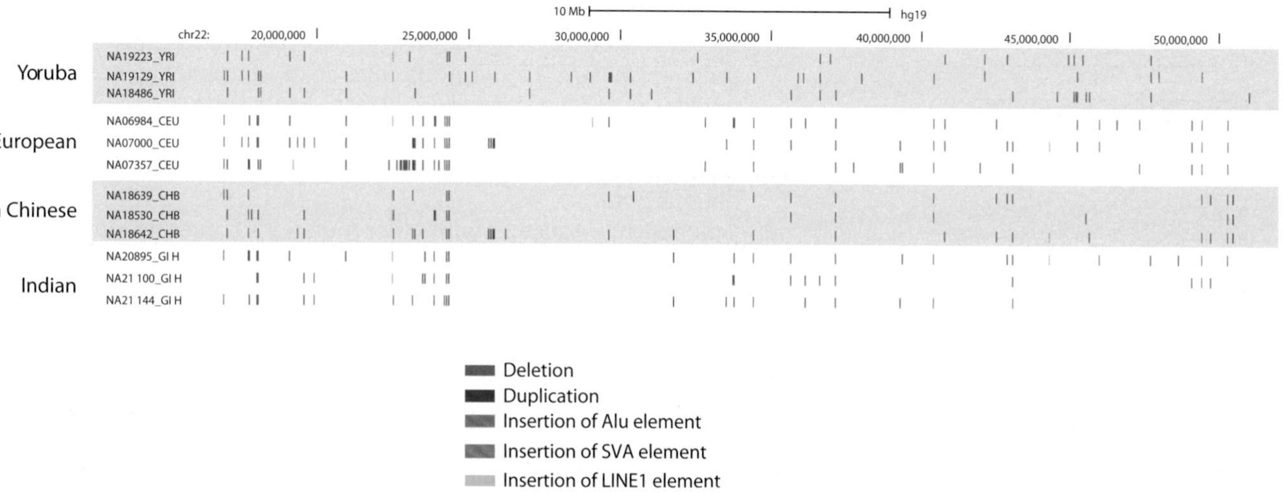

Figure 5.15 *Genomic landscape of structural variation.* The same individuals as seen in Figure 5.13 are shown but for a larger genomic region of chromosome 22. Three of the structural variants, Alu, LINE1, and SVA elements, are insertions of retrotransposons. The total numbers of the five different classes of structural variants shown in this region of the chromosome are deletions (214), duplications (48), ALU element insertions (85), LINE1 insertions (12), and SVA insertions (18). (Based on information from the 1000 Genomes Project.)

PHENOTYPIC IMPLICATIONS OF INDIVIDUAL VARIATION: SINGLE GENE DISORDERS AND COMPLEX DISEASES

After having introduced SNVs and structural variation, we now turn to their functional significance. The variants have different effects, depending on the nature of the genetic change and on its location in the human genome.

Remember that the **genotype** is the genetic makeup (the DNA nucleotide sequence) of an individual, whereas we refer to observable effects and consequences of the DNA sequence as the **phenotype**. Human diseases, as well as **traits** like eye and hair color, are all examples of phenotypes.

Some variants will give rise to disease. What are the diseases with a genetic background? Throughout this book, we will mostly be addressing **single gene** or **monogenic disorders**, as they most clearly illustrate the relationship of a distinct DNA sequence and the phenotype. These disorders are all the result of a mutation in a single gene. They are also often referred to as **Mendelian disorders** because of their inheritance pattern (Chapter 2). How many such disorders are there? The database Online Mendelian Inheritance in Man (OMIM) is in 2018 listing a total of 6,905 single gene disorders—out of these, 5,328 have a known molecular background, that is, we know what gene is affected. However, genomes associated with inherited disorders are right now being analyzed at a high rate. It therefore seems likely that a lot more single gene disorders and their molecular backgrounds will be revealed in the near future.

Many disorders with a genetic component are not single gene disorders. A **complex genetic** or **multifactorial disorder** is caused by genetic changes in a number of different genes. In addition, there are typically environmental and lifestyle factors that are associated with a complex disorder. Many common human disorders are of this nature, such as type II diabetes, schizophrenia, and Alzheimer's disease. For a lot of the complex disorders, a large number of genes have been implicated. For instance, schizophrenia involves more than 200 genes. Another disorder with a distinct genetic background is cancer—to be covered in more detail later in this chapter.

What are the possible phenotypic effects of a variant in the genome? We may list the following, in addition to the common *no observable or known effect*. A variant could:

1. Be responsible for a nondisease phenotypic trait. Examples of human traits are height, blood group, eye color, and intelligence. A limited number of traits are determined by a single gene, whereas a majority of them involve multiple genes.
2. Be responsible for a single gene disorder. Examples are found throughout this book.
3. On its own or together with other mutations confer an altered risk for a complex genetic disorder, including cancer.
4. Influence sensitivity to a pharmacological drug.

When we are to understand the phenotypic consequences of a genetic change, we also need to take into consideration that humans are diploid organisms (Chapters 1 and 2). We previously considered the inheritance of the sickle cell anemia globin gene. Individuals homozygous for the sickle cell allele express sickle cell anemia, whereas heterozygotes express the trait, with blood cells producing approximately equal amounts of normal and variant β-globin. The phenotypic effects of a mutation are further illustrated in **Figure 5.16**. First, the outcome of a mutation depends on whether

(a)

(b)

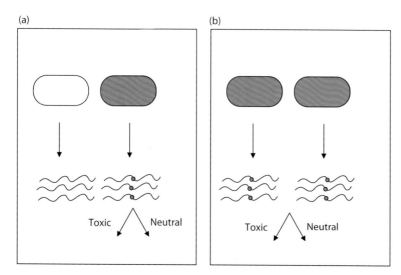

Figure 5.16 *Mutations and phenotypic scenarios.* Maternal and paternal copies of a gene are shown, with mutant genes in red. The effect of the mutation could at one extreme be dysfunctional or toxic to the cell. At another extreme the effect is neutral. (a) A scenario with one normal gene and one mutated gene. When there is a balanced expression approximately equal amounts of protein are produced from the two genes. The level of the normal protein is about half of that in a normal cell. The phenotype is determined by the effect of the mutation. (b) Both genes carry the mutation, and no normal protein is produced. As in (a), the phenotype is dependent on the properties of the mutant protein.

it is present in one or two copies (panels A or B). Second, the expression of a variant gene may have an effect on the phenotype (such as toxic effect, as in sickle cell anemia) or it may be neutral. It should also be kept in mind that for many variants, we do not have any information on the genotype/phenotype relationship.

PHENOTYPIC IMPLICATIONS OF SNPs MAY BE INFERRED FROM A GENOME-WIDE ASSOCIATION STUDY

SNPs are the genetic variants that have been most extensively studied with respect to their phenotypic implications. Different methods are available to elucidate the function of genomic variants. One popular method is a **genome-wide association study** (**GWAS**, pronounced "gee-was"). In such a study, SNPs (or SNVs) may be compared in two groups of human individuals—one with a disease or trait (case group) and one group without (control). The SNPs may be monitored using SNP arrays or with whole genome sequencing. For each of the SNPs, their frequencies in the case and control group are evaluated with a statistical test to calculate the likelihood that the frequency is significantly different between the case and control groups (**Figure 5.17**). In this way the *association* between SNP and disease/trait is estimated. The results of a GWAS are often shown as a **Manhattan plot** in which the genomic localization of the SNP is on the *x*-axis and the association value is on the *y*-axis. GWAS was used successfully the first time in 2005, when it was used to identify genetic factors related to a common eye disorder, age-related macular degeneration (**Figure 5.18**). In 2018, about 3,600 different GWAS studies were carried out that are in the public domain, and about 80,000 trait-associated SNVs in the genome have been mapped in these projects.

Figure 5.17 *Genome-wide association study.* An example of a GWAS is shown where SNPs are compared in two groups of human individuals—one with a disease or trait (case group) and one group without (control). (a) Each of the two matrices with circles represents 100 individuals monitored for a specific SNP, here labeled "SNP 1." The variant is present in the individuals in green, whereas the gray circles represent individuals that have the normal (reference) allele. The left matrix represents a set of individuals with a specific trait or disease. From this matrix we observe that 70% of the individuals have the SNP 1. To the matrix at the right is the corresponding result for a normal population—in this case the frequency of SNP 1 is 25%. (b) The experiment in (a) is repeated for approximately one million SNPs. The first six of these experiments are outlined here.

(a)

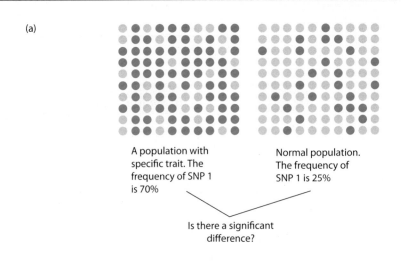

A population with specific trait. The frequency of SNP 1 is 70%

Normal population. The frequency of SNP 1 is 25%

Is there a significant difference?

(b)

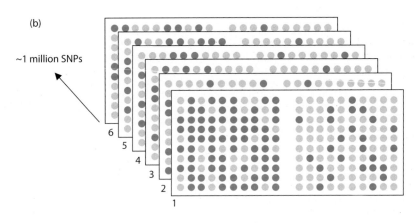

~1 million SNPs

Figure 5.18 *Manhattan plot.* The plot shows the results from a GWAS for the disorder age-related macular degeneration. The genomic localization of the SNP is on the x-axis, and the association level is on the y-axis. Two of the SNPs most strongly associated with the disease are rs10737680, located in an intron of gene *CFH*, and rs10490924, a silent mutation in an exon of the gene *ARMS2/HTRA1*. Age-related macular degeneration is a disorder that gives rise to damage of the macula, a part of the retina. As a result, there is a loss of central vision. It is a common disease, being the leading cause of vision loss in people 50 years or older.

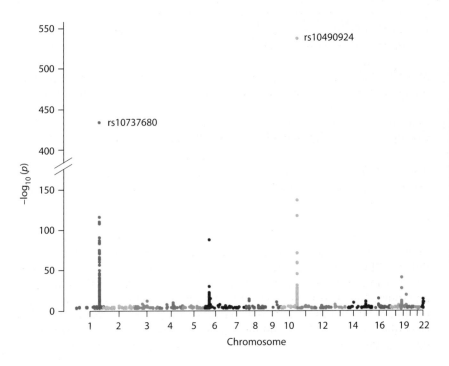

A RESTRICTED NUMBER OF VARIANTS HAVE A KNOWN DISTINCT FUNCTION

The large majority of SNVs do not have a specific phenotypic association. This may be because they are at positions in the genome where a mutation does not make a difference. Alternatively, we may yet have failed to reveal their importance. However, there are a limited number of SNVs that are associated with a known biological function. This association could be strong, as for the sickle cell anemia mutation, or it could be relatively weak, as for many variants identified in GWAS studies. The SNPedia database in 2018 listed about 109,000 SNVs/SNPs with an associated function. Some are responsible for monogenic disorders, some confer an increased risk of developing a specific disease, some are associated with disease resistance, and some are affecting the response to a specific drug. There are also variants associated with a phenotypic trait not related to a particular disease or drug response (**Table 5.1**). It should be noted that for many variants, like those examined by GWAS, they are associated with a particular disease, but their effect is relatively marginal. For instance, there is no cause for alarm if you learn that a certain variant you have increases the risk of developing a particular disorder with say 1%.

SNP tests may be carried out by a hospital to address a clinical question. But such tests are also commercially available (direct-to-consumer [DTC] testing) and may be ordered by an individual from companies like 23andMe, FamilyTreeDNA, and Ancestry. In 2018 it is estimated that more than 12 million individuals have been tested with DTC services, and over 10,000 new DTC kits are purchased daily. Typical applications of such a personal SNP array test include revealing genealogy relationships or SNPs with health and disease implications. The company 23andMe was in 2017 allowed by the US Food and Drug Administration (FDA) to market tests that assess genetic risks for 10 disorders, including Parkinson's and late-onset Alzheimer's diseases. And in 2018 the FDA approved the testing of three different *BRCA1/BRCA2* mutations.

One commonly used SNP array used in DTC is able to monitor about 711,000 SNPs (based on the "HumanOmniExpress" chip, Illumina, Inc.). It should be noted that these SNPs are all previously known variants that are common in the population. Therefore, any rare or individual SNVs will not be revealed by the SNP array. Of the SNPs on the aforementioned chip, about 15,000 of them are listed in SNPedia. Selected SNPs as identified from an analysis of the author's genetic variants are shown in **Table 5.2**.

CANCER: A GENETIC DISEASE

A major focus of this book is the function of individual DNA sequence elements in our genome. The importance of these functions will mainly be illustrated by inherited monogenic disorders. This is because a monogenic disease may be explained by a single mutation occurring in the genome, and for this reason there is a distinct relationship between genotype and phenotype. In the case of complex disorders, more than one mutation contributes to the disease. This is also true for cancer, which is typically caused by multiple changes in the genome. For complex disorders and cancer, there are also environmental factors that play a role. For these reasons it is difficult to identify the contributions of different individual mutations for these disorders. Nevertheless, cancer is by and large a genetic disease, and in this book we occasionally use it to illustrate the functional role of genomic elements.

Table 5.1 Functional implications of selected SNPs

Category of phenotype	dbSNP identifier	Disease/Phenotype	Gene	Variation, reference strand
Monogenic disease	rs334	Sickle cell disease/ trait	HBB	Coding, T > C
	rs121434592	Proteus syndrome	AKT1	Coding, C > T
	rs28941770	Tay-Sachs disease	HEXA	Coding, C > A,G,T
	rs121908745	Cystic fibrosis	CFTR	Coding, TATC > T
	rs113993962	Bloom syndrome	RECQL3 or BLM	Coding, ATCTGA > TAGATTC
Disease risk	rs112176450	Parkinson's disease	EIF4G1	Coding, A > G
	rs35095275	Parkinson's disease	GBA	Coding, A > C/G
	rs429358	Alzheimer's disease	APOE	Coding, C > T
	rs7412	Alzheimer's disease	APOE	Coding, C > T
	rs4444903	Liver cancer	EGF	5' UTR, A > G
	rs80359675	Breast cancer	BRCA2	Coding, ->TCAAA
Disease resistance	rs333	Resistance to HIV infection	CCR5	Intron, GTCAG CA/-
Drug response	rs4149056	Clearance of methotrexate	SLCO1B1	Coding, C > T
	rs2359612	Affects warfarin dose	VKORC1	Intron, A > G
Other traits	rs713598	PTC bitter taste	TAS2R38	Coding, C > G
	rs10246939	PTC bitter taste	TAS2R38	Coding, C > T
	rs4481887	Ability to perceive asparagus odor in urine	(In cluster of olfactory receptor genes)	Intergenic region, A > G
	rs4988235	Lactose intolerance	LCT	Regulatory element upstream of gene, C > T
	rs1800497	Dopamine receptor may influence the sense of pleasure	AKK1	Coding, C > T
	rs1799971	Makes alcohol cravings stronger	OPRM1	Coding, A > G
	rs53576	A silent G to A change in the oxytocin receptor (OXTR) gene. Relationship to empathy and handling of stress?	OXTR	Intron, A > G
	rs6152	May influence baldness	AR	Coding, A > G

Note: Information is from dbSNP (www.ncbi.nlm.nih.gov/SNP/) and SNPedia (www.snpedia.com)

One major characteristic of cancer cells is that they divide in an uncontrolled manner. They may have acquired this property because cell growth is stimulated or because functions that normally restrict growth have been inactivated. The genes that are mutated in cancer may accordingly be classified in either of two groups. **Oncogenes** are genes that encode proteins with the potential to promote cancer development. Whereas an oncogene in normal tissue is expressed at a restricted level, a mutation in the oncogene causes it to be expressed at a higher level. This activation may be caused by a change in gene regulation or by a duplication of the gene. Conversely, **tumor suppressor genes** normally prevent cells from critical steps toward cancer. However, when these genes are lost or when their expression is dramatically lowered, cells may more easily progress toward cancer.

Table 5.2 Functional implications of selected SNPs of the author

Category of phenotype	dbSNP identifier	Implication
Disease risk	rs1108580 and rs1611115	Lower heart attack risk than average
Disease risk	rs5848(G;G)	Lower dementia risk, lower risk of frontotemporal dementia, Alzheimer's disease and Parkinson's disease
Disease risk	rs4143094(G;T)	Slightly (17%) higher risk of colorectal cancer correlated with consumption of processed meats
Disease risk	rs5882(G;G)	Longer life span
Drug response	rs2235015(G;G)	Somewhat less likely to respond to certain antidepressants
Drug response	rs4825476(G;G)	1.9 times higher risk of suicidal thoughts when taking citalopram
Drug response	rs8050894(C;G)	Average warfarin response (~5 mg/day)
Other trait	rs4988235(C;T), rs4988235(T;T), rs182549(C;T), rs182549(T;T)	Probably able to digest milk
Other trait	rs16969968(A;G)	Slightly higher risk for nicotine dependence, lower risk for cocaine dependence
Other trait	rs1426654(A;A)	Probably light-skinned, European ancestry
Other trait	rs1800407(G;G)	Blue/gray eyes more possible

Note: Genomic DNA was examined with a SNP array. SNP data were then analyzed with Promethease (https://promethease.com). Among the 700,000 SNPs on the SNP array, about 11,000 had functional annotation and were listed in the Promethease report.

What is the genetic contribution to cancer? First, there are inherited genomic variants that increase the risk of developing cancer. As one example, there are specific variants in the *BRCA1* and *BRCA2* genes that are associated with an elevated risk of breast cancer. *BRCA1* and *BRCA2* both encode proteins involved in the repair of DNA. There are other links between cancer and DNA repair. Thus, there are inherited disorders linked to cancer that are caused by mutation in other DNA repair genes. Examples are xeroderma pigmentosum, a disorder where the affected individuals are prone to develop skin cancers because of a defect in nucleotide excision repair. Another example is the Li-Fraumeni syndrome. It is caused by mutations in the gene *TP53*, encoding the p53 protein. The normal function of the protein is to take part in the control of cell division and growth. It will make sure that DNA is adequately repaired before proceeding to synthesis of new DNA. Another function of p53 is to initiate apoptosis if DNA damage cannot be repaired sufficiently. When p53 is mutated, however, the normal control of cell growth invoked by this protein is disrupted—this gene is therefore an example of a tumor suppressor.

Although inherited mutations contribute to cancer, a more common cause is genetic changes that occur in somatic cells. Tumors develop by a stepwise process, where a set of mutations are acquired over time (**Figure 5.19**). Some of these changes confer a selective advantage that enables cells to grow faster than the other neighboring cells. The stepwise modifications of cells have been particularly well studied in colorectal cancer. This disease originates from epithelial cells of the colon or rectum. The first mutation typically affects the *APC* gene ("Adenomatosis Polyposis Coli"), a gene part of the Wnt signaling pathway. This is a pathway initiated through the binding of specific molecules to a class

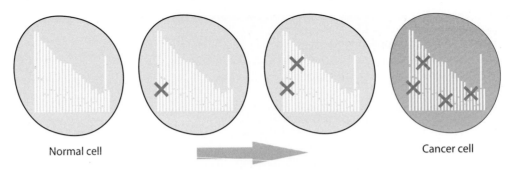

Figure 5.19 *Cancer develops by the accumulation of critical mutations over time.* A normal cell (left) is transformed into a cancer cell (right) in a stepwise manner, where a set of mutations are acquired over time. Within each cell the chromosomes are depicted, with red crosses representing mutations. In this example, once four critical mutations have been introduced a cancer cell has arisen.

of cell surface receptors, a binding that gives rise to specific intracellular responses. Cells mutated in *APC* grow slowly, but a second mutation in another gene, such as *KRAS*, allows a clonal expansion. Additional genes, such as *TP53*, are then mutated, eventually leading to the development of a malignant tumor.

Cancer initiates as a result of mutations in specific genes. Mutations that make possible a growth advantage to a cancer cell are called **driver** mutations. But each tumor has a large number of mutations, and most of them confer no selective growth advantage; they are known as "passenger" mutations. It is difficult to distinguish driver from passenger mutations. However, it has been estimated that there are about 300 genes that are able to "drive" cancer development. The number of driver gene mutations needed to cause cancer varies between tumors, but is within the range of 1–11. The driver genes encode proteins that are all part of a restricted set of signaling pathways related to cell fate, cell survival, and processes responsible for genome integrity, such as DNA repair.

NUMEROUS CANCER GENOMES HAVE BEEN SEQUENCED

Our understanding of genetic changes in cancer development is aided by large-scale projects to sequence DNA and RNA from tumor material. For instance, there is the Cancer Genome Atlas (TCGA), a collaboration between the National Cancer Institute (NCI) and the National Human Genome Research Institute (NHGRI) in the United States. It includes the analysis of 33 types of cancer. It has tumor tissue and matched normal tissues from more than 11,000 patients. The project includes about 1,000 sequenced **cancer genomes**—each such genome is obtained by whole genome sequencing where the DNA is derived from a sample of tumor tissue. TCGA also includes the same number of corresponding normal noncancer genomes. The amount of data in the TCGA project is impressive—2.5 petabytes of data, some of which is publically available. It allows researchers to address a large variety of questions related to cancer biology. Another important resource for cancer researchers is the Catalogue of Somatic Mutations in Cancer (COSMIC) (Wellcome Trust Sanger Institute), a manually curated database of somatic mutation information related to cancer.

Each cancer genome typically has a vast number of mutations—more than 100,000 single nucleotide or indel mutations are sometimes observed. Cancer genomes are also characterized by copy number variation (**Figure 5.20**).

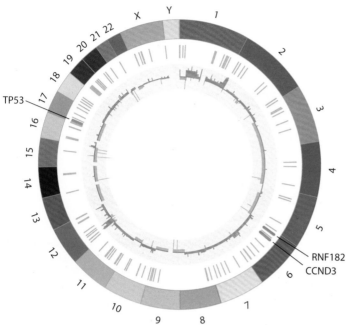

Figure 5.20 *Genetic changes in a breast cancer genome.* A sample of breast invasive carcinoma cancer from the Cancer Genome Atlas project was analyzed by DNA sequencing. In this case only the exons were sequenced (a procedure known as exome sequencing, see also Chapter 13). Individual chromosomes are depicted on the outer circle followed by concentric tracks for SNVs and copy number variation (CNV) data relative to mapping position in the genome. Highlighted in the SNV track are three different genes that are mutated and that are known to be associated with driver mutations in breast cancer. It should be noted that a much larger number of SNVs would have been identified if the whole genome had been sequenced instead of only the exonic regions. In the CNV track are shown portions in red that are enriched in the cancer, and conversely blue regions reveal lower copy number as compared to the normal cell. The CNV track reveals copy number variation in longer continuous regions of the genome, reflecting changes in chromosome number and larger chromosomal rearrangements.

In common solid tumors (such as colon and breast), there are in the order of 50 different genes with mutations that change the protein sequence (but only a subset of these are mutations that are expected to be driver mutations; remember that a tumor has less than 12 such mutations). Some tumors have a much larger number of mutations. For instance, lung cancers from smokers have 10 times more mutations as those from nonsmokers—this is because tobacco smoke contains powerful mutagens.

CANCER: TRANSLOCATIONS AND MORE DRAMATIC GENOMIC ABERRATIONS

Cancer genomes also show examples of chromosomal aberrations, like changes in chromosome number and chromosomal translocations. Some translocations produce a **fusion gene**. One example is a fusion between the genes *BCR* and *ABL* that results from an interchange of material from chromosomes 9 and 22 (**Figure 5.21**). This occurs in human leukemias such as chronic myeloid leukemia, a rare type of blood cancer that gives

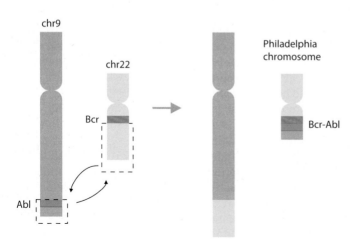

Figure 5.21 *Chromosomal translocation with formation of a fusion gene.* An interchange of material from chromosomes 9 and 22 occurs in chronic myeloid leukemia. The genes *BCR* and *ABL* are fused as a result of the translocation, giving rise to an aberrant activity of ABL. The chromosome with the fusion gene is also known as the Philadelphia chromosome, named so because it was discovered by Peter Nowell and David Hungerford, both affiliated with research facilities in the city of Philadelphia.

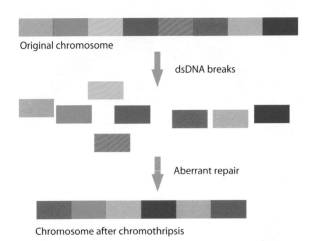

Original chromosome

dsDNA breaks

Aberrant repair

Chromosome after chromothripsis

Figure 5.22 *Chromothripsis.* A single shattering event may give rise to tens to hundreds of fragments. This is followed by the stitching of fragments into new combinations. Many new breakpoints are thus formed, and some portions of the normal chromosome may be missing.

rise to a high production of white blood cells. The fusion of *BCR* and *ABL* results in an activation of the tyrosine kinase activity of the *ABL* gene product. This activation leads to uncontrolled cell growth.

A yet more dramatic change in the cancer genome is a phenomenon known as **chromothripsis** (*chromo* for "chromosomes," and *thripsis* for "shattering into pieces") (**Figure 5.22**). A single shattering event gives rise to a large number of double-strand DNA breaks. Tens to hundreds of fragments may be formed. This is followed by the stitching of fragments into new combinations. Many new fusion points are thus formed, and some portions of the normal chromosome may be missing. There is a significant association of chromothripsis and poor prognosis. For instance, a 62-year-old patient with chronic lymphocytic leukemia from which chromothripsis was first observed rapidly got worse and relapsed quickly even though a drug was administered that is commonly used in the treatment of this cancer.

All cases of chromothripsis seem to involve a limited set of chromosomes, a single chromosome, or even just a small portion of a chromosome. The mechanisms of chromothripsis are poorly understood. The mechanism of stitching seems to be the DNA repair pathway NHEJ.

WHAT ARE THE CAUSES OF CANCER?

We have seen that a large number of different mutations and chromosomal aberrations are observed in cancer genomes. Only a fraction of them are causative (driver mutations). How do these critical genetic changes come about? We can identify three major elements when defining cancer risk. First, there could be a *hereditary* factor, like mutations in *BRCA1* and *BRCA2* genes that are important risk factors for cancer development. But it is estimated that only about 2%–3% of all cancers are linked to inherited mutations. Instead **somatic** mutations are much more important events. These can arise by different mechanisms as discussed previously in this chapter. There are two major important categories of somatic mutations. First, they can occur spontaneously, either during DNA replication or because of normal endogenous DNA damage. In this category, chance plays an important role. Second, mutations may arise because of external influence such as irradiation (for instance, exposure to sunlight) or chemicals (for instance, from smoking). These are examples where there are environmental or lifestyle components in the development of cancer. What are the relative contributions of these different factors causing cancer? In studies by Cristian Tomasetti and Bert Vogelstein, the conclusion is reached that DNA replication mutations are responsible for as many as two-thirds of mutations in human cancers. However, these results do not exclude that cancer may be prevented. As a simple example, consider a certain cancer where there are three driver mutations, and all three of them are required for its development. The cancer can in that case be prevented if two of the mutations are caused by replication errors and the third is caused by environmental factors.

SUMMARY

- The effect of mutations depends on when and where they occur. Germline mutations may give rise to inherited disorders, whereas cancer and a few rare noncancer disorders are caused by somatic mutations.
- It is estimated that in the human germline about 0.3 mutations occur per cell division, whereas the mutation rate of somatic cells may be about one order of magnitude higher.
- Mutations occur naturally as a result of normal cellular biochemical activity, as well as from external influence, such as irradiation and chemicals. Mutations may be caused by (1) errors during DNA replication; (2) DNA damage because of reactions occurring spontaneously in the cell (examples are depurination and deamination of cytosine); (3) the action of external or endogenous chemicals (example is the formation of 8-oxo-guanine); and (4) irradiation with UV light (the formation of pyrimidine dimers) and ionizing radiation (double-strand DNA breaks).
- Common mechanisms to repair DNA include base excision repair, nucleotide excision repair, and mismatch repair. Translesion synthesis is a pathway to repair DNA damage, such as pyrimidine dimers and apurinic/apyrimidinic sites.
- Specific mechanisms exist to repair double-strand breaks, microhomology-mediated end joining (MMEJ), homologous recombination, and NHEJ. Translesion synthesis and NHEJ are error-prone repair mechanisms.
- Mutations may be categorized into (1) substitution and indel mutations (that involve a small number of nucleotides) and (2) structural variation that involves larger regions.
- A SNP or SNV is a variation in a specific site in the genome. The most common of these variants are nucleotide substitutions. SNPs/SNVs may be identified with genome sequencing or with SNP arrays.
- Examples of structural variation are (1) copy number variation that results from larger deletions, insertions, and duplications, and (2) chromosomal translocations.
- In a single gene disorder, the genetic defect is a mutation in a single gene. A complex multifactorial disorder is typically caused by mutations in multiple genes, and there are contributing lifestyle and environmental factors.
- A GWAS is one method used to elucidate the function of genomic variants.
- Out of millions of human SNVs, only a small fraction of these have a known biological function.
- Cancer is a genetic disease. It may be caused by inherited mutations, but the more common cause of cancer is somatic mutation. Tumors

develop by a stepwise process where mutations are acquired over time. Tumors have 1–11 driver mutations required for cancer progression.

- Thousands of cancer genomes have been sequenced within TCGA project. Cancer genomes may have as many as 100,000 substitution or indel mutations as well as extensive copy number variation and chromosomal aberrations.

QUESTIONS

1. Define and discuss the consequences of *germline* and *somatic* mutations, respectively.
2. What are the different molecular causes of mutations?
3. What is a *depurination*, and how may this error be repaired?
4. What are the errors in DNA that result from *irradiation*?
5. What are the mechanisms used to repair double-strand breaks in DNA? How are they different in terms of recreating the original DNA sequence? Are there other DNA repair methods than repair of double-strand breaks that are error prone?
6. Explain the concept of SNP. What is the difference between a SNP and a SNV?
7. How many SNPs and SNVs are there in the human genome?
8. How is a SNV different from *structural variation*?
9. How is *copy number variation* defined?
10. What is characteristic of a *monogenic* disorder?
11. Discuss the possible *phenotypic* outcomes of mutations/genetic variants.
12. What are the sources of information exploited in a GWAS?
13. What is meant by *oncogene* and *tumor suppressor gene*?
14. Cancer may be caused by *inherited* mutations and *somatic* mutations. Which of these two categories are the most common causes of cancer?
15. What are the causes of *somatic* mutations that are responsible for cancer?
16. Are there diseases other than cancer that arise because of *somatic* mutations?
17. What is a *driver* mutation? How many such mutations are typically required to develop cancer?
18. Why does a cancer cell typically contain many mutations as compared to a normal cell?
19. What is a *fusion gene* in the context of cancer, and why could it be harmful?
20. You may hear someone saying that you get cancer as a result of chance, and there is nothing you can do to change the odds. How could you argue against this statement?

URLs

NCBI dbSNP (www.ncbi.nlm.nih.gov/SNP/)

OMIM (www.omim.org)

GWAS data (www.ebi.ac.uk/gwas/; www.gwascentral.org/)

SNPedia database (www.snpedia.com/)

1000 genomes project (www.internationalgenome.org)

Promethease (https://promethease.com)

The TCGA Research Network (http://cancergenome.nih.gov/)

COSMIC (https://cancer.sanger.ac.uk/cosmic)

FURTHER READING

Mutation rates

Jonsson H, Sulem P, Kehr B et al. 2017. Parental influence on human germline de novo mutations in 1,548 trios from Iceland. Nature 549(7673):519–522.

Milholland B, Dong X, Zhang L et al. 2017. Differences between germline and somatic mutation rates in humans and mice. Nat Commun 8:15183.

Shendure J, Akey JM. 2015. The origins, determinants, and consequences of human mutations. Science 349(6255):1478–1483.

Noncancer diseases caused by somatic mutations

Erickson RP. 2010. Somatic gene mutation and human disease other than cancer: An update. Mutat Res 705(2):96–106.

Lindhurst MJ, Sapp JC, Teer JK et al. 2011. A mosaic activating mutation in AKT1 associated with the Proteus syndrome. N Engl J Med 365(7):611–619.

DNA damage

De Bont R, van Larebeke N. 2004. Endogenous DNA damage in humans: A review of quantitative data. Mutagenesis 19(3):169–185.

DNA repair with NHEJ

Lieber MR. 2010. The mechanism of double-strand DNA break repair by the nonhomologous DNA end-joining pathway. Annu Rev Biochem 79:181–211.

The human gene mutation database

Stenson PD, Mort M, Ball EV et al. 2017. The Human Gene Mutation Database: Towards a comprehensive repository of inherited mutation data for medical research, genetic diagnosis and next-generation sequencing studies. Hum Genet 136(6):665–677.

The 1000 Genomes Project

The Genomes Project C. 2015. A global reference for human genetic variation. Nature 526:68.

Variants and human traits

Crouch DJM, Winney B, Koppen WP et al. 2018. Genetics of the human face: Identification of large-effect single gene variants. Proc Natl Acad Sci 115(4):E676–E685.

Marouli E, Graff M, Medina-Gomez C et al. 2017. Rare and low-frequency coding variants alter human adult height. Nature 542(7640):186–190.

Sniekers S, Stringer S, Watanabe K et al. 2017. Genome-wide association meta-analysis of 78,308 individuals identifies new loci and genes influencing human intelligence. Nat Genet 49:1107.

Yang J, Benyamin B, McEvoy BP et al. 2010. Common SNPs explain a large proportion of the heritability for human height. Nat Genet 42(7):565–569.

Canadian Personal Genome Project

Reuter MS, Walker S, Thiruvahindrapuram B et al. 2018. The Personal Genome Project Canada: Findings from whole genome sequences of the inaugural 56 participants. CMAJ 190(5):E126–E136.

SNPs and SNVs

Keinan A, Clark AG. 2012. Recent explosive human population growth has resulted in an excess of rare genetic variants. Science 336(6082):740–743.

Lek M, Karczewski KJ, Minikel EV et al. 2016. Analysis of protein-coding genetic variation in 60,706 humans. Nature 536(7616):285–291.

Nelson MR, Wegmann D, Ehm MG et al. 2012. An abundance of rare functional variants in 202 drug target genes sequenced in 14,002 people. Science 337(6090):100–104.

Sherry ST, Ward MH, Kholodov M et al. 2001. dbSNP: The NCBI database of genetic variation. Nucleic Acids Res 29(1):308–311.

Telenti A, Pierce LC, Biggs WH et al. 2016. Deep sequencing of 10,000 human genomes. Proc Natl Acad Sci U S A.

Tennessen JA, Bigham AW, O'Connor TD et al. 2012. Evolution and functional impact of rare coding variation from deep sequencing of human exomes. Science 337(6090):64–69.

Structural variation and CNVs

Bruder CE, Piotrowski A, Gijsbers AA et al. 2008. Phenotypically concordant and discordant monozygotic twins display different DNA copy-number-variation profiles. Am J Hum Genet 82(3):763–771.

Conrad DF, Pinto D, Redon R et al. 2010. Origins and functional impact of copy number variation in the human genome. Nature 464(7289):704–712.

Kidd JM, Cooper GM, Donahue WF et al. 2008. Mapping and sequencing of structural variation from eight human genomes. Nature 453(7191):56–64.

O'Huallachain M, Karczewski KJ, Weissman SM et al. 2012. Extensive genetic variation in somatic human tissues. Proc Natl Acad Sci USA 109(44):18018–18023.

Pang AW, MacDonald JR, Pinto D et al. 2010. Towards a comprehensive structural variation map of an individual human genome. Genome Biol 11(5):R52.

Sudmant PH, Rausch T, Gardner EJ et al. 2015. An integrated map of structural variation in 2,504 human genomes. Nature 526(7571):75–81.

The Genomes Project C. 2015. A global reference for human genetic variation. Nature 526:68.

Zarrei M, MacDonald JR, Merico D, Scherer SW. 2015. A copy number variation map of the human genome. Nat Rev Genet 16(3):172–183.

OMIM database

Amberger JS, Hamosh A. 2017. Searching Online Mendelian Inheritance in Man (OMIM): A Knowledgebase of Human Genes and Genetic Phenotypes. Curr Protoc Bioinformatics 58:1.2.1–1.2.12.

Age-related macular degeneration

Black JR, Clark SJ. 2016. Age-related macular degeneration: Genome-wide association studies to translation. Genet Med 18(4):283–289.

Cancer

Hanahan D, Weinberg RA. 2000. The hallmarks of cancer. Cell 100(1):57–70.

Hanahan D, Weinberg RA. 2011. Hallmarks of cancer: The next generation. Cell 144(5):646–674.

Nowak MA, Waclaw B. 2017. Genes, environment, and "bad luck." Science 355(6331):1266–1267.

Tomasetti C, Li L, Vogelstein B. 2017. Stem cell divisions, somatic mutations, cancer etiology, and cancer prevention. Science 355(6331):1330–1334.

The Cancer Genome Atlas Network. 2012. Comprehensive molecular portraits of human breast tumours. Nature 490(7418):61–70.

Vogelstein B, Papadopoulos N, Velculescu VE et al. 2013. Cancer genome landscapes. *Science* 339(6127):1546–1558.

The Cancer Genome Atlas (TCGA) project

Ding L, Bailey MH, Porta-Pardo E et al. 2018. Perspective on oncogenic processes at the end of the beginning of cancer genomics. *Cell* 173(2):305–320.e310.

Hoadley KA, Yau C, Hinoue T et al. 2018. Cell-of-origin patterns dominate the molecular classification of 10,000 tumors from 33 types of cancer. *Cell* 173(2):291–304.e296.

Sanchez-Vega F, Mina M, Armenia J et al. 2018. Oncogenic signaling pathways in the Cancer Genome Atlas. *Cell* 173(2):321–337.e310.

Chromothripsis

Liu P, Erez A, Nagamani SC et al. 2011. Chromosome catastrophes involve replication mechanisms generating complex genomic rearrangements. *Cell* 146(6):889–903.

Maher CA, Wilson RK. 2012. Chromothripsis and human disease: Piecing together the shattering process. *Cell* 148(1–2):29–32.

Rode A, Maass KK, Willmund KV et al. 2016. Chromothripsis in cancer cells: An update. *Int J Cancer* 138(10):2322–2333.

The Critical Protein Coding Sequences

6

We discuss throughout this book the effect of mutations in the human genome because they give important clues as to the normal function of different DNA sequence elements. We have seen how a single nucleotide exchange in the genomic DNA causes the inherited disorder sickle cell anemia by changing a glutamic acid codon into a valine codon. In this chapter we see more examples of disorders that give rise to amino acid replacements, but also disorders caused by other changes in the translation product.

MUTATIONS IN CODING SEQUENCES

The molecular mechanisms as discussed in Chapter 5 generate point mutations and indels. What are the consequences of such mutations in a coding region? Remember that the relationship of nucleotide sequences and amino acids is specified by the genetic code. The genetic code informs us about the effect of mutations. There are 61 codons specifying amino acids (also known as **sense** codons) and three codons that mean stop (also known as **nonsense** codons). We first consider mutations that substitute one nucleotide for another (a point mutation), the most common type of mutation.

What is the effect of a point mutation in terms of the protein sequence? First, it may either affect an amino acid (a **nonsynonymous** change) or leave the encoded amino acid intact (a **synonymous** change). One outcome of the nonsynonymous mutation is that one amino acid is replaced by another (also known as a **missense** mutation). This is the more likely event—a random point mutation in a coding sequence gives rise to a change in amino acid in about 71% of cases. Included in the category of nonsynonymous mutations are also those that introduce a stop codon (a nonsense mutation) as well as those that change a stop codon into a codon that specifies an amino acid, as discussed later in this chapter.

As to indels, they will typically be **frameshift** mutations—that is, they cause a shift of reading frame during protein synthesis. Only if the number of inserted or deleted nucleotides is divisible by three is a frameshift avoided, and one or more amino acids will be added to or deleted from the normal amino acid sequence. We examine these different scenarios in more detail.

INHERITED SINGLE GENE DISORDERS ARE OFTEN CAUSED BY NONSYNONYMOUS MUTATIONS

A large number of different mutations have been identified in the context of human inherited disease. How do they distribute in terms of categories such as synonymous/nonsynonymous and coding/noncoding regions in the human genome?

It turns out that many inherited disorders are caused by changes in protein coding sequences. The National Center of Biotechnology Information (NCBI) ClinVar database is collecting clinically significant variants. An analysis of approximately 70,000 mutations labeled as pathogenic in this collection shows that nearly 90% of them are in coding regions (**Table 6.1**). Among these, about 37% are missense changes. The predominance of mutations in protein coding sequences is in marked contrast to the fact that only 1.6% of the human genome is protein coding sequences, as discussed in Chapter 4. Only about 10% of all pathogenic mutations are in intronic regions of protein coding genes. Finally, less than 0.2% are in genes that encode noncoding RNAs, and only 0.4% are in intergenic regions.

It should at this stage be noted that as we learn more about genetic disorders and their molecular background, the distribution as shown in Table 6.1 might be slightly modified. For instance, the current number of mutations in protein genes may be overestimated because by tradition,

Table 6.1 Relative distribution of pathogenic mutations		
Mutation		**Percentage**
Protein coding	Nonsynonymous, missense	36.8
	Nonsense	20.0
	Frameshift deletion	18.9
	Frameshift insertion	8.4
	Synonymous	0.44
	Non-frameshift deletion	1.53
	Frameshift substitution	1.34
	Non-frameshift substitution	0.59
	Non-frameshift insertion	0.32
	Stop-loss	0.11
	Exonic unknown	0.24
UTR	5′ UTR	0.16
	3′ UTR	0.06
Protein gene splicing		8.44
Protein intronic		1.89
ncRNA exonic		0.10
ncRNA intronic		0.04
Intergenic		0.41

Note: Table is based on about 70,000 mutations tagged in the NCBI ClinVar database as pathogenic. A "frameshift substitution" refers to the situation where a sequence of nucleotides is replaced with another sequence and where the difference in length is not divisible by three. In "non-frameshift insertions" and "non-frameshift deletions" the difference in the number of nucleotides is divisible by three.

research has focused on such regions rather than the noncoding parts of the genome.

All point mutations in a coding sequence do not occur with the same frequency. For instance, it has been observed that among nonsynonymous mutations occurring in the context of human genetic disease, arginine and glycine codons are the most frequently mutated. This is explained by the presence of the dinucleotide sequence "CG" in these codons. Such occurrences of CG in the genome are generally referred to as **CpG sites** (the "p" for phosphate is there to distinguish it from the C•G base pair). At such sites there is frequent methylation of the C (see also Chapter 10). In turn, the methyl derivative of C is easily deaminated to thymine. The overall effect is that the C in the CG sequence from an evolutionary perspective tends to be mutated to T with greater frequency (**Figure 6.1**).

MANY DIFFERENT MUTATIONS IN A GENE MAY GIVE RISE TO DISEASE

In certain inherited disorders that are common and that have been well studied, analyzes of patients have revealed a large number of different mutations. As one example, we consider again the β-globin gene (*HBB*). Not only is the mutation replacing glutamate with valine in position 6 of the β-globin protein responsible for sickle cell anemia, but many other mutations cause diseases that result from errors in the β-globin protein or from insufficient amounts of the full-length protein. Many of these disorders have a similar phenotype. A number of mutations responsible for disease are shown in **Figure 6.2**. As a rule, a patient carries only one of these mutations, and this mutation is sufficient to explain the disorder.

The sickle cell mutation was described in Chapters 2 and 3. This is one example of a missense mutation. There is a large collection of such mutations that cause inherited disorders. For now we consider only two more examples. These appear in the context of disorders known as Tay-Sachs disease and hemochromatosis, respectively.

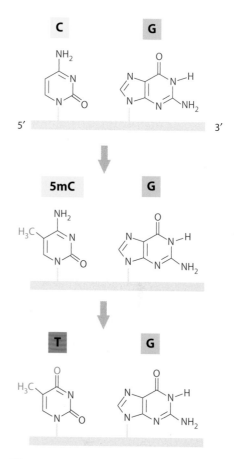

Figure 6.1 CpG sites are prone to mutation. The cytosine in a CG sequence is often modified to form 5-methylcytosine (5 mC). This derivative may in turn be deaminated to thymine, causing a mutation C > T.

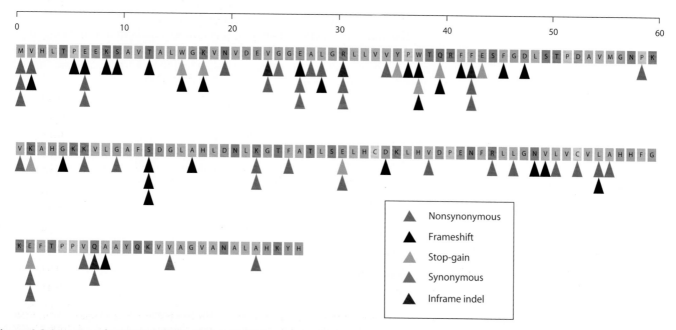

Figure 6.2 Location of mutations in the human β-globin gene (*HBB*). (Based on information in the NCBI ClinVar database and considers mutations believed to be pathogenic.)

THE GENE *HEXA* AND TAY-SACHS DISEASE

Tay-Sachs disease is common among Ashkenazi Jews with a prevalence of about 1 in 3,000. This Jewish population is an example of an ethnic group where certain genetic disorders are more common than in the general population. Different factors may contribute to this situation, such as the *founder* effect, which is to say a loss of genetic diversity that occurs whenever a new population is established by a small number of individuals.

Tay-Sachs disease was first described in the late nineteenth century by physicians Warren Tay and Bernard Sachs. The most common form of the disease affects children. It results in a deterioration of mental and physical abilities and is fatal by age 2–3 years. While the most common cause of the disease is a frameshift mutation, we consider here a less common variant that is the result of a nonsynonymous mutation.

It was revealed in the late twentieth century that Tay-Sachs disease is caused by mutations in the gene *HEXA*, a gene that encodes the α-subunit of the enzyme hexosaminidase. Substrates for this enzyme are gangliosides, carbohydrate components of nerve cell membranes. Hexosaminidase is responsible for the degradation of gangliosides by removing N-acetylgalactosamine (GalNac) residues from the ganglioside GM2. A deficiency of hexosaminidase has the effect that ganglioside GM2 will accumulate, in turn leading to neurodegeneration.

There are two forms of the hexosaminidase enzyme. In HexA there are α- and β-subunits encoded by the genes *HEXA* and *HEXB*, respectively. The other form is HexB with two β-subunits. HexB does not have a known physiological function. However, the HexA enzyme is active in the degradation of GM2, and the α-subunit is critical for this reaction.

We consider here a form of Tay-Sachs disease, known as "B1 variant," that results from an amino acid replacement in the *HEXA* gene product, the α-chain. A nucleotide substitution G > A changes a codon CGC in *HEXA*, encoding the amino acid Arg178, to codon CAC (histidine) (**Figure 6.3**). (In the literature of genetics an alteration at the level of protein sequence is commonly referred to with an expression like "Arg178His" or with one-letter amino acid codes, "R178H.") An heterodimer is formed with normal β-subunits and mutant α-subunits, but as a result of the α-chain mutation, the enzyme cannot hydrolyze GM2. The Arg178 is just adjacent to the active site, and a change in this amino acid is therefore likely to affect enzyme activity. In addition, Arg178 is close to residues of the β-subunit, and the mutation may therefore also affect the interaction between the two subunits (**Figure 6.4**).

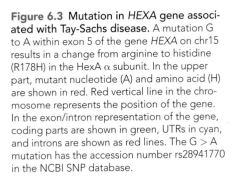

Figure 6.3 Mutation in *HEXA* gene associated with Tay-Sachs disease. A mutation G to A within exon 5 of the gene *HEXA* on chr15 results in a change from arginine to histidine (R178H) in the HexA α subunit. In the upper part, mutant nucleotide (A) and amino acid (H) are shown in red. Red vertical line in the chromosome represents the position of the gene. In the exon/intron representation of the gene, coding parts are shown in green, UTRs in cyan, and introns are shown as red lines. The G > A mutation has the accession number rs28941770 in the NCBI SNP database.

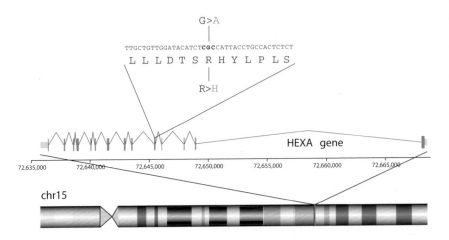

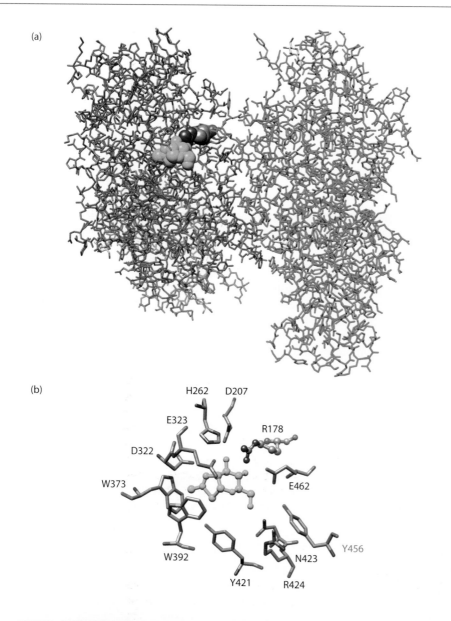

Figure 6.4 Three-dimensional structure of hexosaminidase. (a) The α- and β-subunits of hexosaminidase are shown in red and green, respectively. A synthetic molecule, NAG-thiazoline (cyan) binds to the active site of the α-subunit. The α-subunit Arg178 residue (shown with spheres colored by element) is adjacent to the active site and also close to the β-subunit. (b) Detailed view of the active site amino acid residues. All amino acids shown are of the α-subunit, except Y456 which is from the β-chain. NAG-thiazoline (cyan) binds to the active site, and one of the amino acids in close proximity to the active site is Arg(R)178. A mutation of this amino acid is likely to disturb the catalytic activity of the HexA enzyme as well as interactions with the β-subunit. (Adapted from Lemieux M et al. 2006. *J Mol Biol* 359:913–929. PBD ID: 2GK1)

THE GENE *HFE* AND HEREDITARY HEMOCHROMATOSIS

Hereditary hemochromatosis (HHC) is thought to be the most common genetic disorder in Caucasian populations. It is most frequent among people with Northern European ancestry, particularly those of Celtic descent. The prevalence in individuals of Northern European descent may be as high as 1 in 227 individuals. It is a recessive disorder, and as much as 10% of the Caucasian population is estimated to carry a gene mutation for HHC.

HHC is characterized by a dysregulation of the iron level in the body. An abnormal uptake of dietary iron will result in a severe accumulation of iron in tissues and organs, disturbing their normal functions.

Many HHC patients have mutations in the gene *HFE*. This gene encodes a protein known as *HFE* ("High Iron Fe") or hemochromatosis protein. It is a membrane protein thought to regulate iron absorption. One common mutation is a nucleotide change G > A in exon number 3 so that a cysteine codon (UGC, or TGC at the level of DNA sequence) is changed into a tyrosine codon (UAC) (**Figure 6.5**). As a result, the cysteine in position 282 of

Figure 6.5 Mutation in *HFE* gene associated with hereditary hemochromatosis (HHC). A G > A mutation in exon number 3 of the gene *HFE* causes the amino acid substitution cysteine (C) to tyrosine (Y). In the upper part, mutant nucleotide (A) and amino acid (Y) are shown in red. Red vertical line in the chromosome represents the position of the gene. In the exon/intron representation of the gene, coding parts are shown in green, UTRs in cyan, and introns as red lines.

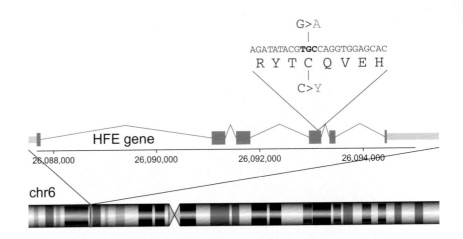

the protein (position 260 in the numbering system of the mature protein) is altered to tyrosine. The cysteine is important for the structural integrity of the protein, as it forms a disulfide bond with another cysteine (**Figure 6.6**). It should be noted that it presumably takes more than the C282Y mutation to develop HHC, as not all individuals who are homozygous for C282Y show any symptoms of the disease.

(a)

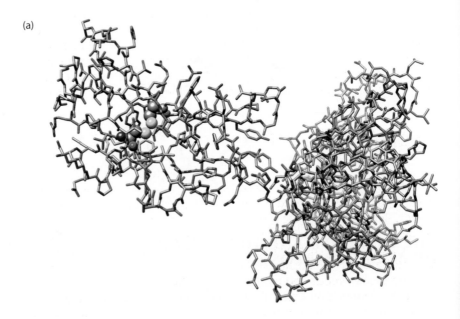

Figure 6.6 Structure of *HFE* protein. (a) *HFE* protein is shown as a ball-and-stick model. The cysteine in position 260 is commonly mutated in HHC. This residue together with cysteine 203 are highlighted with atoms shown as spheres. (b) Detailed view of the cysteines in positions 203 and 260 that are connected through a disulfide bond (sulfur atoms in yellow). This disulfide bond is critical for the structural integrity of the protein. Cys260 corresponds to Cys282 in the precursor form of the protein—the precursor also contains an N-terminal signal peptide eventually cleaved off. (Adapted from Bennett MJ et al. 2000. *Nature* 403:46–53. PBD ID: 1DE4)

(b)

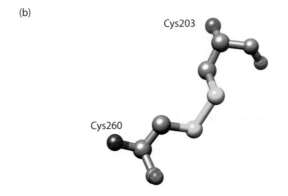

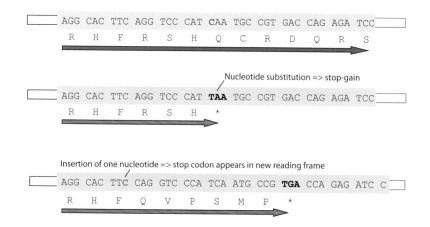

Figure 6.7 Mutations may give rise to new stop codons. Two scenarios are shown whereby stop codons could come about by mutation. First, a point mutation may convert a sense codon to a stop codon. Second, an indel mutation may give rise to a new reading frame in which a stop codon is eventually encountered.

MUTATIONS THAT GENERATE PREMATURE STOP CODONS

We have seen in previous examples that one possible outcome of a point mutation in a coding sequence is an amino acid substitution. However, three codons of the genetic code are stop codons. Therefore, another possible result of a point mutation is a **nonsense** or **stop-gain** mutation, a conversion of a sense codon to a stop codon (**Figure 6.7**).

It is important to note that premature stop codons do not only arise as a result of a mutation converting a sense codon to a stop codon. Also indels causing frameshift mutations are likely to generate stop codons as shown in Figure 6.7. In general, a stop codon is encountered early in the novel reading frame as there is no specific mechanism to select against stop codons in the two alternative reading frames of an mRNA sequence.

NONSENSE MUTATIONS RESULT IN mRNA DEGRADATION

The change to a stop codon causes protein synthesis to terminate at that point, leading to a truncated protein product. Unless the mutation is close to the C-terminal end of the protein, the product is unlikely to have normal biological activity. Truncated proteins are likely to be misfolded, and there is a protein degradation mechanism to remove such proteins. In addition, there is a mechanism to eliminate mRNAs with premature stop codons, referred to as **nonsense-mediated mRNA decay** (**NMD**). For this reason, mRNAs with premature stop codons are not likely to generate a protein product. One reason that NMD, as well as protein degradation mechanisms, have evolved could be that many truncated proteins are likely to be toxic to the cell.

How does the cell know how to distinguish premature stop codons from normal stop codons? Remember from Chapter 4 that exons are combined during splicing in the cell nucleus. It turns out that when mRNAs have reached the cytosol where protein synthesis takes place, they have a memory of where splicing previously has occurred in the nucleus. Whenever a stop codon occurs upstream of a splice site, this situation may be recognized by the NMD machinery (**Figure 6.8**).

Can the stop codon be avoided as a cure for this type of mutation? The drug Ataluren (Translarna in Europe) promotes read-through at stop codons, thereby suppressing the phenotype caused by the premature stop codon. This drug has been tested as therapy for inherited disorders resulting from nonsense mutations, such as cystic fibrosis and Duchenne muscular dystrophy.

Figure 6.8 Nonsense-mediated decay. An mRNA having a premature termination codon is likely to be degraded using the mechanism nonsense-mediated decay (NMD). During splicing in the nucleus, specific proteins attach to exon-exon junctions and form an exon-junction complex (EJC). The proteins of this complex stay attached to the mRNA when it is transported to the cytoplasm. Once in the cytoplasm, the protein UPF2 is recruited to EJC. When a ribosome translating an mRNA reaches a premature termination codon (PTC), it is stalled at this position and a complex of proteins (named SURF) binds to the ribosome. A component of the SURF complex then binds UPF2 of the EJC, thereby bridging EJC to the ribosome. Following this interaction, mRNA is degraded. In essence, the proteins of the splicing junction are a marker that the coding sequence extends beyond this site. If exon-junction complexes that are formed during pre-mRNA splicing still are present after translation termination, this is a signal that termination is premature.

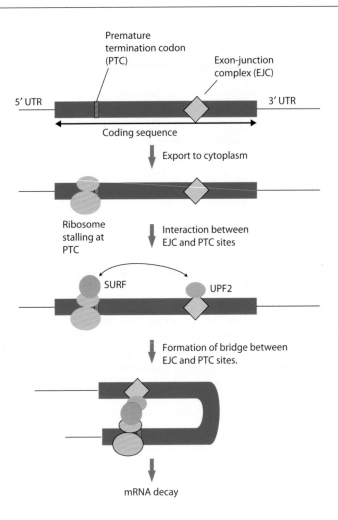

A NONSENSE MUTATION THAT GIVES RISE TO AGGRESSIVE BEHAVIOR

Remember that in the case of a nonsense mutation, the premature stop codon is likely to result in degradation of mRNA and/or protein. Furthermore, any truncated protein remaining in spite of degradation mechanisms is not likely to be functional. There are a number of disorders that may be caused by nonsense mutations, such as cystic fibrosis, Duchenne muscular dystrophy (DMD), β-thalassemia, and Hurler syndrome. Here we examine two other disorders in more detail. The first is a result of a CAG (Gln) to UAG (stop) conversion in the gene MAOA.

In 1978, a woman presented an unusual case to clinical geneticists at the University Hospital in Nijmegen, the Netherlands. In the woman's family there were a significant number of men showing violent behavior. In 1962, the woman's granduncle had prepared a family tree that identified nine other males with the same trait. The trait could be traced back to the nineteenth century. The affected males showed aggressive behavior usually triggered by anger and that was often out of proportion to the provocation. Aggressive behavior tended to cluster in periods of 1–3 days. Some of the affected males were responsible for arson, attempted rape, and exhibitionism. One male raped his sister at the age of 23. He then was at an institution for psychopaths, and there he was sometimes fighting with other inmates. When he was 35 years old, while working in the fields, he stabbed one of the wardens with a pitchfork. He had been told to get on with his work. Another male was working at a sheltered workshop and because he was told by his boss that his work was not satisfactory, he tried to run over him with a car.

The genetic background to this exceptional case of aggressive behavior was examined by Han Brunner and coworkers at the Nijmegen hospital. It was clear that all of the affected individuals were males and that the trait was carried from a mother to her son. This is evidence of **X-linked recessive inheritance**, which is to say that there is a mutation in a gene on the X chromosome that causes the disease to be expressed specifically in males. Remember that males have only one X chromosome, and for this reason, they are dependent on having functional copies of genes on that chromosome.

Brunner eventually managed to map the disorder to the short arm of the X chromosome in the region Xp11-21. Within that region there was one particularly interesting gene, *MAOA*, a gene that encodes the enzyme monoamine oxidase A. This enzyme is responsible for degradation of three different neurotransmitters, norepinephrine, serotonin, and dopamine (for the reaction using dopamine as substrate, see **Figure 6.9**).

Balanced metabolism of the neurotransmitters is required to maintain a normal mental state. Dysregulation of serotonin and dopamine is frequently observed in the context of psychiatric disorders, and it was therefore of interest to examine further such dysregulation in the Dutch family. By analyzing urine samples of affected males, the Brunner group found evidence of disturbances in the metabolism of neurotransmitters.

In a collaboration with Xandra Breakefield, Brunner was able to pinpoint the exact location of the mutation in the *MAOA* gene. In this study, five males with aggressive behavior were examined. They all had a point mutation in exon 8 of the *MAOA* gene, which changes a glutamine codon (CAG) to a termination codon (UAG) (**Figure 6.10**). This mutation was specific to the affected males—it did not occur in other members of the family.

This was the first time that a gene was linked to aggression. The disorder of the Dutch family is now referred to as Brunner syndrome. It is an extremely rare condition and has since been identified in only a few additional families.

The *MAOA* gene later attracted attention because of its potential role in aggression, antisocial behavior, and criminal behavior. It has been named the "warrior gene." In Brunner syndrome, the expression of *MAOA* is completely shut down as a result of the nonsense mutation. However, there are also fairly common variants of the *MAOA* gene that do not turn off the gene completely but that are responsible for different levels of expression. The variation is the result of repeat elements upstream of the gene that influence transcription. In individuals with a low number of repeats, there is less protein produced as compared to individuals with a high number of repeats. A low expression of *MAOA* makes an individual more likely to be

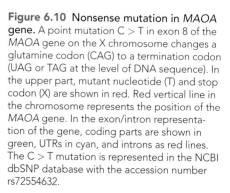

Figure 6.9 Dopamine breakdown catalyzed by enzyme monoamine oxidase A (MAOA).

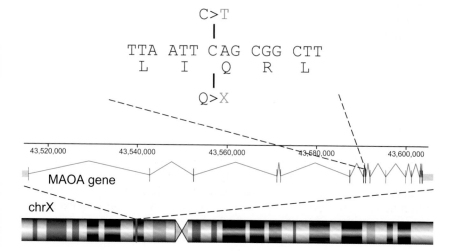

Figure 6.10 Nonsense mutation in *MAOA* gene. A point mutation C > T in exon 8 of the *MAOA* gene on the X chromosome changes a glutamine codon (CAG) to a termination codon (UAG or TAG at the level of DNA sequence). In the upper part, mutant nucleotide (T) and stop codon (X) are shown in red. Red vertical line in the chromosome represents the position of the *MAOA* gene. In the exon/intron representation of the gene, coding parts are shown in green, UTRs in cyan, and introns as red lines. The C > T mutation is represented in the NCBI dbSNP database with the accession number rs72554632.

aggressive. At the same time, low expression of *MAOA* cannot alone predict antisocial behavior; there is also an environmental component. Thus, it has been shown that boys with low *MAOA* expression that in addition have been abused are much more likely to develop antisocial behavior as compared to boys who were not abused. We see here an example of how the genotype and environment act together to produce a certain behavior.

A SPEECH AND LANGUAGE DISORDER

As a second example of a nonsense mutation that is responsible for a genetic disorder, we examine a change from CGA (Arg) to UGA in a gene named *FOXP2*. This gene first caught attention when it was identified as being involved in a language and speech disorder. The story of *FOXP2* begins with a specific multigenerational British family, known as KE, that was starting to be investigated in the late 1980s (**Figure 6.11**). It was noted that some of the family members had a linguistic impairment resulting from articulation problems ("verbal dyspraxia"). Their speech was quite incomprehensible to the inexperienced listener. In one scientific study, the affected individuals were shown to have problems with repetition of words and with carrying out certain oral movements on command. It was demonstrated that there is a neurological background to the disorder.

During the behavioral studies of the disorder, researchers also made attempts to identify the responsible gene. It was first learned that the gene was present in chromosomal band 7q31. Further analysis identified the gene *FOXP2*, and the affected members of the KE family were all characterized by a mutation G > A in this gene, corresponding to the amino acid substitution arginine to histidine (R553H). In 2005, another family with a speech and language disorder was examined and shown to have a mutation in *FOXP2* leading to a conversion of an arginine to a stop codon ("R328X", where X means a stop codon) (**Figure 6.12**). The inherited defects related to *FOXP2* are **dominant**. Both copies of a healthy *FOXP2* gene are required for normal speech and language development.

The FOXP2 protein is involved in the regulation of transcription. The acronym FOX stems from the fact that it has a "forkhead box" DNA-binding domain. The protein is expressed in the brain during development, and it is likely to regulate the expression of a number of genes of importance for brain function. One known target gene is *CNTNAP2* (contactin-associated protein-like 2). This protein has also been found to be associated with language defects. Thus, individuals have been identified with mutations in the *CNTNAP2* gene and as a result have difficulties in repeating nonsense words.

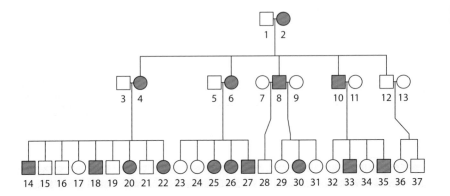

Figure 6.11 Pedigree of KE family. Males are shown as squares and females as circles. Family members affected with speech and language disorder are indicated in red.

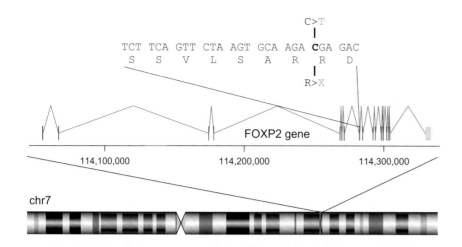

Figure 6.12 Nonsense mutation in *FOXP2* gene. A mutation C > T in the *FOXP2* gene leads to a conversion of an arginine to a stop codon (R328X). In the upper part, mutant nucleotide (T) and stop codon (X) are shown in red. Red vertical line in the chromosome represents the position of the *FOXP2* gene. In the exon/intron representation of the gene, coding parts are shown in green, UTRs in cyan, and introns as red lines.

The FOXP2 protein has been identified in all vertebrates, and its amino acid sequence is fairly conserved. FOXP2 in other animals might have a physiological role related to human speech, although other animals do not have the advanced speech of humans. In humans, speech learning is dependent on hearing other individuals. Also birds learn to sing from a tutor, such as an adult male bird. Experiments have been carried out with zebra finches where the expression of FOXP2 was reduced to 50% of its normal value. Interestingly, these birds were not as efficient as normal birds when it came to imitating tutor songs.

A TERMINATION CODON CHANGED TO A SENSE CODON: A LESS COMMON MUTATION

Whereas a mutation may convert a sense codon into a stop codon, the opposite change, the conversion of a stop codon into a sense codon (also known as a **stop-loss** mutation) is also possible. However, this mutation is much less frequent than nonsense mutations. We find one example in the context of an inherited disorder Hurler-Scheie syndrome. Individuals with this disorder have mutations in the gene *IDUA*, a gene encoding the enzyme α-L-iduronidase. This enzyme is responsible for degradation of glycosaminoglycans, such as dermatan sulfate and heparan sulfate. In one patient with Hurler-Scheie syndrome, a T > G mutation was detected that changes a termination codon UGA to GGA (glycine). This change produces an extended protein with an additional 38 amino acids at the carboxyl terminus of α-L-iduronidase (**Figure 6.13**).

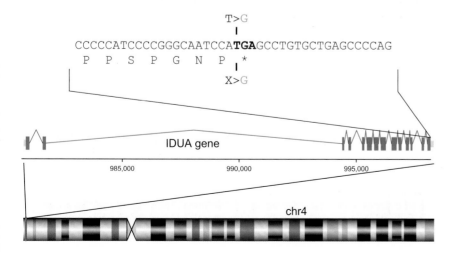

Figure 6.13 Stop-loss mutation in *IDUA* gene. A mutation T > G in the coding sequence of the gene *IDUA* converts a stop codon (TGA) to a glycine codon (GGA). In the upper part, mutant nucleotide (G) and amino acid (G) are shown in red. Red vertical line in the chromosome represents the position of the *IDUA* gene. In the exon/intron representation of the gene, coding parts are shown in green, UTRs in cyan, and introns as red lines.

Figure 6.14 Frameshift mutation in *HEXA* gene. An insertion of TATC in the coding sequence of *HEXA* causes Tay-Sachs disease. In the upper part, the insertion sequence and resulting novel amino acid sequence are shown in red. The small arrow in the codon TCC indicates the site of insertion. Red vertical line in the chromosome represents the position of the *HEXA* gene. In the exon/intron representation of the gene, coding parts are shown in green, UTRs in cyan, and introns as red lines. The insertion mutation is represented in the NCBI SNP database with the accession number rs387906309.

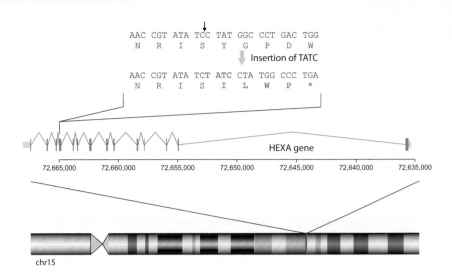

INDEL MUTATIONS OF CODING SEQUENCES: TAY-SACHS DISEASE REVISITED

We now turn to indel mutations in coding sequences. Typically such mutations have a more severe impact on the protein product of the gene. They are likely to disrupt the reading frame during translation, unless the number of inserted or deleted nucleotides is divisible by three. A frameshift error is also likely to generate a premature stop codon, initiating NMD, as discussed above.

We already mentioned Tay-Sachs disease when dealing with point mutations in the coding sequence of the *HEXA* protein gene. There is a much more common mutation in patients with this disorder. About 70% of the Ashkenazi Jewish population with Tay-Sachs disease has a 4 nucleotide insertion in exon 11 of *HEXA*. As illustrated in **Figure 6.14**, the result of this mutation is that a new reading frame is entered where a stop codon is encountered after a few codons downstream of the insertion site.

FRAMESHIFT ERROR AND RESISTANCE TO HIV INFECTION

A second example of an indel mutation is a 32 nucleotide deletion that causes a frameshift error in the gene named *CCR5* (short for "CC chemokine receptor 5") (**Figure 6.15**). The allele with the 32 nucleotide deletion is found in up to 20% of Europeans but in less than 1% in African and eastern

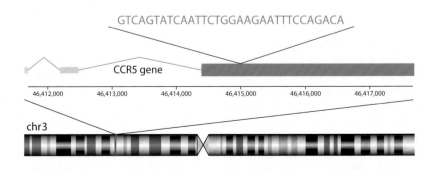

Figure 6.15 Gene CCR5 and 32 nucleotide deletion. A deletion of 32 nucleotides (sequence in red) results in a translational frameshift.

Asian populations. It does not cause disease—rather it protects against infection by the human immunodeficiency virus (HIV).

The product of the *CCR5* gene is a protein found on the surface of white blood cells. Most forms of HIV make use of this protein to enter and infect host cells. The mutant protein is not functional, and as a result individuals with this mutation are resistant to infection by HIV. It may have been selected because it protected against smallpox in Europe. Individuals with the deletion allele of *CCR5* are healthy, suggesting that *CCR5* is dispensable. Homozygotes are strongly resistant to HIV-1 infection, whereas heterozygotes are much less resistant. A therapy against HIV infection is being tried where patient cells are modified *ex vivo* to carry the deletion allele, and the modified cells are then reintroduced into the body (for more on gene therapy, see Chapter 14).

AN INFRAME DELETION IS A COMMON CAUSE OF CYSTIC FIBROSIS

We have seen that the effect of the shift of reading frame is that an erroneous protein sequence will be produced right after the mutation site and that a stop codon is likely to occur in the novel reading frame. However, in the event the indel is of a size divisible by three, the outcome is less dramatic for the protein product in the sense that the protein will not be truncated. As an example, we consider a mutation that is a frequent cause of the disorder cystic fibrosis.

Cystic fibrosis (CF) is one of the most common inherited disorders. It affects about 70,000 people worldwide. Although it has been identified in different human populations, it affects mainly individuals of European descent, where 1 in 2,500 newborns are affected. CF is characterized by abnormally viscous secretions in the airways of the lungs and in the ducts of the pancreas. These lead to inflammation and damage to both lungs and pancreas. However, treatment of CF has markedly improved over the years.

CF is a recessive disease. In 1989, the responsible gene was identified. It encodes a protein, CFTR, short for "cystic fibrosis transmembrane conductance regulator." The function of the protein is to transport chloride ions across the cell membrane in a regulated manner. When the CFTR protein is defective, there is a buildup of thick and sticky mucus in the lungs, pancreas, and other organs. In the lungs, inhaled bacteria are cleared inefficiently, with inflammation as a result.

The CFTR protein is made up of different functional domains as shown in **Figure 6.16**. In addition to membrane-spanning domains and the actual chloride channel, there are two nucleotide-binding domains and one regulatory domain.

The most common mutation in CF patients is a deletion of three nucleotides, CTT, from exon 11. The reading frame stays unaffected downstream of the mutation, and the effect on the protein is a deletion of a single amino acid, phenylalanine in position 508 in the polypeptide chain. The mutation is known as F508del (**Figure 6.17**).

What are the molecular consequences of the F508del mutation? First, it results in aberrant folding of CFTR, and the misfolded protein is degraded. The minor fraction of the mutant protein that reaches the cell membrane has reduced activity. What is the effect of the mutation on the structure of the protein? Evidence has been presented that the amino acid F508 mediates an interaction between the surfaces of two different domains of CFTR—NBD1 and a cytoplasmic portion of the C-terminal membrane-spanning

Figure 6.16 The structure of CFTR protein.
(a) The different domains of CFTR are displayed schematically. They include membrane-spanning domains (MSD1 and MSD2, green), NBD1 (nucleotide binding domain 1), NDB2 (nucleotide binding domain 2), and a regulatory subunit (R). In between MSD1 and MSD2 is the crucial chloride channel of the protein. (b) A linear representation of the protein highlights the interaction between a region CL4 and F508 located in the domain NBD1. MSD1 and MSD2 are the two membrane-spanning domains as shown in A. (c) A model of the three-dimensional structure of CFTR shows a close interaction of F508 and CL4. Colors are as in the other panels. ([c] Serohijos AW et al. Phenylalanine-508 mediates a cytoplasmic-membrane domain contact in the CFTR 3D structure crucial to assembly and channel function. *Proc Natl Acad Sci USA* 105(9):3256–3261, Copyright 2008 National Academy of Sciences, U.S.A.)

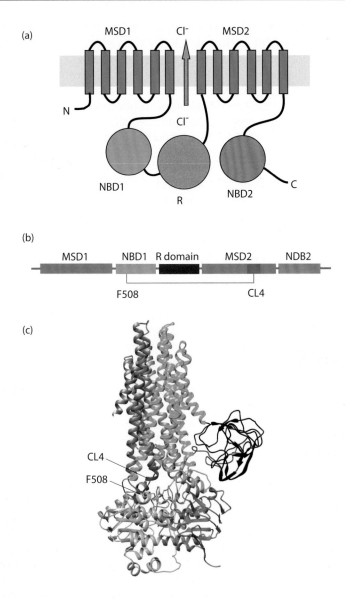

domain (MSD2) (see **Figure 6.16b** and **c**). This interaction is disturbed in the F508del mutant protein, preventing correct folding of CFTR. Finally, in addition to the different effects on protein structure, another effect of the F508del mutation is that it causes a structural change in the mRNA encoding CFTR, leading to a reduction in translation efficiency.

About 90% of CF sufferers have the F508del mutation in at least one allele. There are also many other mutations in CF patients, of which the most common are G542X (5%), G551D (4%), W1282X (4%), N1303K (2.5%), R553X (2.5%), and R117H (1%). These mutations, as well as a selection of other, less common pathogenic mutations are shown in **Figure 6.18**.

Figure 6.17 A common mutation in cystic fibrosis. The most common mutation, F508del (SNP rs113993960), in CF patients is a deletion of a three nucleotide sequence, CTT, from the coding sequence. The effect on the protein is a deletion of phenylalanine (F).

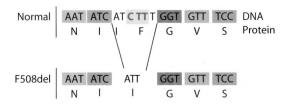

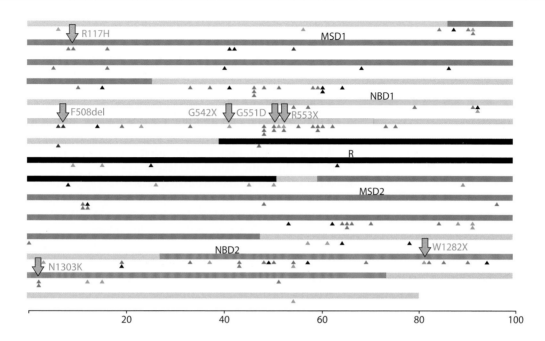

Figure 6.18 Mutational spectrum of CFTR. Pathogenic mutations in CFTR are shown with symbols as in Figure 6.2. Individual protein domains are colored as in Figure 6.16 with MSD1 and MSD2 in green, NBD1 in pink, NDB2 in red, and the R domain in black. The most common mutations in cystic fibrosis, including F508del, are indicated with orange arrows.

SUMMARY

- In this chapter, substitution and indel mutations in coding sequences of proteins were presented. The substitution mutations may confer;
  1. No change in the protein (synonymous mutation);
  2. Amino acid replacement (nonsynonymous mutation, missense);
  3. Conversion to stop codon (nonsense or stop-gain mutation); or
  4. Conversion of stop codon to amino acid (stop-loss mutation).
- An indel mutation is likely to cause a translation frameshift error. This change is typically more deleterious than a point mutation, as it will often result in a truncated protein. In addition, the protein product will typically be degraded or the mRNA subject to nonsense mediated decay. However, when an indel involves a small number of nucleotides divisible by three, it leads to an insertion or deletion of a small number of amino acids.
- Mutations in coding sequences are responsible for a large number of monogenic disorders. In this chapter we encountered Tay-Sachs disease, hereditary hemochromatosis, Brunner syndrome, a speech disorder, and cystic fibrosis.
- A nonsynonymous mutation is the most common cause of monogenic disorders.
- Many different mutations in one gene may give rise to disorder, as illustrated by the many variants identified in the β-globin gene.
- The phenotypes of different mutations in the same coding sequence are often similar but need not be identical.

QUESTIONS

1. Consider the following short 18 nucleotide DNA sequence specifying a protein and with codons separated by dashes:

   5′ ATG-CAA-GTA-TTA-GCG-TGA 3′

   Also consider mutations A > G in position 1, A > G in position 5, T > A in position 11, A > G in position 12, T > C in position 16, and G > A in position 17 of the DNA sequence. Which of these mutations may be classified into *nonsynonymous*, *synonymous*, *missense*, *stop-gain*, or *stop-loss* (*nonsense*) mutations, respectively?

2. Consider different insertion mutations in the sequence in the previous question. What is the effect on the translation product with insertions T, TA, TAT, and TATA, respectively, in between positions 3 and 4?

3. Consider the different categories *synonymous*, *missense*, *stop-gain*, and *stop-loss* mutations. What is the most common type of mutation in terms of inherited monogenic disorders?

4. Why do glycine and arginine codons mutate more rapidly than most other codons?

5. Why are certain inherited disorders more common in specific populations such as the Ashkenazi Jews?

6. What kinds of mutations may give rise to Tay-Sachs disease?

7. A common mutation in hereditary hemochromatosis is one that changes a cysteine codon to a tyrosine codon. What is the effect on the structure and function of the protein encoded?

8. Referring to the previous question—use the OMIM database (www.omim.org) to find out whether hereditary hemochromatosis is associated with changes in any of the other amino acids adjacent to the active site depicted in Figure 6.4b.

9. How does a nonsense (stop-gain) mutation affect translation and the translation product?

10. Why do most frameshift mutations give rise a premature stop codon?

11. What is the function of the protein encoded by the *MAOA* gene? How is the gene *MAOA* related to aggressive behavior?

12. Explain how gene *FOXP2* is linked to a language disorder.

13. What is the effect of a stop-loss mutation in a gene?

14. Give an example from the chapter where a variant/mutation may be beneficial.

15. What is the most common mutation in patients with cystic fibrosis? What is the effect of the mutation at the level of amino acid sequence?

URLs

NCBI ClinVar database (www.ncbi.nlm.nih.gov/clinvar/)

Cystic Fibrosis Mutation Database (www.genet.sickkids.on.ca/app)

Most common CFTR mutations in the world (http://www.genet.sickkids.on.ca/resource/Table1.html)

FURTHER READING

Mutation frequency of glycine and arginine codons

Cooper DN, Krawczak M. 1990. The mutational spectrum of single base-pair substitutions causing human genetic disease: Patterns and predictions. *Hum Genet* 85(1):55–74.

Vitkup D, Sander C, Church GM. 2003. The amino-acid mutational spectrum of human genetic disease. *Genome Biol* 4(11):R72.

Tay-sachs disease and hexosaminidase

Myerowitz R, Costigan FC. 1988. The major defect in Ashkenazi Jews with Tay-Sachs disease is an insertion in the gene for the alpha-chain of beta-hexosaminidase. *J Biol Chem* 263(35):18587–18589.

HEXA structure

Lemieux MJ, Mark BL, Cherney MM et al. 2006. Crystallographic structure of human beta-hexosaminidase a: Interpretation of Tay-Sachs mutations and loss of GM2 ganglioside hydrolysis. *J Mol Biol* 359(4):913–929.

Mark BL, Mahuran DJ, Cherney MM et al. 2003. Crystal structure of human β-hexosaminidase B: Understanding the molecular basis of Sandhoff and Tay-Sachs disease. *J Mol Biol* 327(5):1093–1109.

Structure of HFE

Bennett MJ, Lebron JA, Bjorkman PJ. 2000. Crystal structure of the hereditary haemochromatosis protein HFE complexed with transferrin receptor. *Nature* 403(6765):46–53.

Brunner syndrome

Brunner HG, Nelen M, Breakefield XO et al. 1993. Abnormal behavior associated with a point mutation in the structural gene for monoamine oxidase A. *Science* 262(5133):578–580.

Brunner HG, Nelen MR, van Zandvoort P et al. 1993. X-linked borderline mental retardation with prominent behavioral disturbance: Phenotype, genetic localization, and evidence for disturbed monoamine metabolism. *Am J Hum Genet* 52(6):1032–1039.

Piton A, Redin C, Mandel J-L. 2013. XLID-causing mutations and associated genes challenged in light of data from large-scale human exome sequencing. *Am J Hum Genet* 93(2):368–383.

FOXP2 and a language disorder

Fisher SE, Scharff C. 2009. *FOXP2* as a molecular window into speech and language. *Trends Genet* 25(4):166–177.

Lai CS, Fisher SE, Hurst JA et al. 2000. The SPCH1 region on human 7q31: Genomic characterization of the critical interval and localization of translocations associated with speech and language disorder. *Am J Hum Genet* 67(2):357–368.

Lai CS, Fisher SE, Hurst JA et al. 2001. A forkhead-domain gene is mutated in a severe speech and language disorder. *Nature* 413(6855):519–523.

MacDermot KD, Bonora E, Sykes N et al. 2005. Identification of *FOXP2* truncation as a novel cause of developmental speech and language deficits. *Am J Hum Genet* 76(6):1074–1080.

Nudel R, Newbury DF. 2013. *FOXP2*. Wiley interdisciplinary reviews. *Cogn Sci* 4(5):547–560.

Vernes SC, Newbury DF, Abrahams BS et al. 2008. A functional genetic link between distinct developmental language disorders. *N Engl J Med* 359(22):2337–2345.

HIV and *CCR5* gene

Faure E, Royer-Carenzi M. 2008. Is the European spatial distribution of the HIV-1-resistant CCR5-Delta32 allele formed by a breakdown of the pathocenosis due to the historical Roman expansion? *Infect Genet Evol: J Mol Epidemiol Evol Genet Infect Dis* 8(6):864–874.

A therapy against HIV infection

Tebas P, Stein D, Tang WW et al. 2014. Gene editing of CCR5 in autologous CD4T cells of persons infected with HIV. *N Engl J Med* 370(10):901–910.

Cystic fibrosis and *CFTR*

Castellani C, Cuppens H, Macek M Jr. et al. 2008. Consensus on the use and interpretation of cystic fibrosis mutation analysis in clinical practice. *J Cyst Fibros* 7(3):179–196.

Cutting GR. 2015. Cystic fibrosis genetics: From molecular understanding to clinical application. *Nat Rev Genet* 16(1):45–56.

Davies JC, Alton EW, Bush A. 2007. Cystic fibrosis. *BMJ* 335(7632):1255–1259.

Kerem B, Rommens JM, Buchanan JA et al. 1989. Identification of the cystic fibrosis gene: Genetic analysis. *Science* 245(4922):1073–1080.

Lewis HA, Zhao X, Wang C et al. 2005. Impact of the deltaF508 mutation in first nucleotide-binding domain of human cystic fibrosis transmembrane conductance regulator on domain folding and structure. *J Biol Chem* 280(2):1346–1353.

MacDonald KD, McKenzie KR, Zeitlin PL. 2007. Cystic fibrosis transmembrane regulator protein mutations: "Class" opportunity for novel drug innovation. *Paediatr Drugs* 9(1):1–10.

Riordan JR, Rommens JM, Kerem B et al. 1989. Identification of the cystic fibrosis gene: Cloning and characterization of complementary DNA. *Science* 245(4922):1066–1073.

Rommens JM, Iannuzzi MC, Kerem B et al. 1989. Identification of the cystic fibrosis gene: Chromosome walking and jumping. *Science* 245(4922):1059–1065.

Serohijos AW, Hegedus T, Aleksandrov AA et al. 2008. Phenylalanine-508 mediates a cytoplasmic-membrane domain contact in the CFTR 3D structure crucial to assembly and channel function. *Proc Natl Acad Sci USA* 105(9):3256–3261.

Watson MS, Cutting GR, Desnick RJ et al. 2004. Cystic fibrosis population carrier screening: 2004 revision of American College of Medical Genetics mutation panel. *Genet Med* 6(5):387–391.

Zhang Z, Chen J. 2016. Atomic structure of the cystic fibrosis transmembrane conductance regulator. *Cell* 167(6):1586–1597.e1589.

Triplet Repeats and Neurodegenerative Disorders

7

In Chapter 6, we became familiar with changes in DNA that affect the protein-coding regions of proteins. We considered point mutations that change the amino acid sequence of a protein and indel mutations that gave rise to a shift in reading frame. We now turn to another type of mutation that may alter the coding sequence. It affects DNA regions of a repetitive nature. For instance, a common type of repeat is a **trinucleotide tandem repeat**. These are trinucleotide sequences that are repeated multiple times and where the trinucleotide sequences are directly adjacent to each other. Such repetitive DNA is relatively likely to be mutated to achieve an expansion in the number of the repeats. First we consider a disorder caused by such a mutation.

THE DEATH OF PHEBE HEDGES

Phebe Hedges, born in 1764, lived in East Hampton, Long Island, New York. In June 1806 when she was 40 years old, she decided to take her life. The *Suffolk Gazette* published a notice in June 30, 1806:

> It has become our duty to publish the following melancholy circumstance, which took place at Easthampton a few weeks since :— Capt. David Hedges returned to his home in the evening, and found Mrs. Hedges ironing clothes, and apparently in health—he retired to bed and left her at that employment, but on awaking in the morning she was not to be found. After considerable search and inquiry her footsteps were traced from the house thro' fields of grain to the shore. Mrs. Hedges was about 40 years of age, and was much esteemed by her neighbors. This extraordinary step is attributed to her extreme dread of the disorder called St. Vitus' dance, with which she began to be affected, and which her mother now has to a great degree. From some arrangements of her clothing it appears she had for some time contemplated her melancholy end.

It was thought that Phebe Hedges suffered from a disease then known as St. Vitus' dance. We now refer to it as Huntington's disease (HD) or Huntington's chorea. A chorea is a disorder characterized by abnormal involuntary movements. The term *chorea* is from the Greek word *choreia*, "dance," as the movements of feet and hands are reminiscent of dancing. The English physician Thomas Sydenham described chorea already in 1696:

> He that is affected with this Disease, can by no means keep his hand in the same Posture for one Moment, if it be brought to the Breast or any other part, but it will be distorted to another position or Place by a certain Convulsion, let the Patient do what he can. If a Cup of Drink be put into his Hand, he represents a thousand Gestures, like Juglers, before he brings it to his Mouth—for whereas he cannot carry it to his Mouth in a Rightline, his hand being drawn hither and thither by the Convulsion, he turns it often about for some time, till at length happily reaching his Lips, he flings it suddenly into his Mouth, and drinks it greedily, as if the poor Wretch designed only to make sport.

GEORGE HUNTINGTON

The type of chorea that Phebe Hedges suffered from was more fully described in a medical text in 1842. The author was Charles Oscar Waters, a 26-year-old who had recently graduated from the Jefferson Medical College in Philadelphia, Pennsylvania. Another American physician, George Huntington (**Figure 7.1**), later examined the disorder more carefully. He was only 22 years old when he described his first results in a paper published in 1872. This was incidentally the only contribution of Huntington to academic medical science.

George Huntington was born in 1850 in East Hampton. He followed a family tradition as he became a physician in East Hampton. Thus, George's grandfather Abel started as a young man a medical practice in East Hampton in 1797 and also his son was in the same profession. George Huntington was early in his life fascinated by the disorder. He recalls:

> Driving with my father through a wooded road leading from East Hampton to Amagansett, we suddenly came upon two women, mother and daughter, both tall, thin, almost cadaverous, both bowing, twisting, grimacing. I stared in wonderment, almost in fear. What could it mean? My father paused to speak with them and we passed on [...] From this point on my interest in the disease has never wholly ceased.

Huntington later describes the symptoms of the disease. The symptoms begin extremely gradually, "by the irregular and spasmodic action of certain muscles, as of the face, arms, etc." The movements grow progressively worse "until the hapless sufferer is but a quivering wreck of his former self." The patient cannot hope for remission. "I have never known a recovery or even an amelioration of symptoms in this form of chorea: when once it begins it clings to the bitter end."

INHERITANCE OF HUNTINGTON'S DISEASE

Both Waters and Huntington realized that the disorder was hereditary. Huntington was probably helped in this conclusion by the fact that his father and grandfather were both medical practitioners in East Hampton. As Abel

Figure 7.1 George Huntington. 1850–1916. (Published under CC BY 4.0.)

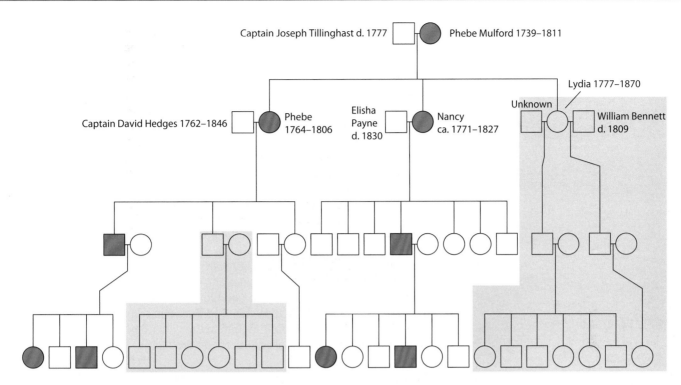

Figure 7.2 Family tree of Phebe Hedges. Tree shows the descendants of Phebe Mulford and Captain Joseph Tillinghast. Individuals affected by Huntington's disease are indicated in red. Males are shown as squares and females as circles. Descendents of children who do not receive the mutated gene are shown with a gray background.

had Phebe Hedges as one of his patients, he must have informed his son about her case.

Studies of the descendents of Phebe's mother reveal a pedigree as shown in **Figure 7.2**. Phebe's mother was the first verified carrier of the disease, and it was transmitted to later generations, showing the characteristics of a **dominant** disease. Remember that in such disorders only one copy of the mutated gene is required for the disorder to be expressed (Chapter 2). If one of the parents in a family has the disease, this individual has one normal gene and one mutated gene. In such a case, the probability that a child will inherit the disease is 50% (for probabilities of inheritance, see Chapter 2). The children who do not receive the mutated gene will be healthy and will not pass the disease on to coming generations. (These branches of the tree are shown with a gray background in Figure 7.2.)

A NEUROPSYCHIATRIC DISORDER

Huntington's disease (HD) is not only a movement disorder; the first symptoms are changes in behavior and cognition, and the disease may cause the patient to be aggressive, violent, anxious, and depressed. The folk singer Woody Guthrie (1912–1967) suffered from HD. He made notes describing his various symptoms. In a relatively early phase of the disease, he felt "terribly restless always. I get here and I want to be yonder. I get over yonder and I want to be back over here ... I don't trust anybody I see."

Woody inherited the disorder from his mother, Nora Guthrie. Nora was eventually admitted to a mental facility in Oklahoma. Woody described his mother's odd behavior in an earlier period of her life:

She would be alright for awhile, and treat us kids as good as any mother, and all at once it would start in something bad and awful something would start coming over her, and it would come by slow degrees. Her face would twitch and her lips would snarl and her teeth would show. Spit would run out of her mouth and she would start out in a low grumbling voice and gradually get to talking as loud as her throat could stand it—and her arms would draw up at her sides, then behind her back and swing in all kinds of curves. Her stomach would draw up into a hard ball, and she would double over into a terrible-looking hunch and turn into another person, it looked like, standing right there before Joy and me.

We have thus seen that HD is a dominantly inherited neuropsychiatric disorder that progresses slowly. HD affects mainly a group of nerve cells at the base of the brain known as the basal ganglia. Functions of the basal ganglia include control of movements. Symptoms generally start between the ages of 30 and 50 years. There are behavioral and cognitive symptoms in addition to motor dysfunction. The life expectancy after onset is approximately 10–20 years. The younger the onset of the disease occurs, the more severe are the symptoms. Depression is common in HD, and suicide attempts are more common than in the general population. HD affects about 1 in 10,000 individuals of European descent.

THE RESPONSIBLE GENE IS IDENTIFIED

For a long time after the characterization of HD by George Huntington, the genetic basis of the disorder was completely unknown. However, in 1983 the gene of interest was localized to chromosome 4. A breakthrough was made in 1993 when that precise gene was identified. The achievement relied on an analysis of DNA samples from families in the Lake Maracaibo region of Venezuela, an area with a high density of HD and relatively high level of inbreeding. The method used was **genetic linkage analysis,** and this was the first time this methodology was used to identify an autosomal[1] disease locus. The method is based on events during **meiosis,** which is the process where sperm and egg cells are generated. During meiosis a single cell divides twice to produce four cells that are haploid—that is, they each contain half the original amount of genetic information (**Figure 7.3**). Genetic linkage analysis is based on the fact that during meiosis two different loci on a chromosome have a certain probability of being separated through the natural process of chromosomal crossing over. The closer the two loci are along the length of the chromosome, the more likely they are to stay together—they are said to be genetically linked. Conversely, two loci that are distant are more likely to be separated into different chromatids during crossing over (see Figure 7.3). The location of a disease gene may be inferred from an analysis of genotypes in a family affected by the disease as outlined in **Figure 7.4**.

With the help of linkage analysis, it was shown that the gene responsible for HD is a gene now known as *HTT*. Further examination of this gene showed that it encodes the protein **huntingtin,** which is 3,144 amino acids in length.

---

[1]An autosomal disease involves a mutation in any of the nonsex chromosomes 1–22.

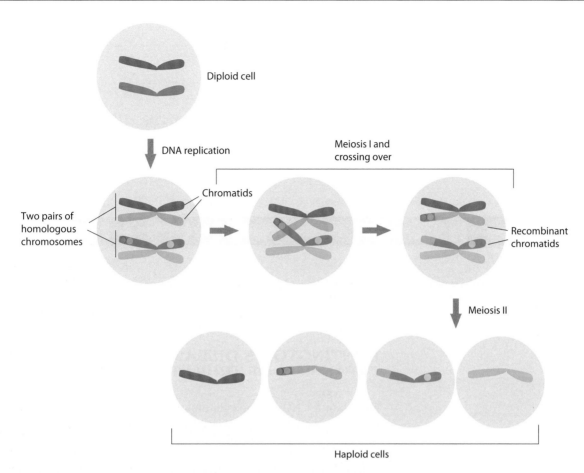

Figure 7.3 Closely located sites on a chromosome are likely to stay together after meiosis recombination. During meiosis, DNA of a diploid cell is first replicated. During meiosis I, two pairs of homologous chromosomes are aligned and then recombination causes an exchange of material between the chromosomes (crossing-over). In this process, two adjacent sites (shown with red and orange circles) are less likely to be separated than two distant sites (such as those shown with red and yellow circles). During meiosis II, four haploid cells are formed.

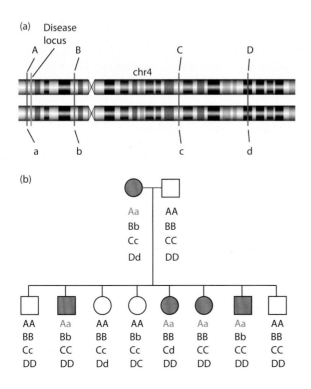

Figure 7.4 Genetic linkage analysis. For an example of linkage analysis, we imagine here a monogenic disease locus on chromosome 4 and assume a dominant mode of inheritance. (a) The two copies of chromosome 4 of one individual are shown with four polymorphic sites. Each site has two different alleles, for instance, one site has the "A" and "a" alleles. You could think of these as the variation of a single nucleotide polymorphism. This particular individual has the genotype Aa/Bb/Cc/Dd. The four polymorphisms are at different distances from the disease locus. (b) We consider a family with parents and eight children where the mother and four children (symbols filled with red) are affected by the disease. Males are shown as squares and females as circles. The genotypes of all family members have been determined. The mother has the genotype as in (a), whereas the father has the genotype AA/BB/CC/DD. It is noted that all affected family members have the genotype "Aa" (marked in green). At the same time, those affected with the disease do not all have the Bb genotype. Neither do they all have Cc or Dd. If we previously did not know about the location of the disease gene, it seems likely from the pedigree analysis that the disease locus is closer to the Aa site than any of the other polymorphic sites in chromosome 4.

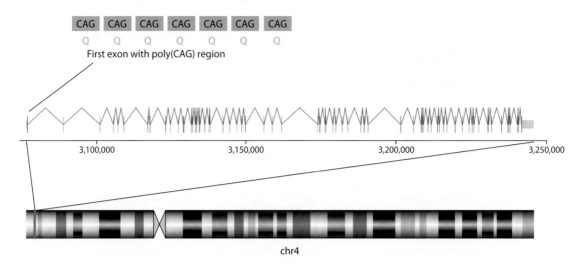

Figure 7.5 Gene *HTT* encoding huntingtin. Red vertical line in the chromosome represents the position of the gene. In the exon/intron representation of the gene, coding parts are shown in green, UTRs in cyan, and introns as red lines.

HUNTINGTON'S DISEASE IS CAUSED BY AN EXPANSION OF TRIPLET REPEATS

Important for HD is a polyglutamine (polyQ) region where the glutamine residues are encoded by CAG triplets in the mRNA (**Figure 7.5**). The polyQ region in huntingtin is located close to the N-terminus of the protein and is encoded by the first exon of the *HTT* gene (see Figure 7.5). The length of the polyQ region is subject to variation among human individuals. In healthy individuals, the CAG sequence is repeated 9–35 times, with an average median of between 17 and 20 triplets. When the number of repeats exceeds 35, this gives rise to HD.

Persons with a number of repeats in the lower 30s are nonsymptomatic but carry a **premutation**, which is to say they are at risk of having children with a larger number of repeats and, therefore, symptomatic disease. This means that repeats above a certain size range are "unstable" and likely to be expanded in the next generation. It is interesting to note that in the premutation state, an expansion in repeat length is apparently more likely than reduction.

Individuals with 36–39 repeats are less likely to show symptoms of the disease, and if they are affected by the disease it is likely to develop later in life. The age of onset of HD is actually inversely proportional to the number of repeats. For instance, onset during childhood is associated with 75 or more repeats. The phenomenon that the symptoms of some genetic conditions tend to appear at an earlier age and/or become more severe as the disorder is passed from one generation to the next is known as **anticipation**. The pedigree in **Figure 7.6** illustrates changes in the number of repeats, age of onset, and anticipation in a family affected by HD.

THE STRUCTURE OF HUNTINGTIN

Based on the amino acid sequence of the huntingtin protein, it is predicted to contain a number of HEAT repeats. (The acronym "HEAT" is derived from four proteins in which the repeat is found—**h**untingtin, elongation factor 3 [**E**F3], protein phosphatase 2A [PP2**A**], and the yeast kinase **T**OR1.) Each HEAT repeat is about 40 amino acids and is composed of two

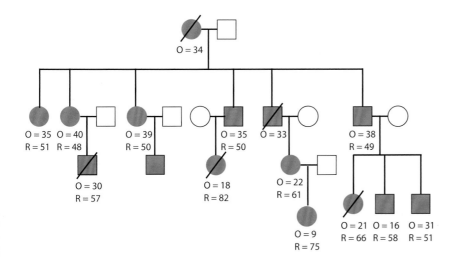

Figure 7.6 Pedigree illustrating anticipation in the context of Huntington's disease. Symbols filled with red are individuals affected by HD, and diagonals indicate deceased family members. Males are shown as squares and females as circles. The majority of affected individuals are indicated by the age of disease onset (O = N) and the number of CAG repeats in the *HTT* gene (R = N). The pedigree shows that from one generation to the next, the number of repeats tends to increase, and disease onset is earlier in life. (Adapted from *Am J Hum Genet* 57(3), Ranen NG et al. Anticipation and instability of IT-15 (CAG)n repeats in parent-offspring pairs with Huntington disease. 593–602. Copyright 1995, with permission from Elsevier.)

α-helices joined by a short loop (**Figure 7.7a**). The prediction of HEAT repeats is verified by structural studies using cryo-electron microscopy. **Figure 7.7b** shows a structure where huntingtin is in a complex with the protein HAP40. Huntingtin is here composed of three domains: the N-HEAT and C-HEAT domains with a bridge domain in between. The N-HEAT and C-HEAT domains form solenoid structures, superhelical structures formed by the HEAT repeats.

The polyQ region in huntingtin is close to the N-terminal end of the protein. However, significant parts of it are not visible using x-ray

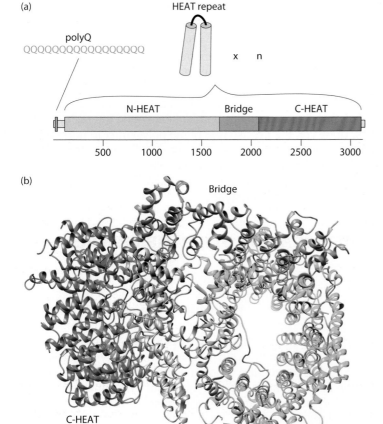

Figure 7.7 Huntingtin structure. (a) The major part of huntingtin is composed of a large number of HEAT repeats, where each repeat is built from two α-helices that are linked by a short sequence of amino acids. The positions of three subdomains of the protein, N-HEAT, bridge domain, and C-HEAT, along the primary structure of huntingtin are indicated. (b) Structural model of huntingtin as elucidated with cryo-electron microscopy. The domains N-HEAT (cyan), C-HEAT (red), and the intervening bridge domain (orange) are indicated, as well as the protein HAP40 (gray). The unordered polyQ region of huntingtin is not visible in the structure. ([b] Adapted from Guo Q et al. 2018. *Nature* 555:117–120, PDB ID:6 EZ8.)

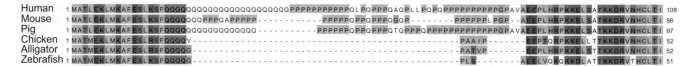

Figure 7.8 Huntingtin is conserved among vertebrates. Amino acid sequences of the N-terminal portion of huntingtin from six different vertebrates are shown. Mammals have more extensive polyglutamine and polyproline regions as compared to the lower animals. The background colors reflect the physicochemical properties of the amino acids.

crystallography. Based on this, we may conclude that the polyQ tract is flexible and does not have a distinct conformation. These structural properties are characteristic also of polyQ regions that are present in many other proteins.

MOLECULAR FUNCTIONS OF HUNTINGTIN

The precise role of the human huntingtin protein is not known. The protein is present in all vertebrates. It is conserved in this group of animals, meaning that similar sequences are found in many different animals (**Figure 7.8**). During evolution of the huntingtin protein in vertebrates, the size of the polyQ region was increased in mammals, suggesting that thereby the properties of huntingtin were changed. This may seem a hazardous means of modifying its activity, as a too extensive repeat expansion causes HD.

Huntingtin is ubiquitously expressed in human cells, although it is more abundant in the nervous system than in other tissues. It is essential for normal embryonic development. There are a variety of processes within the cell that involve huntingtin. For instance, there is evidence that it is important during transport processes. Furthermore, it is associated with **apoptosis**, and it may also be involved in transcriptional control as it is known to bind a number of different transcription factors.

A polyQ tract is present in many other proteins than huntingtin, and in general, they take part in protein-protein interactions. They are known to be present in many proteins related to transcriptional regulation. The size of the polyQ repeat determines both solubility and interactions with other proteins. However, the precise role of the polyQ tract in normal huntingtin remains unknown. Furthermore, it is not clear how the increase in polyQ length gives rise to Huntington's disease. We here see an example of how much more research is required to understand the detailed relationship between genotype and phenotype.

DNA REPLICATION SLIPPAGE ERRORS

Why does the number of CAG repeats in DNA sometimes change? When a region of DNA containing short tandem repeats is replicated, there is an increased probability of **replication slippage** (**Figure 7.9**). This is a mispairing that occurs between the two DNA strands because of ambiguity, where one strand is shifted with respect to the other. For instance, a repeat expansion may occur because the newly synthesized strand loops out during replication (known as backward slippage). Conversely, if a portion of the template strand loops out, the number of repeats will decrease (forward slippage), a mechanism much less common than backward slippage.

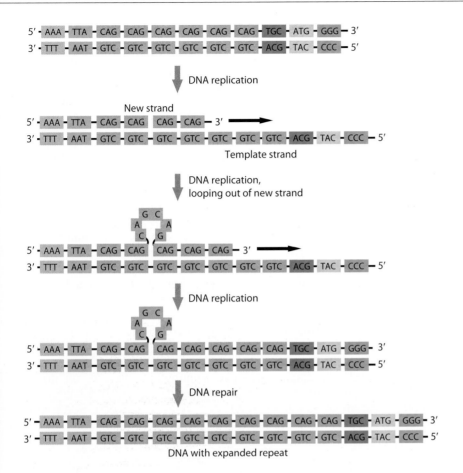

Figure 7.9 Errors that arise by slippage during DNA replication. Replication of DNA with a tandem repeat region (orange) is illustrated in which backward slippage occurs. The newly synthesized strand first detaches from the template strand and pairs with another repeat region upstream. Then DNA polymerase reassumes replication, but the enzyme repeats the insertion of sequences previously added. The result is that a portion of the template strand is replicated twice into the newly synthesized strand. Thereby, the template strand and the newly synthesized strands are not identical. A DNA repair mechanism then realigns the template strand with the new strand, and the loop is straightened out. The example here shows a change involving two repeats, but looping out may involve any number of repeats.

OTHER REPEAT DISORDERS

Huntington's disease is not the only inherited disorder that involves a protein with polyQ tracts. A number of the diseases known as **spinocerebellar ataxias** (**SCAs**) are associated with triplet repeats. Many of these are a result of CAG triplet expansion as discussed in more detail in the following section. However, there are other examples, such as a GCG repeat in the coding region of the protein encoded by the *PABPN1* gene. The GCG repeat encodes a polyalanine tract. A normal size of the repeat is 6 codons, but when it is expanded to 7–13 it gives rise to a disease named *oculopharyngeal muscular dystrophy*.

Many of the triplet repeats resulting in disorders are not in the coding region of a protein. For instance, there are rare SCAs that are caused by tri-, tetra-, or penta-nucleotide repeats in the noncoding regions of genes. One example is SCA12 with a CAG repeat in a region toward the 5′ end of the gene, presumably involved in the regulation of transcription. In addition, in the case of a hereditary form of the disease amyotrophic lateral sclerosis (ALS), a hexanucleotide (GGGGCC) repeat expansion is in an intronic region. The effect of this expansion may be at the level of RNA, or possibly at the level of amino acid sequence, because there is evidence that the repeat region is translated using an unusual mode of translation that is not dependent on a AUG start codon. Finally, there is the fragile X syndrome, another example of a disease resulting from repeat expansion. This disorder is associated with CGG triplets upstream of the coding sequence, as further discussed in Chapter 10. But for the remaining part of this chapter, we focus on coding sequence CAG repeats that result in different forms of spinocerebellar ataxias.

SPINOCEREBELLAR ATAXIAS CAUSED BY CAG TRIPLET EXPANSION

The spinocerebellar ataxias form a large group of inherited disorders. Ataxia is derived from a Greek word meaning "lack of order," and in the medical literature ataxia refers to difficulties of coordinating movements. The diseases are associated with defects in the cerebellum ("little brain"), and pons and inferior olives (both part of the brainstem), as well as the spinal cord, hence the term *spinocerebellar* ataxia.

In patients with inherited SCA, the first symptoms are gait imbalance followed by problems with carrying out voluntary, planned movements by the hands, arms, and legs. Slurring of speech is another common symptom. There may also be visual problems, such as difficulty with focusing, double vision, and problems with saccades (rapid eye movements).

The different forms of inherited spinocerebellar ataxias have been assigned numbers according to the order in which the disease was identified (SCA1—SCA43). The gene behind SCA1 was identified in 1994.

Spinocerebellar ataxias are—like Huntington's disease—all dominantly inherited disorders. It is estimated that 1–5 in 100,000 have SCA. Among these about 65% have SCA1, 2, 3, 6, 7, or 8, where SCA3, also known as Machado-Joseph's disease, is the most common. Among the most common SCAs, SCA1, 2, 3, and 7 are caused by an expansion of CAG repeats in the coding sequence of the respective gene, *ATXN1*, *ATXN2*, *ATXN3*, and *ATXN7*. SCA6 is caused by a CAG repeat in a 5′ UTR region and SCA8 by an expansion of a CTG repeat in a noncoding RNA. (SCA8 is here classified as a "CAG" repeat disorder although the trinucleotide of the repeat is expressed as 5′-CTG-3′. This is because the CAG is the complement of CTG and thus present at the level of the double-stranded DNA.) For more information on the different SCAs associated with CAG repeats, see **Table 7.1**.

In the versions of SCA related to polyQ expansion, several organs in the body are affected. The damage to the brainstem is more significant than in the cerebellum. There is a significant loss of cells in different parts of the central nervous system. The severity of disease is often dependent on the length of the repeat, as we have already noted for HD. For more details on repeats lengths in spinocerebellar ataxias and in HD, see **Table 7.2**.

It is interesting to note that apart from the polyQ domain, the proteins responsible for SCA1, 2, 3, and 7 are completely unrelated in terms of sequence and structure. Thus, a number of ataxias are caused by proteins that share a polyQ region, although they are otherwise structurally and functionally unrelated. One is led to think that the actual polyQ tract plays a central role in the pathogenic action of the protein and that there could be molecular mechanisms involving polyQ that are common between the different ataxias. Evidence was recently presented that a common function of extended polyQ tracts, including those of huntingtin, is to interfere with an early step in an intracellular protein degradation pathway known as autophagy. At the same time, there are other polyQ-containing proteins that are not subject to expansion and that do not cause any inherited disorders. How common are CAG repeats in human proteins? There are about 264 proteins with a tandem repeat of at least 6 Qs, corresponding to about 1% of all human proteins.

Table 7.1 Spinocerebellar ataxias associated with CAG repeats

Disorder	Gene	Symptoms	Frequency (%) among all SCAs	Protein function (from UniProt/Genetics home reference)	Location of mutation
SCA1[a]	ATXN1	Ataxia. Signs of widespread cerebellar and brainstem dysfunction with relatively little supratentorial involvement. Neuropathologic findings include neuronal loss in the cerebellum and brainstem, and degeneration of spinocerebellar tracts.	6	Chromatin-binding factor. Regulates transcription.	CDS
SCA2[b]	ATXN2	Ataxia, dysarthria, slow saccades, and peripheral neuropathy.	15	May regulate trafficking of Epidermal Growth Factor Receptor (EGFR).	CDS
SCA3 (MJD)[c]	ATXN3	Also called Machado–Joseph disease (MJD). Most often begins as a progressive ataxia accompanied by lid retraction and infrequent blinking (creating "staring eyes"), ophthalmoparesis, and impaired speech and swallowing. Neuropathological findings include widespread degeneration of cerebellar afferent and efferent pathways, pontine and dentate nuclei, and the cell bodies of the substantia nigra, subthalamic nucleus, globus pallidus, cranial motor nerve nuclei and anterior horns.	21	Enzyme involved in a system for protein degradation that removes damaged or excess protein. Also regulates transcription.	CDS
SCA6[d]	CACNA1A	In contrast to the other relatively common SCAs (1, 2, 3 and 7), SCA6 represents a "milder" disease, most often manifesting as a "pure" cerebellar ataxia accompanied by dysarthria and gaze-evoked nystagmus. Compared to other SCAs, noncerebellar symptoms occur much less frequently in SCA6. The disease progresses more slowly than in other SCAs and is usually compatible with a normal life span.	15	Part of a channel that transports calcium ions across cell membranes. Such calcium channels play a key role in a cell's ability to generate and transmit electrical signals. Calcium ions are involved in many different cellular functions, including cell-to-cell communication, muscle contraction and the regulation of certain genes.	CDS
SCA7[e]	ATXN7	Is distinguished from other SCAs by the presence of retinal degeneration. In other respects, SCA7 resembles the other SCAs characterized by ataxia and brainstem findings.	5	Involved in transcriptional control.	CDS
SCA12[f]	PPP2R2B	CAG repeat expansion in 5′ UTR. First symptom is typically an action tremor of the arms. The tremor is eventually accompanied by head tremor and ataxia.	<1	A regulatory subunit of protein phosphatase 2 (PP2). PP2 is an enzyme implicated in the negative control of cell growth and division.	5′ UTR
SCA17[g,h,i]	TBP	More than any other SCA, SCA17 manifests with widespread cerebral as well as cerebellar dysfunction. Affected persons typically present in young-adulthood or mid-adulthood with progressive gait and limb ataxia that is usually accompanied by dementia, psychiatric symptoms, and varying extrapyramidal features, including parkinsonism, tremor, dystonia, and sometimes chorea. Ataxia may not be the predominant feature. In some cases, SCA17 even resembles Huntington disease. Consistent with this more widespread neurological phenotype, the MRI findings in SCA17 include diffuse cerebral and cerebellar atrophy.	<1	General transcription factor that functions at the core of the DNA-binding multiprotein factor TFIID. Binding of TFIID to the TATA box is the initial transcriptional step of the pre-initiation complex (PIC), playing a role in the activation of eukaryotic genes transcribed by RNA polymerase II.	CDS

Source: Symptoms are as described in Paulson HL. 2009. *J Neuroophthalmol* 29(3):227–237.

a   Zuhlke C et al. 2002. Spinocerebellar ataxia type 1 (SCA1): Phenotype-genotype correlation studies in intermediate alleles. *Eur J Hum Genet* 10(3):204–209.
b   Sanpei K et al. 1996. Identification of the spinocerebellar ataxia type 2 gene using a direct identification of repeat expansion and cloning technique, DIRECT. *Nat Genet* 14(3):277–284.
c   Kawaguchi Y et al. 1994. CAG expansions in a novel gene for Machado-Joseph disease at chromosome 14q32.1. *Nat Genet* 8(3):221–228.
d   Li L, Saegusa H, Tanabe T. 2009. Deficit of heat shock transcription factor 1–heat shock 70 kDa protein 1A axis determines the cell death vulnerability in a model of spinocerebellar ataxia type 6. *Genes Cells* 14(11):1253–1269.
e   David G et al. 1997. Cloning of the SCA7 gene reveals a highly unstable CAG repeat expansion. *Nat Genet* 17(1):65–70.
f   Holmes SE et al. 1999. Expansion of a novel CAG trinucleotide repeat in the 5′ region of PPP2R2B is associated with SCA12. *Nat Genet* 23(4):391–392.
g   Nakamura K et al. 2001. SCA17, a novel autosomal dominant cerebellar ataxia caused by an expanded polyglutamine in TATA-binding protein. *Hum Mol Genet* 10(14):1441–1448.
h   Zuhlke C et al. 2001. Different types of repeat expansion in the TATA-binding protein gene are associated with a new form of inherited ataxia. *Eur J Hum Genet* 9(3):160–164.
i   Koide R et al. 1999. A neurological disease caused by an expanded CAG trinucleotide repeat in the TATA-binding protein gene: A new polyglutamine disease? *Hum Mol Genet* 8(11):2047–2053.

Table 7.2 Repeat lengths in spinocerebellar ataxias		
Gene	Repeat range of healthy individuals	Pathogenic repeat range
ATXN1	6–35	49–88
ATXN2	14–32	33–77
ATXN3	12–40	55–86
ATXN7	7–17	38–120
TBP	25–44	47–63
HTT	9–36	37–100
CACNA1A	6–17	21–30

Note: The normal repeat ranges and pathogenic ranges are shown for a number of genes involved in spinocerebellar ataxias and Huntington's disease.

SUMMARY

- The locus of Huntington's disease (HD) was identified with genetic linkage analysis. The method is based on the fact that during meiosis two different loci on a chromosome have a certain probability of being separated through the process of chromosomal crossing over.
- The gene responsible for HD is named *HTT* and encodes the protein huntingtin.
- HD is caused by an expansion of CAG triplet repeats in the *HTT* gene. The repeats are located in the coding sequence and specify a polyglutamine stretch at the N-terminus of the protein.
- Individuals with a *premutation* carry an intermediate number of repeats not quite sufficient for the disease to be expressed, but the offspring are more likely to have more repeats and clinically apparent disease.
- *Anticipation* is the phenomenon that the symptoms of some genetic conditions tend to appear at an earlier age and/or become more severe as the disorder is passed from one generation to the next.
- A mechanism for generating a different number of tandem repeats is *slippage* during DNA replication. An expansion of repeats is more common than a reduction.
- In addition to HD, a number of *spinocerebellar ataxias* (SCAs) are caused by changes in the number of trinucleotide repeats. The most common SCAs are caused by CAG triplet repeat expansion. The genes involved in SCAs depend on the particular disease, but typically involve a polyglutamine region of the gene product.

QUESTIONS

1. Provide a sequence (5′ to 3′) of 12 nucleotides representing a region of a trinucleotide *tandem repeat*. Also provide the complementary sequence.
2. Why are short repeat sequences prone to mutation?
3. What is the molecular background to Huntington's disease?
4. Draw a simple pedigree to illustrate how a disorder is dominantly inherited.
5. Consider a normal allele A and a disease variant allele a. If one of the parents has the genotype Aa and the other AA, what is the probability that a child of these parents is affected by the disease in case it is a

recessive and *dominant* disorder, respectively? (Make use of the Punnett square method in Chapter 2.)

6. What was the method used to identify the genomic locus responsible for Huntington's disease? What are the basic principles of this method?
7. What is characteristic of the three-dimensional structure of huntingtin? What is a HEAT repeat? What part of the protein is related to Huntington's disease and what is characteristic of its structure?
8. Explain the concepts of *premutation* and *anticipation*.
9. What disorders other than Huntington's disease are related to an expansion of poly(CAG) regions?
10. Consider *dinucleotide* and *tetranucleotide* repeat regions within a coding sequence. What types of amino acid sequences are encoded by such repeat regions, and what would be the effect of an expansion?

FURTHER READING

Early history of Huntington's disease

Wexler A. 2008. *The woman who walked into the sea: Huntington's and the making of a genetic disease.* Yale University Press, New Haven, CT.

Wexler AR. 2002. Chorea and community in a nineteenth-century town. *Bull Hist Med* 76(3):495–527.

On Woody Guthrie

Klein J. 1980. *Woody Guthrie: A life.* A.A. Knopf, New York.

Paulson HL, Albin RL. 2011. Huntington's disease: Clinical features and routes to therapy. In: Lo DC, Hughes RE, eds. *Neurobiology of Huntington's Disease: Applications to Drug Discovery.* CRC Press/Taylor and Francis Group, Boca Raton, FL.

Identification of the huntingtin gene

Gusella JF, Wexler NS, Conneally PM et al. 1983. A polymorphic DNA marker genetically linked to Huntington's disease. *Nature* 306(5940):234–238.

The Huntington's Disease Collaborative Research Group. 1993. A novel gene containing a trinucleotide repeat that is expanded and unstable on Huntington's disease chromosomes. *Cell* 72(6):971–983.

Repeat expansion by DNA repair mechanisms

Gomes-Pereira M, Hilley JD, Morales F et al. 2014. Disease-associated CAG·CTG triplet repeats expand rapidly in non-dividing mouse cells, but cell cycle arrest is insufficient to drive expansion. *Nucleic Acids Res* 42(11):7047–7056.

Su XA, Freudenreich CH. 2017. Cytosine deamination and base excision repair cause R-loop-induced CAG repeat fragility and instability in *Saccharomyces cerevisiae. Proc Natl Acad Sci USA* 114(40):E8392–E8401.

Huntington's disease and anticipation/premutation

Ranen NG, Stine OC, Abbott MH et al. 1995. Anticipation and instability of IT-15 (CAG)n repeats in parent-offspring pairs with Huntington disease. *Am J Hum Genet* 57(3):593–602.

Huntingtin protein

Andrade MA, Bork P. 1995. HEAT repeats in the Huntington's disease protein. *Nat Genet* 11(2):115–116.

Gemayel R, Chavali S, Pougach K et al. 2015. Variable glutamine-rich repeats modulate transcription factor activity. *Molecular Cell* 59(4):615–627.

Kim MW, Chelliah Y, Kim SW et al. 2009. Secondary structure of huntingtin amino-terminal region. *Structure* 17(9):1205–1212.

Li W, Serpell LC, Carter WJ et al. 2006. Expression and characterization of full-length human huntingtin, an elongated HEAT repeat protein. *J Biol Chem* 281(23):15916–15922.

Palidwor GA, Shcherbinin S, Huska MR et al. 2009. Detection of alpha-rod protein repeats using a neural network and application to huntingtin. *PLOS Comput Biol* 5(3):e1000304.

Saudou F, Humbert S. 2016. The biology of huntingtin. *Neuron* 89(5):910–926.

Schaefer MH, Wanker EE, Andrade-Navarro MA. 2012. Evolution and function of CAG/polyglutamine repeats in protein-protein interaction networks. *Nucleic Acids Res* 40(10):4273–4287.

Seong IS, Woda JM, Song JJ et al. 2010. Huntingtin facilitates polycomb repressive complex 2. *Hum Mol Genet* 19(4):573–583.

Vijayvargia R, Epand R, Leitner A et al. 2016. Huntingtin's spherical solenoid structure enables polyglutamine tract-dependent modulation of its structure and function. *eLife* 5:e11184.

Huntingtin protein structure

Guo Q, Bin H, Cheng J et al. 2018. The cryo-electron microscopy structure of huntingtin. *Nature* 555:117.

Amyotrophic lateral sclerosis and repeat expansion

Dolgin E. 2017. The hexanucleotide hex. *Nature* 550:S106.

Freibaum BD, Lu Y, Lopez-Gonzalez R et al. 2015. GGGGCC repeat expansion in C9orf72 compromises nucleocytoplasmic transport. *Nature* 525(7567):129–133.

Herdewyn S, Zhao H, Moisse M et al. 2012. Whole-genome sequencing reveals a coding non-pathogenic variant tagging a non-coding pathogenic hexanucleotide repeat expansion in C9orf72 as cause of amyotrophic lateral sclerosis. *Hum Mol Genet* 21(11):2412–2419.

Jovicic A, Mertens J, Boeynaems S et al. 2015. Modifiers of C9orf72 dipeptide repeat toxicity connect nucleocytoplasmic transport defects to FTD/ALS. *Nat Neurosci* 18(9):1226–1229.

Paul JW, Gitler AD. 2014. Cell biology. Clogging information flow in ALS. *Science* 345(6201):1118–1119.

Zhang K, Donnelly CJ, Haeusler AR et al. 2015. The C9orf72 repeat expansion disrupts nucleocytoplasmic transport. *Nature* 525(7567):56–61.

Translation modes, including repeat associated non-AUG (RAN) translation

Gao FB, Richter JD, Cleveland DW. 2017. Rethinking unconventional translation in neurodegeneration. *Cell* 171(5):994–1000.

GCG repeat expansions in the *PABPN1* gene

Brais B, Bouchard JP, Xie YG et al. 1998. Short GCG expansions in the *PABP2* gene cause oculopharyngeal muscular dystrophy. *Nat Genet* 18(2):164–167.

Spinocerebellar ataxias

Bird TD. Hereditary ataxia overview. 1998 Oct 28 [Updated 2018 Sep 27]. In: Adam MP, Ardinger HH, Pagon RA et al., eds. *GeneReviews*® [Internet]. University of Washington, Seattle; 1993–2018. Available from: https://www.ncbi.nlm.nih.gov/books/NBK1138/

Orr HT, Chung MY, Banfi S et al. 1993. Expansion of an unstable trinucleotide CAG repeat in spinocerebellar ataxia type 1. *Nat Genet* 4(3):221–226.

Paulson HL. 2009. The spinocerebellar ataxias. *J Neuroophthalmol* 29(3):227–237.

Schols L, Bauer P, Schmidt T et al. 2004. Autosomal dominant cerebellar ataxias: Clinical features, genetics, and pathogenesis. The Lancet. *Neurology* 3(5):291–304.

The Untranslated Parts of a Message

8

In the previous chapters we considered the protein coding regions in the human genome. These regions cover only about 1.6% of the entire genome, but a large fraction of inherited disorders are the result of mutations in protein coding sequences (see Table 6.1). However, at this point we are to abandon the amino acid sequence coding regions of the genome and instead examine all other regions of interest. The first of these, to be covered in this chapter, are all other functional elements of the mRNA that are not the protein coding sequence. Thus, there are sequences both upstream and downstream of the coding region, the portions referred to as the 5′ and 3′ untranslated regions (UTRs). These UTRs contain a number of important functional elements, typically responsible for the regulation of protein synthesis. Some of these elements are responsible for interaction with proteins, and some of them with other RNAs.

ORGANIZATION OF mRNA AND FUNCTIONS WITHIN UTRs

A schematic view of an mRNA with UTRs is depicted in **Figure 8.1**. The 5′ UTR region is on average shorter than the 3′ UTR—the mean length of a 5′ UTR is about 500 nucleotides and that of a 3′ UTR is about 2000 nucleotides. In comparison, the mean length of the coding sequence is about 2100 nucleotides. While coding sequences cover about 1.6% of the human genome, the 5′ and 3′ UTRs make up about 0.3% and 1.4%, respectively.

Unlike the coding sequence of mRNA, the UTRs have a number of regulatory functions. The amount of protein formed is controlled in different ways, through control of mRNA translation initiation, translation rate, mRNA stability, efficiency of export of mRNA from the nucleus to the cytosol, and subcellular localization of the mRNA.

The 5′ UTR is involved in translational regulation, in particular, the process of translation initiation. As an example, we consider here an RNA structure element with a regulatory role. We also examine shorter 5′

Figure 8.1 Structure of mRNA and functional elements. The structure of an mRNA is shown with coding sequence (CDS), 5' and 3' untranslated regions (UTRs), and a poly(A) tail. Boxes below the mRNA show general functions, and boxes above show specific functional elements. Further abbreviations used are IRE (iron responsive element), uORF (upstream open reading frame), SECIS (selenocysteine insertion sequence), and SeCys (selenocysteine).

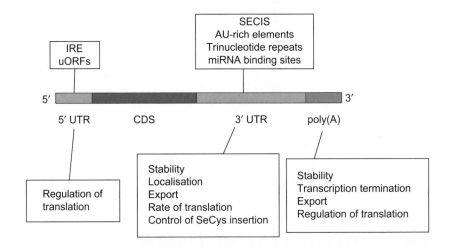

UTR **open reading frames**[1] that control translation of the main coding sequence of the mRNA.

As the 3' UTR is much longer than the 5' UTR, it has potential for more regulatory signals. We discuss two different examples showing how the 3' UTR is important for the stability of the mRNA; both of the examples are connected to specific inherited disorders. There are also sequence elements that guide the mRNA to the right part of the cell where synthesis of the protein is to occur. In addition, the 3' UTR is involved in the transport of mRNA from the nucleus to the cytosol. We also account for an example of sequence elements in the 3' UTR that are necessary for the decoding of codons UGA as the amino acid selenocysteine. Finally, 3' UTRs contain sequence elements recognized by a class of RNAs known as microRNAs—these elements take part in a mechanism of mRNA degradation and translational repression.

All eukaryotic mRNAs also contain a **poly(A) tail**, a sequence of As located at the 3' end of the molecule. The poly(A) tail is not encoded in the genome but is added enzymatically after transcription. Poly(A) tails initially have a size of about 250 residues, but the length may be reduced by enzymatic deadenylation. Essential for efficient initiation of translation is an interaction of the poly(A) tail with the 5' end of the mRNA. This circularization of the mRNA is mediated by proteins, and its efficiency is dependent on the length of the poly(A) tail. If the tail is shorter than 50 residues, translation is typically repressed. Thus, translational efficiency may be controlled by enzymatic polyadenylation and deadenylation. Polyadenylation also has a crucial role in control of transcription termination (Chapter 10).

IRON-RESPONSIVE ELEMENT REGULATES TRANSLATION AND mRNA STABILITY

Remember that many noncoding RNAs (ncRNAs), such as tRNA (Chapter 3), adopt characteristic structures based on internal base-pairing in the molecule. Not only in ncRNAs are such structures formed—they are present in almost any RNA sequence, including mRNA. Therefore, when considering an mRNA sequence, we must not only pay attention to its **primary structure**—the actual nucleotide sequence—but also its **RNA secondary structure**. Many important regulatory elements have a specific secondary structure. One such well-characterized element in mRNA UTRs is the **iron**

---

[1]An "open reading frame" is a continuous stretch of codons that contain a start codon at the 5' end and a stop codon at the 3' end.

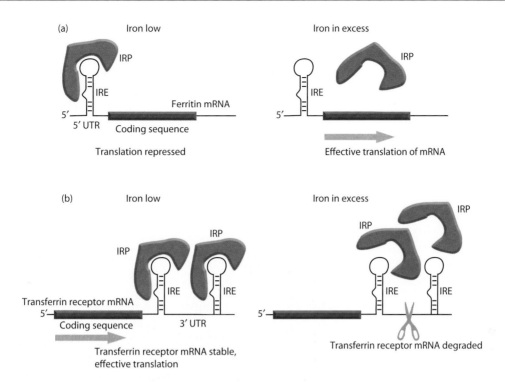

Figure 8.2 Regulation of translation by iron responsive elements. (a) In ferritin mRNA an iron responsive element (IRE) is located in the 5′ untranslated region (5′ UTR). When the level of iron is low, the IRE is bound by iron regulatory proteins (IRPs). Translation is blocked as a result of this protein binding. When the iron level is high, the IRP proteins have a lower affinity for IRE, and translation of the mRNA is efficient. (b) In the transferrin receptor (Tfr) mRNA, there are five copies of IRE located in the 3′ UTR—only two of these IREs are shown here. The binding of the IRP proteins results in stabilization of Tfr mRNA. Conversely, when iron is abundant, IRP proteins tend to have lower affinity to IREs, and as a result, Tfr mRNA degradation is stimulated.

responsive element (**IRE**) that plays a significant role in the regulation of iron metabolism.

A number of proteins are involved in the control of iron metabolism. We previously encountered one of them, the protein encoded by the hemochromatosis gene (HFE-1) with mutations like C282Y resulting in iron overload (Chapter 6). Another important protein is ferritin, whose function is to bind iron ions, ions that would otherwise be toxic to the cell. The production of ferritin is carefully controlled to maintain a suitable level of free iron ions. Ferritin is composed of two subunits, the light and heavy chains. The synthesis of both chains of ferritin is regulated through an intricate mechanism involving translation of its mRNA (**Figure 8.2a**). In the 5′ UTR of the ferritin mRNA is located a copy of the IRE. In a situation where cells are starved for iron, proteins called *iron regulatory protein* (IRP) tend to bind the IRE. As a result of this binding, translation is repressed, and ferritin production is lowered. Conversely, when iron is abundant, the IRP has a lower affinity to the IRE, and consequently, ferritin translation is more efficient.

Many other proteins of iron metabolism are regulated at the level of translation, such as the transferrin receptor (Tfr). Transferrin is a protein with a role to transport iron ions in the blood and to deliver these ions to cells. As part of this mechanism, cells have Tfrs located at the cell surface. The production of Tfr is subject to a control mechanism involving IRE elements (**Figure 8.2b**). In this case, the IREs are not located in the 5′ UTR, but instead the 3′ UTR of the Tfr mRNA contains five copies of the IRE. Their function is not to regulate the initiation of translation like for ferritin mRNA, but instead they affect the stability of the Tfr mRNA. When the level of iron is comparatively low, the binding of IRP tends to stabilize the mRNA. When iron is abundant, these proteins do not bind to the UTR. The result is that degradation of Tfr mRNA is stimulated, and less Tfr protein is produced.

The importance of the IRE is illustrated with cases of a genetic disorder, the **hereditary hyperferritinemia cataract syndrome**. It is a rare, dominant disorder with an estimated frequency of 1 in 200,000 individuals. Hyperferritinemia cataract syndrome is the result of an excess of ferritin in the blood (hence the term *hyperferritinemia*) and tissues. The protein starts to accumulate early in life, and one effect is a cataract—a clouding of the eye lenses. The first cases of this disorder were described in 1995. One of the reports was concerned with two families that had lived in Northern Italy for many generations. The affected family members experienced already in childhood poor visual acuity and difficulties seeing in the presence of very bright light. They also had high levels of ferritin in the blood. It turned out that the molecular basis of the disease is a mutation affecting the mRNA encoding the ferritin light chain. This mutation confers a change in an essential part of the IRE.

STRUCTURE OF THE IRON RESPONSIVE ELEMENT

The IRE is an RNA hairpin structure. Its context with respect to the ferritin light chain mRNA and important structural elements of the ferritin and Tfr IREs are shown in **Figure 8.3**. Consensus structures emerge when comparing IREs from different human genes and the corresponding genes in other

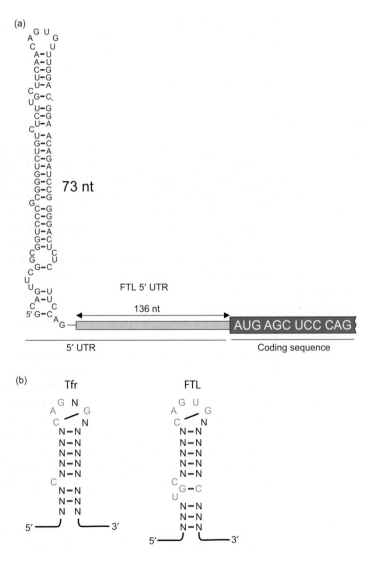

Figure 8.3 Structure of IRE. (a) The 5′ end of the ferritin light chain mRNA (encoded by the FTL gene) and the structure of the 73 nucleotide (nt) long IRE are shown. The IRE is at the 5′ end of the mRNA. The first four codons of the protein coding sequence are shown. (b) Important elements of the IREs may be obtained by comparing the nucleotide sequence of such elements from different mammalian mRNAs. Here the consensus structures of IREs from the transferrin receptor and the ferritin light chain mRNAs are compared. Conserved bases are highlighted in red. Both IREs have a conserved hexanucleotide apical loop and a bulging unpaired C between two helical regions. These elements are important for the specific interaction with IRP proteins.

(a)

(b)

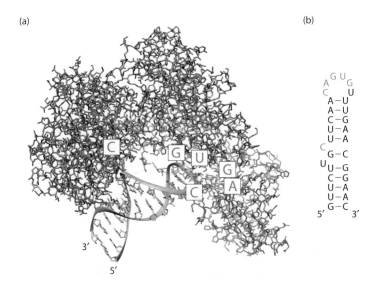

Figure 8.4 Interaction between ferritin IRE RNA and iron regulatory protein 1 (IRP1). (a) The structure of IRP1 (green) in complex with the IRE (blue) is shown. The essential sequence CUGUG and the bulging C are highlighted. (b) Secondary structure model of the IRE sequence used in the structural study. ([a] Adapted from Walden WE et al. 2006. *Science* 314:1903–1908. PBD ID: 3SNP.)

animals. Thus, characteristic of the IRE is a CAGUGN loop sequence (where N is any nucleotide). In the Tfr IRE, there is a C residue six positions upstream of the CAGUGN sequence, creating a bulge in the hairpin, whereas the IREs present in ferritin mRNAs instead have a conserved bulge/loop UGC/C.

The three-dimensional structure of ferritin and Tfr IREs in complex with IRP protein has been studied with x-ray crystallography. This structure reveals that the conserved positions in the RNA—that is, the CUGUG sequence as well as the C bulge—are specifically recognized by the protein. This is shown for the ferritin IRE in **Figure 8.4**.

Returning to the hereditary defect in the Italian family referred to above, the mutation identified was a substitution G > C in position 41, a position that is part of the highly conserved CAGUGN loop sequence.

Intriguingly, it was later found that other families with the same type of ferritinemia disorder have other mutations in the same loop sequence. Since the molecular characterization of this disease, many other cases of inherited hyperferritinemia cataract syndrome have been analyzed. A large number of substitutions and deletions were identified. **Figure 8.5** shows

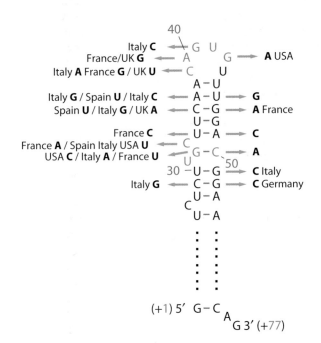

Figure 8.5 Hyperferritinemia cataract syndrome is caused by mutations in IRE. A selection of point mutations in the ferritin light chain (gene FTL) IRE that cause hereditary hyperferritinemia cataract syndrome are indicated. Strongly conserved bases are in red, and nucleotide numbering is in green. Mutations were identified in families with country affiliations as shown. (Adapted from Luscieti S et al. 2012. *Orphanet J Rare Dis* 8:30. Published under CC BY 2.0.)

a summary of substitution mutations in the most highly conserved part of the IRE. The large majority of these mutations are likely to interfere with the binding of IRPs to the IRE. This will give rise to an overproduction of ferritin, because when IRPs are poorly bound to the IRE, translation will be constitutive and poorly regulated.

AN RNA ELEMENT NECESSARY FOR THE INCORPORATION OF SELENOCYSTEINE IN PROTEINS

IREs are not unique as RNA regulatory elements. We know more examples where local RNA structures in untranslated regions of mRNAs play an important role in translational control. One of them is the **selenocysteine insertion sequence** (SECIS), which is present in all domains of life, including eukaryotes. We have previously seen how the genetic code encodes 20 different amino acids. However, there is yet another amino acid, selenocysteine, that is encoded by mRNA. This amino acid is relatively rare and present in some 25 human proteins only. Many of these are enzymes involved in oxidation and reduction reactions where the selenocysteine is critical for the catalytic activity.

Selenocysteine is specified by a UGA codon, a codon that is normally signifying termination of protein synthesis. However, in the context of a SECIS element, UGA instead specifies selenocysteine. In bacteria, the SECIS element is just downstream of the UGA codon and within the coding sequence, but in eukaryotes it is located in the downstream 3' UTR.

There are a number of macromolecules that contribute to the eukaryotic mechanism of UGA recoding. First, there is a selenocysteine-specific tRNA (Sec-tRNASec) with an **anticodon** recognizing UGA. Second, a dedicated protein, an elongation factor EFsec, carries the tRNA. Finally, there are two proteins that bind the SECIS element, SBP2 (SECIS binding protein 2) and ribosomal protein L30 (**Figure 8.6**).

There are two conserved motifs in the SECIS that are essential for the faithful encoding of UGA as selenocysteine. First, there is the SECIS "core"

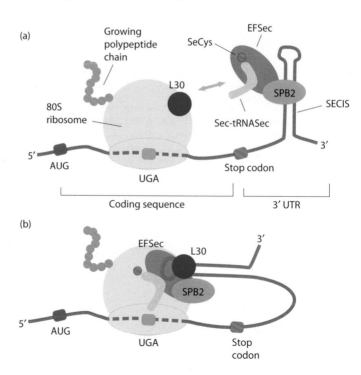

Figure 8.6 Model for UGA recoding in eukaryotes. (a) A SECIS element is located in the 3' UTR, downstream of the coding sequence. It binds a protein SBP2, which recruits another protein EFsec and a specific tRNA, Sec-tRNASec, carrying selenocysteine (SeCys, red). (b) After association with the ribosome, the ribosomal protein L30 replaces SBP2 on the SECIS element. The binding of L30 may aid in the reading of a UGA codon by Sec-tRNASec. (Adapted from Chavatte L et al. 2005. *Nat Struct Mol Biol* 12:408–416. With permission from Macmillan Publishers.)

with a quartet of non-Watson-Crick base pairs (two tandem G•A pairs) that are required for the binding of SBP2 and L30. Second, there is a sequence AAA in the SECIS apical loop of unknown function (**Figure 8.7**).

REGULATION OF TRANSLATION BY 5′ UTR OPEN READING FRAMES

How does the translation machinery find the start site in a human mRNA? The start codon is always AUG. Originally it was thought that the first AUG encountered from the 5′ end of the mRNA was the translation start site, and a model was proposed in which the ribosome would scan the mRNA from its 5′ end and start translation at the first AUG to be identified. However, it was eventually discovered that many mRNAs contain additional AUGs upstream of the actual protein coding sequence. These upstream AUGs may be recognized by the translation machinery and form the initiation site for an **upstream open reading frame** (**uORF**). These uORFs give rise to protein products that are distinct from those of the main coding sequence. As a general rule, the uORF inhibits translation of the main coding sequence. A number of different mechanisms could contribute to this effect. For instance, a uORF could block the translational machinery. However, there are also situations, such as stress conditions, where this inhibitory function of the uORF may be alleviated to achieve more efficient translation of the main coding sequence.

WHEN UPSTREAM OPEN READING FRAMES ARE MUTATED

It is estimated that about 35%–49% of all human mRNAs contain at least one uORF. The biological significance of uORFs is demonstrated by a number of inherited disorders. One example is the Marie Unna hereditary hypotrichosis (MUHH). This is a rare autosomal dominant disorder that was first described by the German dermatologist Marie Unna in 1925. It is characterized by sparse or absent hair at birth, the growth of coarse and wiry hair during childhood, and progressive hair loss at puberty.

The genetic basis of the disease was traced to a gene named *HR* (acronym is derived from "hairless"). The protein encoded by the gene takes part in negative regulation of transcription of multiple genes that ultimately affect hair growth. While the Marie Unna disorder is caused by mutations affecting a uORF, there are other disorders that are also characterized by hair loss—known as alopecia and atrichia—that result from mutations elsewhere in the *HR* gene.

In the human *HR* gene there are four different uORFs. The mutation responsible for MUHH was shown to be a substitution T > C in the second position of the uORF named U2HR (**Figure 8.8**). This ORF is highly

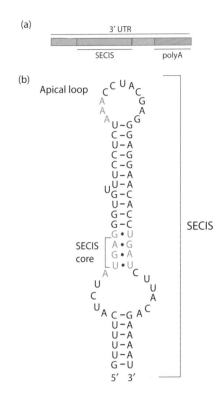

Figure 8.7 The structure of the SECIS element in the 3′ UTR of *GPX1*. The gene *GPX1* encodes the enzyme glutathione peroxidase 1. This is one of the genes where selenocysteine is encoded by a UGA codon. (a) The location of the SECIS element in the 3′ UTR of *GPX1*. (b) Structural details of the SECIS element in *GPX1*, highlighting two portions important for its function (nucleotides in red). First, there is the SECIS "core" with two tandem G•A pairs that are required for the binding of SBP2 and L30. Second, there is a sequence AAA of unknown function.

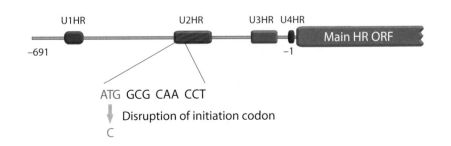

Figure 8.8 A point mutation in a uORF in Marie Unna hereditary hypotrichosis. The human *HR* gene has four different uORFs: U1HR, U2HR, U3HR, and U4HR. The mutation responsible for the disorder is a T > C substitution in the second position of U2HR, disrupting the initiation codon ATG (AUG at the level of RNA sequence). Numbers refer to positions relative to the transcription start site, where the start site has number +1 and positions upstream have negative numbers counting back from −1.

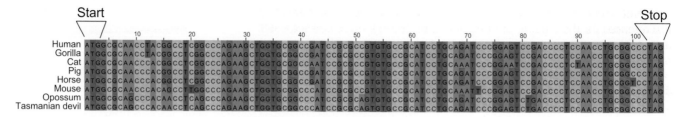

Figure 8.9 *HR* uORF is conserved during evolution. The second uORF, U2HR, of the human gene *HR* is compared to the corresponding uORF in other mammals, including the two marsupials opossum and Tasmanian devil. The background coloring is based on nucleotide identity. The uORF is initiated with an ATG start codon and ends with a TAG stop codon.

conserved among mammals (**Figure 8.9**). The mutation in MUHH destroys the initiation codon ATG, and thereby reduces significantly the initiation of translation of this uORF. This, in turn, has the effect of increasing the translation of the main coding sequence.

AU-RICH ELEMENTS ARE INVOLVED IN REGULATION OF mRNA STABILITY

While there are many different biologically significant sequence elements in the 3′ UTR of human mRNAs, the most common are **AU-rich elements (AREs)**. An ARE is a region important for mRNA stability and translational regulation with an abundance of the bases A and U. AREs are typically found in short-lived mRNAs, such as those involved in transient regulatory responses. We do not know exactly how many human mRNAs are regulated by AREs, but it has been estimated that such elements are present in about 5%–8% of all genes.

A characteristic sequence in AREs is AUUUA, which often occurs in a U-rich environment and in multiple copies in 3′ UTR. For such an example, see the mRNA of the *FOS* gene in **Figure 8.10a**. There are also more degenerate ARE sequence motifs, such as WUUUW (where W is A or U)—for more examples, see **Figure 8.10b**.

What is the role of AREs? A number of proteins, such as HuA, HuB, HuC, HuD, and HuR (where Hu refers to "human antigen"), bind to AU-rich elements and have the effect of stabilizing the mRNA. At the same time, other proteins that bind to AREs instead have the opposite effect and cause degradation of the mRNA.

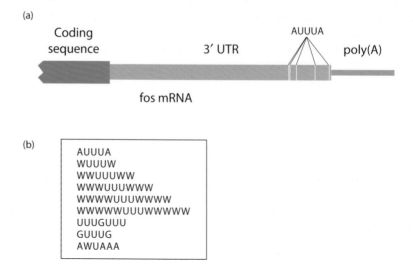

Figure 8.10 AU-rich elements in mRNAs.
(a) In FOS mRNA, the sequence AUUUA is present in five copies at the 3′ end of the 3′ UTR, close to the poly(A) tail. (b) A number of sequence patterns specifying AU-rich elements in human genes. W represents A or U.

HuR contains three RNA-recognition motif (RRM) domains. An RRM is a sequence of about 90 amino acids that binds a single-stranded RNA. It is a fairly common domain in human proteins—between 200 and 250 human proteins have one or more RRM domains. In HuR, there are two N-terminal tandem RRM domains that can selectively bind AREs, while a third RRM domain (RRM3) is involved in binding to the poly(A) tail of mRNAs. The structure in **Figure 8.11** shows the first two RRM domains (RRM1/2) of HuR and how they specifically interact with AU-rich sequences characteristic of an ARE.

Inherited disorders associated with AREs are not common. One example is a case of band keratopathy, a disorder where a white band appears across the central cornea, formed by the precipitation of calcium salts. Members of three related Saudi families were found to be affected by this disorder, and the causative mutation is a G > A substitution in the 3′ UTR of the gene SLC4A4. This mutation creates a novel ARE (**Figure 8.12**). The mutation presumably destabilizes the mRNA and causes it to be present at a lower level in patient cells.

A 3′ UTR REPEAT EXPANSION AND MYOTONIC DYSTROPHY

In Chapter 7, we saw how trinucleotide repeats may give rise to a defective protein and inherited disease. Examples are Huntington's disease and many spinocerebellar ataxias, both associated with an expansion of CAG repeats, corresponding to polyglutamine regions in proteins. Repeat expansion may, however, occur most anywhere in the genome, for instance in mRNA UTRs. One example of an inherited disorder associated with repeat expansion in a noncoding region is **myotonic dystrophy**, a disease characterized by progressive muscle wasting and weakness. Affected individuals often have prolonged muscle contractions and are not able to effectively release muscles after use.

Two types of myotonic dystrophy have been described. Type 1 (DM1, also known as Steinert disease) is caused by mutations in the *DMPK* gene and type 2 (DM2) in the *CNBP* gene. In DM2, the repeat expansion is in the first intron of the gene, but we here focus on DM1, where the expansion affects the 3′ UTR.

DM1 is a dominantly inherited multisystem disease. Thus, it affects not only muscles but also other organs. Important consequences are muscle weakness, muscle stiffness, and cataract. It is the most common adult-onset muscular dystrophy and affects about 1 in 8,000 individuals. It is even more common in specific areas such as Quebec, Canada, and Basque Country, Spain.

The genetic basis of the disorder was identified in 1992 by different research groups. The responsible gene *DMPK* is located at chromosome 19 and encodes a protein with the same acronym, short for "dystrophica myotonica protein kinase." An expansion of a CUG repeat located in the 3′ UTR is responsible for the disease (**Figure 8.13**).

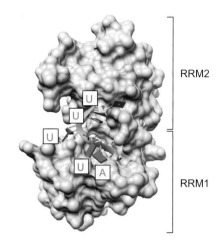

Figure 8.11 Binding of protein HuR to AU-rich elements. The protein HuR is shown as a surface model, and a portion of an RNA molecule is shown bound to a cleft in the protein. The sequence of the RNA present in the structure is 5′- AU<u>UUUUA</u>UUUU -3′, where the underlined sequence is highlighted in the structure. RRM1 and RRM2 are two different RNA binding domains in HuR. (Adapted from Wang H et al. 2013. *Acta Crystallogr [Section D]* 69:373–380. PDB ID: 4ED5.)

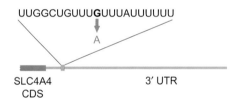

Figure 8.12 A point mutation in the 3′ UTR of the gene *SLC4A4* gives rise to a novel AU-rich element. The coding sequence (CDS) and 3′ UTR of the SLC4A4 mRNA, as well as the mutation G > A are shown.

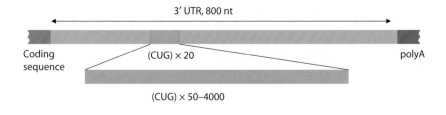

Figure 8.13 Expansion of a CUG repeat in the 3′ UTR of *DMPK* causes myotonic dystrophy. A number of repeats (20) characteristic of a healthy individual is compared to repeat expansions (50–4,000) of a myotonic dystrophy patient.

Healthy individuals typically have between 5 and 37 CUG repeats. Those with between 38 and 49 repeats are not symptomatic but carry a premutation. In DM1, the number of copies is much larger, and just as in Huntington's disease (Chapter 7), the more copies of the repeat, the more severe are the symptoms of the disease. In the most critical form of the disease, there are more than 800 copies and symptoms appear after birth.

A POSSIBLE MECHANISM OF MYOTONIC DYSTROPHY TYPE 1

The normal function of DMPK is not fully understood. It is a serine-threonine kinase known to phosphorylate a number of different proteins. It is not clear how a repeat expansion gives rise to the myotonic disease. One possible mechanism is shown in **Figure 8.14**. Specific proteins that take part in transcription and RNA processing may be sequestered because of binding to the repeat region. As a consequence, their normal functions will be affected. The CUG repeat forms specific hairpin structures (dsCUG RNA) by G•C base pairing (see Figure 8.14). A protein involved in **splicing** (for more on splicing see Chapter 9), MBNL1 (muscleblind-like 1), is one of the proteins binding to these hairpins.

The normal function of MBNL1 is to bind to specific sequence elements in the mRNA precursor. Sequence-specific binding of MBNL1 to a short RNA sequence 5'-CGCUGU-3' is illustrated in **Figure 8.15**. It is one of many examples in this book of how a protein may recognize a specific

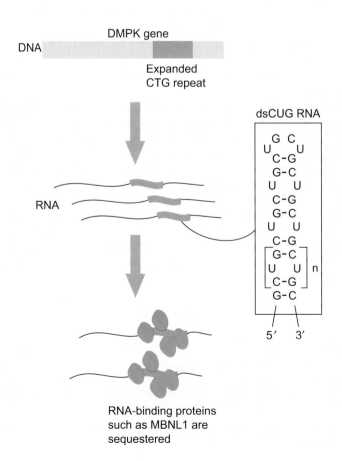

Figure 8.14 A possible mechanism contributing to myotonic dystrophy type 1. An expanded CTG repeat in the 3' UTR of the *DMPK* gene is responsible for myotonic dystrophy type 1. The CUG repeat forms specific structures (dsCUG RNA) that may sequester proteins that take part in transcription and RNA processing, such as MBNL1 (blue).

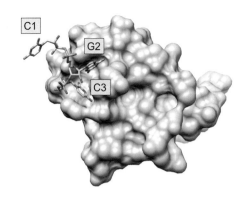

Figure 8.15 Sequence-specific binding to RNA by protein MBNL1. The MBNL1 protein is a splicing regulator that binds to an intronic consensus sequence 5′-YGCUKY-3′ (where Y is C or U, K is G or U). MBNL1 has four different domains that each bind to RNA. Here is illustrated the specific recognition of the sequence 5′-CGCUGU-3′ by one of these domains, shown as a surface model. The first three bases, C1, G2, and C3, of the RNA hexamer are indicated as a ball-and-stick model. Proteins make direct interactions with G2 and C3. (Adapted from Teplova M, Patel DJ. 2008. *Nat Struct Mol Biol* 15:1343–1351. PDB ID: 3D2S.)

nucleotide sequence. Remember that many DNA and RNA sequence elements of the human genome have a function to bind specifically to proteins. We see many more examples of this when discussing transcriptional control in Chapter 10.

MBNL1 does not bind to the CUG repeats present in the normal *DMPK* mRNA, but it does bind to the expanded dsCUG RNA in DM1. The structural details of the binding of MBNL1 to excessive CUG repeats are not known, but MBNL1 is sequestered, and as a result, the splicing of many genes is misregulated, and splicing defects are observed in DM1.

A UNIVERSE OF SILENCING RNAs

We now turn to a final category of significant sequence elements located in the 3′ UTR of mRNAs. These elements play a role in **RNA interference** (or **RNA silencing**) and are targets for a group of small RNAs. Critical experiments leading to the discovery of small silencing RNAs were carried out by Andrew Fire and Craig Mello in the 1990s. These researchers injected double-stranded RNA (dsRNA) into cells of the worm *Caenorhabditis elegans*. Fire and Mello observed that if the dsRNA contained a sequence matching a specific gene, the expression of that particular gene was dramatically reduced. In 2006 they received the Nobel Prize in Physiology or Medicine for their pioneering work on this regulatory mechanism.

The most well-studied silencing RNAs are **microRNA (miRNA)** and **small interfering RNA (siRNA)**. The function of siRNA is to downregulate exogenous sequences, such as those that enter the cell during a viral infection. They do this by inducing cleavage of the foreign RNA. The function of miRNA is to control the expression of endogenous genes. The first miRNAs were discovered in *C. elegans* by Victor Ambros and Gary Ruvkun.

The syntheses of miRNA and siRNA occur by different pathways. All miRNA is encoded in the genome. A pathway for its synthesis is shown in **Figure 8.16**. One miRNA precursor RNA may have more than one miRNA. In Figure 8.16 is shown such a case where two different miRNAs are encoded within the precursor.

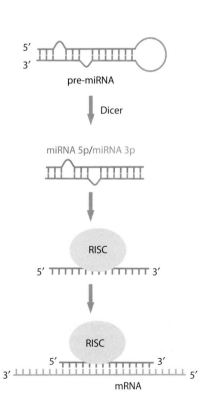

Figure 8.16 Synthesis and mRNA targeting of miRNA. Transcription first gives rise to a pri-miRNA (not shown), which may contain one or more miRNAs. The pri-miRNA is then processed to give pre-miRNA. The pre-miRNA is in turn transported to the cytosol and processed by the enzyme Dicer to produce a small duplex RNA. Here is shown such a duplex containing two miRNAs referred to as 5p (green) and 3p (orange). The duplex structures contain bulges as the 5p and 3p sequences are not perfectly complementary. Each of the miRNAs may associate with proteins to form an RNA-induced silencing complex (RISC). Here is shown such a complex with the 5p miRNA. This complex will finally target an mRNA (blue) with a sequence complementary to the miRNA.

Figure 8.17 Regulation of translation and mRNA stability by miRNAs. A RISC complex of miRNA bound to an Argonaute (AGO) protein (light green) specifically targets an mRNA (gray bar) by partial base complementarity. The target sequence of the miRNA is typically located in the 3′ UTR of the mRNA. The RISC complex then recruits the protein TNRC6 (cyan), which interacts with the poly(A)-binding protein (PABPC, orange) and recruits either of two deadenylation enzymes, PAN2/PAN3 or a complex that includes NOT proteins and CCR4. The deadenylation in turn gives rise to decapping (removal of the specific modified nucleotide present at the 5′ end of the mRNA, gray circle). Decapping results in rapid degradation of the mRNA from its 5′ end. In addition, binding of TNRC6 causes the mRNA to be less efficiently translated, presumably by influencing the assembly and/or activity of a complex with the translation factors eIF4G, eIF4E, and eIF4A. (Adapted from Jonas S, Izaurralde E. 2015. *Nat Rev Genet* 16:421–433. With permission from Macmillan Publishers.)

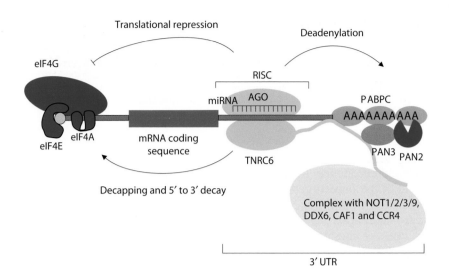

A final product in the pathway of miRNA synthesis is a complex between miRNA and proteins called an RNA-induced silencing complex (RISC). The miRNA in the RISC complex will target an mRNA with a sequence complementary to the miRNA. Interaction of the miRNA-RISC complex with the 3′ UTR leads to the inactivation of the mRNA. This inactivation occurs through a combination of translational repression, deadenylation, decapping (removal of a modified 5′ end nucleotide), and degradation of the mRNA starting from its 5′ end (**Figure 8.17**).

The miRNAs and siRNAs are both small RNAs—an miRNA is typically 21–23 nucleotides, and siRNAs are often in the size range of 20–24 nucleotides. Both RNAs will target mRNAs based on sequence complementarity. In the case of siRNA, the complementarity is perfect, whereas for miRNA there are mismatches in the hybrid formed. However, positions 2–8 in the miRNA, referred to as the seed region, must typically be perfectly complementary to the mRNA target (**Figure 8.18**).

The exact number of miRNAs encoded by the human genome is not known, but about 2,600 are reported in the database miRBase. The miRNAs typically have names like "miR-" followed by a number and extensions like "−5p" or "−3p" when more than one miRNA is derived from a specific precursor duplex RNA (see Figure 8.16). The genes of miRNAs are distributed throughout the genome, and their localization is not related to their target genes. An miRNA may have one or more binding sites in the 3′ UTR of a specific mRNA (**Figure 8.19**). Each of the miRNAs is able to bind not just a single mRNA, but on average an miRNA targets about 400 different mRNAs. Several computational methods have been developed to predict mRNA targets based on the miRNA sequence. Because of the large number of mRNA targets identified, it would seem that a considerable portion of all human protein genes is regulated with miRNAs. This type of regulation is particularly important during development. With this in mind, it is not surprising

Figure 8.18 Principle of interaction between an miRNA and its target mRNA. An miRNA (here 21 nt) recognizes its target mRNA by partial base complementarity. Typically a region spanning nucleotide positions 2–8 of the miRNA, the seed region, must be perfectly complementary to the mRNA. In the other regions of the miRNA, mismatches with the mRNA are possible.

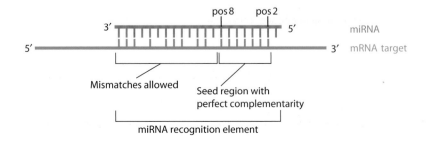

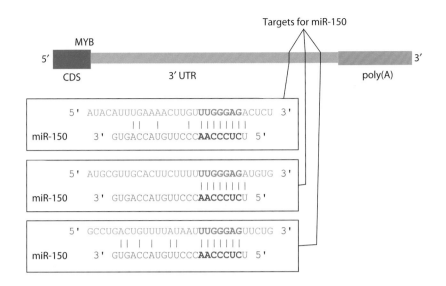

Targets for miR-150

Figure 8.19 Recognition of 3′ UTR sequences by a miRNA. The mRNA of human gene *MYB* has three recognition sites for the miRNA miR-150. The positions of the miRNA seed region are shown in bold. Information is from the database miRDB (www.mirdb.org).

that misregulation of miRNA gene expression is associated with a number of disorders, including cancer. Some miRNAs function as **oncogenes** and others as **tumor suppressors** (Chapter 5).

DISEASES ASSOCIATED WITH MUTATIONS IN miRNAs OR TARGET 3′ UTRs

There are a few examples of inherited disorders involving miRNAs. For instance, there is one case of a disorder known as *keratoconus*, a thinning of the central cornea that leads to progressively poor vision. A number of genes have been identified as associated with the disorder. The cause of the disease was examined in a Northern Irish family in which 18 of 38 individuals from three generations were affected by keratoconus. One the variants identified was a mutation C to T in the miRNA named miR-184 (**Figure 8.20a**). This is a mutation in a central position of the essential seven nucleotide seed region.

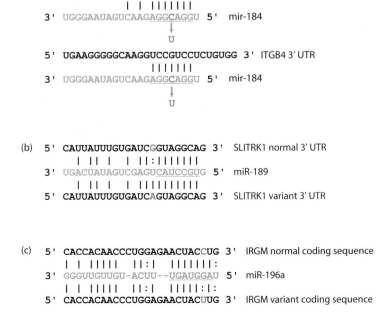

Figure 8.20 Disorders associated with miRNA or target mRNAs. The miRNAs are shown in orange, and seed regions are underlined. Conventional base-pairing is indicated with vertical lines and a G•U pair with a colon. (a) A mutation C > T in miR-184 (arrows) occurs in a central position of the essential seed region in a case of keratoconus and is expected to disrupt the function of this miRNA. Two predicted mRNA target 3′ UTRs, those of INPPL1 and ITGB4, of miR-184 are shown. (b) A variant G > A in the SLITRK1 3′ UTR increases the affinity to the miRNA miR-189. (c) A variant C > U in the coding sequence of gene *IRGM* gives rise to a lower affinity to miR-196a.

The RNA miR-184 is the most abundant miRNA in the corneal and lens epithelia. There are several predicted mRNA targets of miR-184. Evidence has been presented that one gene that may be of interest is INPPL1 (inositol polyphosphate phosphatase-like 1), producing a protein with a role in the regulation of apoptosis. A second possible mRNA target is that of gene *ITGB4*. However, it cannot be excluded that the disease phenotype is caused by other mRNAs targeted by miR-184.

There are also examples of mutations in mRNA that affect miRNA binding and that are associated with disease. Many known SNPs in 3′ UTR regions may actually be functionally important as part of target sequences for miRNAs. For instance, a variant G > A in the 3′ UTR of the *SLITRK1* gene is associated with Tourette's syndrome, a neuropsychiatric disorder. The variant increases the affinity to the miRNA miR-189 and thereby lowers expression of *SLITRK1* expression (**Figure 8.20b**). Similarly, there are polymorphisms in the gene *IRGM* associated with Crohn's disease, an inflammatory bowel disease. One of the variants, C > U, is a synonymous change in a coding region. The variant gives rise to a lower affinity to miR-196a, leading to increased expression of the IRGM protein (**Figure 8.20c**).

SUMMARY

- There are sequence elements in the 5′ and 3′ UTRs of mRNA that are of regulatory significance. They control translation, stability, transport, and subcellular localization of mRNAs. Their significance is emphasized by diseases caused by mutations in these elements.
- The IRE is an RNA structure located in specific mRNAs related to iron metabolism. IREs take part in the regulation of translation initiation and mRNA stability.
- The SECIS element is an RNA structure that is part of a mechanism to decode UGA as selenocysteine.
- The uORFs, located in the 5′ UTR, regulate translation. A mutation in a uORF in the *HR* gene causes Marie Unna hereditary hypotrichosis.
- AREs are located in the 3′ UTR and are important in the control of mRNA stability.
- A trinucleotide repeat expansion in the 3′ UTR of the *DMPK* gene causes myotonic dystrophy type 1.
- The translation and stability of a large number of mRNAs are regulated by miRNAs that interact in a sequence-specific manner with the mRNA 3′ UTRs. In this interaction, mismatches are allowed, but in positions 2–8 in the miRNA, referred to as the seed region, there is typically perfect complementary to the mRNA target. One miRNA can target hundreds of mRNAs.
- A mutation in the miRNA miR-184 is likely to reduce the expression of one or more mRNAs and causes keratoconus. Conversely, there are variants in mRNA 3′ UTRs that are associated with disease resulting from altered miRNA binding.

QUESTIONS

1. What important functions reside in the 5′ and 3′ untranslated regions of an mRNA?
2. Explain the mechanisms related to translational control of mRNAs encoding ferritin and the transferrin receptor. An RNA hairpin structure is crucial in that mechanism. What are the important features of the hairpin structure?

3. How is the incorporation of selenocysteine in proteins different from that of the other 20 amino acids specified by the genetic code?

4. What is meant by an *upstream open reading frame*? Give an example of a disease where such a reading frame is involved.

5. What is the mutation accounted for in this chapter responsible for Marie Unna hereditary hypotrichosis? What are the consequences of this mutation?

6. What are AREs? What is the most characteristic nucleotide sequence?

7. What are the proteins that recognize AREs, and what is the biological significance of this interaction?

8. What kinds of mutations in mRNA 3′ UTRs are responsible for *myotonic dystrophy type 1*? What specific RNA structure could be involved in the mechanism of disease?

9. Explain the difference between *siRNAs* and *miRNAs*.

10. What is meant by the *seed region* of an miRNA?

11. There are diseases associated with variants/mutations in mRNAs that affect miRNA regulation. What are these diseases, and which miRNAs are involved?

URLs

MicroRNA databases: miRBase (www.mirbase.org) and miRDB (www.mirdb.org)

FURTHER READING

UTRs and regulation

Barrett LW, Fletcher S, Wilton SD. 2012. Regulation of eukaryotic gene expression by the untranslated gene regions and other non-coding elements. *Cell Mol Life Sci* 69(21):3613–3634.

Matoulkova E, Michalova E, Vojtesek B, Hrstka R. 2012. The role of the 3′ untranslated region in post-transcriptional regulation of protein expression in mammalian cells. *RNA Biol* 9(5):563–576.

Pichon X, Wilson LA, Stoneley M et al. 2012. RNA binding protein/RNA element interactions and the control of translation. *Curr Protein Pept Sci* 13(4):294–304.

Iron metabolism

Hentze MW, Muckenthaler MU, Galy B, Camaschella C. 2010. Two to tango: Regulation of mammalian iron metabolism. *Cell* 142(1):24–38.

Outten FW, Theil EC. 2009. Iron-based redox switches in biology. *Antioxid Redox Signal* 11(5):1029–1046.

Hyperferritinemia cataract syndrome

Beaumont C, Leneuve P, Devaux I et al. 1995. Mutation in the iron responsive element of the L ferritin mRNA in a family with dominant hyperferritinaemia and cataract. *Nat Genet* 11(4):444–446.

Girelli D, Corrocher R, Bisceglia L et al. 1995. Molecular basis for the recently described hereditary hyperferritinemia-cataract syndrome: A mutation in the iron-responsive element of ferritin L-subunit gene (the "Verona mutation"). *Blood* 86(11):4050–4053.

Girelli D, Corrocher R, Bisceglia L et al. 1997. Hereditary hyperferritinemia-cataract syndrome caused by a 29-base pair deletion in the iron responsive element of ferritin L-subunit gene. *Blood* 90(5):2084–2088.

Girelli D, Olivieri O, Gasparini P, Corrocher R. 1996. Molecular basis for the hereditary hyperferritinemia-cataract syndrome. *Blood* 87(11):4912–4913.

Kato J, Fujikawa K, Kanda M et al. 2001. A mutation, in the iron-responsive element of H ferritin mRNA, causing autosomal dominant iron overload. *Am J Hum Genet* 69(1):191–197.

Structure and evolution of IRE

Casey JL, Hentze MW, Koeller DM et al. 1988. Iron-responsive elements: Regulatory RNA sequences that control mRNA levels and translation. *Science* 240(4854):924–928.

Piccinelli P, Samuelsson T. 2007. Evolution of the iron-responsive element. *RNA* 13(7):952–966.

SECIS

Chavatte L, Brown BA, Driscoll DM. 2005. Ribosomal protein L30 is a component of the UGA-selenocysteine recoding machinery in eukaryotes. *Nat Struct Mol Biol* 12(5):408–416.

Latrèche L, Jean-Jean O, Driscoll DM, Chavatte L. 2009. Novel structural determinants in human SECIS elements modulate the translational recoding of UGA as selenocysteine. *Nucleic Acids Res* 37(17):5868–5880.

Mix H, Lobanov AV, Gladyshev VN. 2007. SECIS elements in the coding regions of selenoprotein transcripts are functional in higher eukaryotes. *Nucleic Acids Res* 35(2):414–423.

Walczak R, Westhof E, Carbon P, Krol A. 1996. A novel RNA structural motif in the selenocysteine insertion element of eukaryotic selenoprotein mRNAs. *RNA* 2(4):367–379.

Zinoni F, Birkmann A, Leinfelder W, Bock A. 1987. Cotranslational insertion of selenocysteine into formate dehydrogenase from *Escherichia coli* directed by a UGA codon. *Proc Natl Acad Sci USA* 84(10):3156–3160.

Zinoni F, Birkmann A, Stadtman TC, Bock A. 1986. Nucleotide sequence and expression of the selenocysteine-containing polypeptide of formate dehydrogenase (formate-hydrogen-lyase-linked) from *Escherichia coli*. *Proc Natl Acad Sci USA* 83(13):4650–4654.

uORFs and disease

Barbosa C, Onofre C, Romão L. 2014. Upstream open reading frames and human genetic disease. In: eLS. John Wiley and Sons, Chichester, www.els.net [doi: 10.1002/9780470015902.a0025714].

Wen Y, Liu Y, Xu Y et al. 2009. Loss-of-function mutations of an inhibitory upstream ORF in the human hairless transcript cause Marie Unna hereditary hypotrichosis. *Nat Genet* 41(2):228–233.

Prediction of AU-rich elements

Fallmann J, Sedlyarov V, Tanzer A et al. 2016. AREsite2: An enhanced database for the comprehensive investigation of AU/GU/U-rich elements. *Nucleic Acids Res* 44(D1):D90–D95.

Halees AS, El-Badrawi R, Khabar KS. 2008. ARED Organism: Expansion of ARED reveals AU-rich element cluster variations between human and mouse. *Nucleic Acids Res* 36(Database issue):D137–D140.

AU-rich elements and disease

Patel N, Khan AO, Al-Saif M et al. 2017. A novel mechanism for variable phenotypic expressivity in Mendelian diseases uncovered by an AU-rich element (ARE)-creating mutation. *Genome Biol* 18(1):144.

Suhl JA, Chopra P, Anderson BR et al. 2014. Analysis of FMRP mRNA target datasets reveals highly associated mRNAs mediated by G-quadruplex structures formed via clustered WGGA sequences. *Hum Mol Genet* 23(20):5479–5491.

Myotonic dystrophy

Aslanidis C, Jansen G, Amemiya C et al. 1992. Cloning of the essential myotonic dystrophy region and mapping of the putative defect. *Nature* 355(6360):548–551.

Brook JD, McCurrach ME, Harley HG et al. 1992. Molecular basis of myotonic dystrophy: Expansion of a trinucleotide (CTG) repeat at the 3′ end of a transcript encoding a protein kinase family member. *Cell* 68(4):799–808.

Buxton J, Shelbourne P, Davies J et al. 1992. Detection of an unstable fragment of DNA specific to individuals with myotonic dystrophy. *Nature* 355(6360):547–548.

Conne B, Stutz A, Vassalli J-D. 2000. The 3′ untranslated region of messenger RNA: A molecular "hotspot" for pathology? *Nature Med* 6(6):637–641.

Conne B, Stutz A, Vassalli JD. 2001. *3′ UTR mutations and human disorders*. eLS, John Wiley and Sons, New York.

Fu YH, Pizzuti A, Fenwick RG Jr. et al. 1992. An unstable triplet repeat in a gene related to myotonic muscular dystrophy. *Science* 255(5049):1256–1258.

Mahadevan M, Tsilfidis C, Sabourin L et al. 1992. Myotonic dystrophy mutation: An unstable CTG repeat in the 3′ untranslated region of the gene. *Science* 255(5049):1253–1255.

Mateos-Aierdi AJ, Goicoechea M, Aiastui A et al. 2015. Muscle wasting in myotonic dystrophies: A model of premature aging. *Front Aging Neurosci* 7:125.

MBNL1

Ho TH, Savkur RS, Poulos MG et al. 2005. Colocalization of muscleblind with RNA foci is separable from mis-regulation of alternative splicing in myotonic dystrophy. *J Cell Sci* 118(Pt 13):2923–2933.

Teplova M, Patel DJ. 2008. Structural insights into RNA recognition by the alternative-splicing regulator muscleblind-like MBNIL1. *Nat Struct Mol Biol* 15(12):1343–1351.

miRNA

Abelson JF, Kwan KY, O'Roak BJ et al. 2005. Sequence variants in SLITRK1 are associated with Tourette's syndrome. *Science* 310(5746):317–320.

Bartel DP. 2018. Metazoan microRNAs. *Cell* 173(1):20–51.

Brest P, Lapaquette P, Souidi M et al. 2011. A synonymous variant in IRGM alters a binding site for miR-196 and causes deregulation of IRGM-dependent xenophagy in Crohn's disease. *Nat Genet* 43(3):242–245.

Fire A, Xu S, Montgomery MK et al. 1998. Potent and specific genetic interference by double-stranded RNA in *Caenorhabditis elegans*. *Nature* 391(6669):806–811.

Hughes AE, Bradley DT, Campbell M et al. 2011. Mutation altering the miR-184 seed region causes familial keratoconus with cataract. *Am J Hum Genet* 89(5):628–633.

Jonas S, Izaurralde E. 2015. Towards a molecular understanding of microRNA-mediated gene silencing. *Nat Rev Genet* 16(7):421–433.

Kawahara Y. 2014. Human diseases caused by germline and somatic abnormalities in microRNA and microRNA-related genes. *Congenital Anom* 54(1):12–21.

Maziere P, Enright AJ. 2007. Prediction of microRNA targets. *Drug Discov Today* 12(11–12):452–458.

Peng Y, Croce CM. 2016. The role of microRNAs in human cancer. *Signal Transduct Targeted Ther* 1:15004.

Exons, Introns, and a Royal Bleeding Disorder

9

In Chapter 8, we saw that a human gene is much more than the protein coding sequence, as its mRNA contains untranslated regions with regulatory importance. In addition, genes are interspersed with introns. The formation of a mature mRNA involves the removal of these introns in the process of splicing. In this chapter, we look at the consequences of the exon-intron structure, both good and bad, starting with a famous case of an inherited bleeding disorder.

QUEEN VICTORIA PASSED HEMOPHILIA ON TO MANY OF HER DESCENDENTS

In 1853, Queen Victoria of the United Kingdom (**Figure 9.1**) gave birth to her eighth child—Leopold. Already in childhood it was realized that the boy suffered from hemophilia, a blood clotting disorder. None of his siblings showed symptoms of this disease. When Leopold was 30 years old, he suffered from joint pain, worsened by the British winter climate. It was recommended by his doctors that he go to Cannes. However, on March 27, 1884, in his Cannes residence, he slipped and fell. As a result, he injured his knee and hit his head. In a healthy individual, these would have been benign injuries. But Leopold died the next day, presumably from a brain hemorrhage.

When Leopold was born, Victoria was given chloroform as an anesthetic, and it was speculated at the time that this may have caused the disease. However, it later turned out that many descendents of Queen Victoria suffered from hemophilia, strongly suggesting a genetic component of the disease.

THE ROYAL HEMOPHILIA IS INHERITED THROUGH THE X CHROMOSOME

Hemophilia is a disease where blood clotting cannot proceed normally because there is a defect in a blood protein, a coagulation factor, required for this process. How is the royal variant of hemophilia inherited? The

Figure 9.1 Queen Victoria, 1887.

Figure 9.2 Transmission principles of hemophilia gene. (a) The son inherits the X chromosome that carries the mutated gene (red bold X) from his mother and has hemophilia ("affected"). (b) Like in (a) but the son inherits the normal X chromosome from his mother and is healthy. (c) The daughter inherits the mutated X chromosome from her father who is hemophilic. The daughter becomes a carrier of the disease gene but is healthy. (d) Son inherits Y chromosome from his hemophilic father and gets normal X and Y chromosomes.

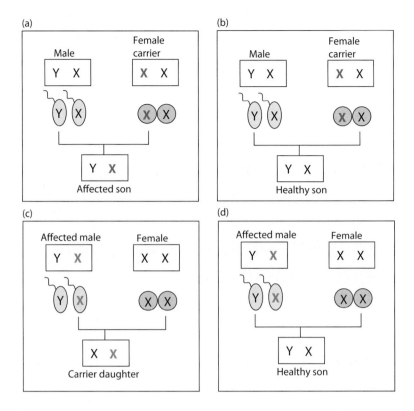

gene for the coagulation factor with the disease mutation is located on the X chromosome. Such a disorder shows the pattern of inheritance as illustrated in **Figure 9.2**. Remember that females carry two X chromosomes, whereas males have one X and one Y chromosome. Hemophilia is a **recessive** disorder. As females always have two copies of the X chromosome, they do not show symptoms of hemophilia because they have at least one normal version of the gene. Therefore, females only pass the mutation on to the next generation. However, if males inherit the mutated gene from their mother, they only have a mutated copy of the gene and develop, as a result, hemophilia (see Figure 9.2a). The sons of a father with hemophilia will not carry the mutated gene, as they receive the Y and not the X chromosome from their father (see Figure 9.2d). In the language of genetics, hemophilia is a sex-linked recessive disease. "Sex-linked" because the mutated gene is on the X chromosome.

Queen Victoria was clearly a carrier of the disease gene, but was she the first carrier? We do not know for certain, but there is no trace of the disease in the preceding generations. It therefore seems likely that the mutation appeared for the first time with the conception of Victoria. Her father Edward, Duke of Kent, was 52 when Victoria was born. As mutations in sperm cells are more likely to occur in older fathers, it is possible that the hemophilia mutation first arose in the X chromosome of Edward's germ cells. Another possibility as to the origin of the hemophilia is that Queen Victoria was an illegitimate child, but there is no convincing evidence to support this idea.

The disease of Victoria was further transmitted to other European royal families (**Figure 9.3**). It was also passed on to the Romanovs in Russia as Alix, a granddaughter of Queen Victoria, married tsar Nicholas II. We will return later to this branch of the tree as we discuss the actual mutation responsible for the disease. It would now appear that the royal bleeding disorder is extinct, and that the last carrier of the disease in the royal family was Prince Waldemar of Prussia, who died in 1945. It is therefore an example of a genetic disorder staying within the population only for a limited

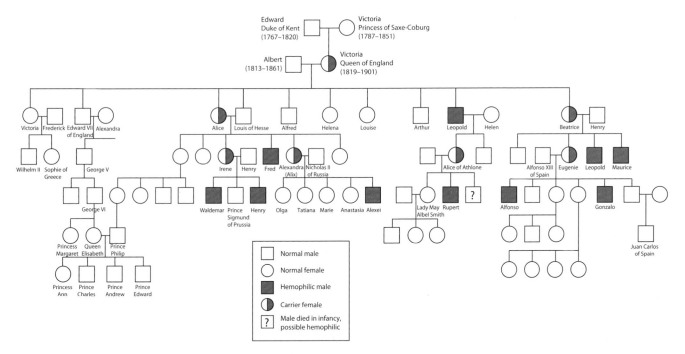

Figure 9.3 Royal hemophilia pedigree. (Adapted from Aronova-Tiuntseva Y, Herreid CF Figure 9.1 [Queen Victoria of England's family tree highlighting the appearance of haemophilia] in: "Hemophilia: The Royal Disease" ©National Center for Case Study Teaching in Science [NCCSTS], University at Buffalo, State University of New York.)

amount of time. However, it should be noted that we do see other cases of inherited hemophilia today. These are comparatively rare conditions.

BLOOD CLOTTING DEPENDS ON A CASCADE OF ENZYMATIC REACTIONS

An outline of reactions taking place during blood coagulation is shown in **Figure 9.4**. Blood clotting or coagulation is a highly complex and regulated process preventing bleeding as a result of an injury. In an injury where a blood vessel is damaged, blood will eventually form a clot. The formation of the clot is a process involving multiple enzymatic steps. Why are so many

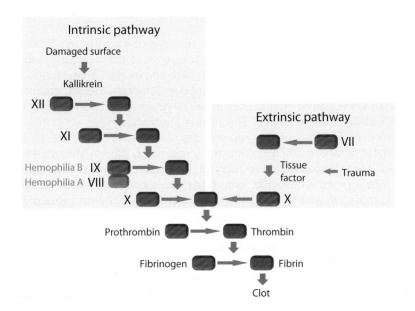

Figure 9.4 Blood clotting pathways. The blood clotting pathways involve many factors acting in a coordinated cascade to produce a blood clot. A recurring theme in the blood coagulation cascade is that an inactive form of an enzyme (red rectangles) is activated by another enzyme to an active form (green rectangles). For the initiation of clot formation, the extrinsic pathway is the most important. It is caused by damage to a blood vessel, which exposes tissue factor. This protein activates factor X, which in turn converts prothrombin to thrombin through proteolytic cleavage. Thrombin converts fibrinogen to fibrin, which is the molecule forming the actual clot. In the intrinsic pathway the conversion of prekallikrein to kallikrein activates factor XII, and a series of steps leads to the activation of factor X. Finally, factor X acts on prothrombin as described for the extrinsic pathway. Hemophilia A and B are caused by mutations in the genes encoding factor VIII and factor IX, respectively. Factor VIII is a cofactor to factor IX.

steps involved? One reason is that a strong amplification of a molecular signal is achieved in each step. The purpose is to achieve an efficient and rapid response to a trauma where the initial molecular signal is quite weak. Blood clotting is also highly regulated. For instance, if clotting formation is too strong, dangerous clots may be formed in blood vessels, preventing normal flow of blood. There are two major pathways of blood clotting—intrinsic and extrinsic.

There are two variants of hemophilia. Hemophilia A is caused by a deficiency in the activity of coagulation factor VIII, whereas hemophilia B is a result of changes in the factor IX protein (see Figure 9.4). The hemophilia B disease is also called Christmas disease, because Stephen Christmas was the first patient described with this disease and the first report describing it was published in the Christmas edition of the *British Medical Journal*. It was early realized that the royal hemophilia gene should be located on the X chromosome because of the sex-linked inheritance pattern. However, the pattern of inheritance cannot be used to distinguish between hemophilias A and B, because both factors VIII and IX are specified by genes on the X chromosome.

DETAILED INFORMATION OF THE ROYAL DISORDER WAS OBTAINED BY DNA ANALYSIS OF THE ROMANOVS

What exactly was the genetic defect carried by Queen Victoria and many of her descendents? For a long time we had no clue, and it was not until 2009 that more detailed information appeared, thanks to modern DNA technology.

We previously referred to Alix, a granddaughter of Queen Victoria. She came to marry Nicholas—or Nikolai Alexandrovich Romanov—who became the Russian czar. Alix changed her name to Alexandra to adapt to her new Russian environment. She bore Nicholas four daughters and one son. It turned out that the son named Alexei, born in 1904, had the bleeding disorder. He survived childhood but was a sickly child.

The Bolshevik revolution in 1917 led to the execution of Nicholas and his family. The dead bodies were transported to a field where nine members of the group were buried in one grave and two of the children in a separate grave. In 1979 a local geologist was able to locate a mass grave near Yekaterinburg suspected to contain remains of the Romanov family. DNA of the skeletons was isolated and analyzed in 1991. By comparing this DNA to other members of the royal pedigree who are alive today, and by using remains of a brother to Nicholas II, rather strong evidence was provided that the grave contained the remains of Nicholas, his wife, and three of their daughters. To provide stronger evidence of their identities, a search for the second grave was initiated, and in 2007 it was found by a group of amateur archeologists, about 70 meters from the first grave. The Russian government invited a team of scientists to perform independent DNA testing of the remains. The results showed without doubt that the grave contained the two missing children, Alexei and one of his sisters.

DNA analysis did not stop at identification of the remains in the grave. It was also possible to identify the responsible gene and the precise location of the mutation within that gene. Either of the two coagulation factors VIII and IX was likely to be involved in the disease. In 2009 researchers analyzed DNA samples from Alexei and his mother with respect to both of these two genes. Remember that many genetic disorders are the result of **nonsynonymous** mutations in coding sequences. However, in this case no such mutation was identified in any of the factor VIII and IX genes. But another mutation was found in the IX gene in a region not specifying the amino acid sequence. How could this affect the factor IX protein? To understand this, we need to know more about how mature mRNAs come about through processing of the original RNA transcribed from DNA.

EUKARYOTIC GENES ARE MOSAICS OF EXONS AND INTRONS

As discussed in Chapter 4, one basic feature of a human gene is that at the level of DNA, the coding sequence is disrupted by several regions that do not code for protein. This basic organization of genes was first demonstrated by Richard Roberts and Philip Sharp in 1977. Roberts and Sharp examined a eukaryotic DNA virus and the mRNAs transcribed from the viral genome. They wanted to find out what portion of the genome encoded a specific mRNA. They used the technique of DNA-RNA hybridization (**Figure 9.5**). Through the principle of base-pairing, the mRNA should pair to the complementary strand of the corresponding DNA region. Roberts and Sharp used electron microscopy to identify nucleic acids and any hybrids formed. Unexpectedly, they found that different segments of the mRNA paired to DNA, whereas there were intervening sequences in the DNA that did not match the mRNA. This was visible with electron microscopy as pieces of DNA looping out from the DNA-RNA hybrid structure. A model was therefore proposed where a gene is a mosaic of **exons**, the parts found in the mature RNA, and **introns**, parts excluded from the mRNA. Roberts and Sharp studied a eukaryotic virus system, but other researchers soon showed that the exon-intron structure applies to all eukaryotes, including human. For this important discovery Roberts and Sharp were awarded the Nobel Prize in 1993.

What is the frequency of introns in human genes? Most genes of the human genome have a large number of introns. Only 3% of human

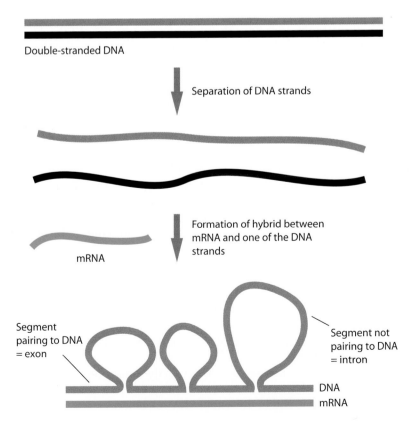

Double-stranded DNA

Separation of DNA strands

mRNA

Formation of hybrid between mRNA and one of the DNA strands

Segment pairing to DNA = exon

Segment not pairing to DNA = intron

DNA
mRNA

Figure 9.5 Critical experiment to reveal introns. Introns were revealed by an experiment of Richard Roberts and Philip Sharp. As a first step in this experiment, the two strands of DNA are separated by raising the temperature. The temperature is then lowered in the presence of mRNA with base complementarity to the DNA. As a result, RNA-DNA hybrids will be formed, which are observable with electron microscopy. The hybrids reveal segments of DNA that do not base-pair to the mRNA—the introns; the segments that pair are the exons.

Figure 9.6 Splicing is a two-step reaction. Splicing occurs by two separate reactions. Here is shown the removal of an intron between two exons labeled 1 and 2. In the first reaction, the 2′ hydroxyl of an adenosine (green A) close to the 3′ end of the intron attacks the phosphodiester bond that is at the 5′ splice site (5′ SS). As a result, the adenosine becomes part of a branched structure. In the second reaction, the free 3′ end of exon 1 attacks with its hydroxyl group the phosphodiester bond of the 3′ splice site (3′ SS). The end products are the covalently joined exons and the intron as a branched structure (lariat).

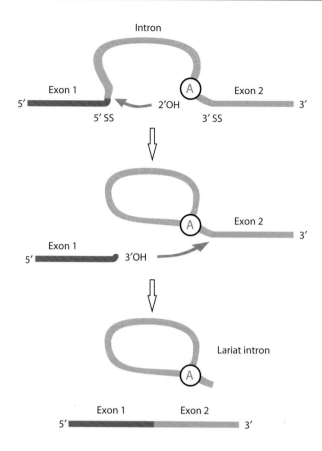

genes are lacking introns entirely. Exons and introns have very different length distributions, as introns are typically much longer than the exons (Figures 4.17 and 4.18). Thus, whereas protein gene introns make up about 38% of the human genome, the exons correspond to only 3% (Figure 4.21). We should also note that introns are not subject to the strong selection pressure characterized by protein-coding exons. For this reason, introns are much less conserved during evolution.

During the process of transcription, the enzyme RNA polymerase produces an RNA copy of the information in DNA. In this process, the introns are transcribed along with the exons so at the level of transcription there is no discrimination of these two kinds of sequences. So how are all of the introns eliminated in the processes leading to the mature mRNA? It turns out that they are removed in a specific two-step process known as **splicing**. It involves reactions at the **5′ splice site** (the site at the 5′ end of the intron, also referred to as the **donor** site) and at the **3′ splice site** (at the 3′ end of the intron, also known as the **acceptor** site) (**Figure 9.6**).

The final result of the splicing reaction is that the two ends of the exons are joined. The intron is displaced as a branched structure (a **lariat**)—the branch point is an adenosine close to the 3′ end of the intron that is bonded to three different nucleotides. There is often no use for the intron, and it is typically degraded to its nucleotide components.

The **spliceosome** is the biochemical machinery responsible for splicing. Important components of the spliceosome are RNA-protein complexes referred to as snRNPs or "snurps." RNAs of the spliceosome have critical functions—for instance, there are specific RNAs that are responsible for its catalytic activity.

How does the spliceosome know exactly what portions to remove from the transcribed RNA? It has been demonstrated that much of the information for splicing is within the actual RNA sequence.

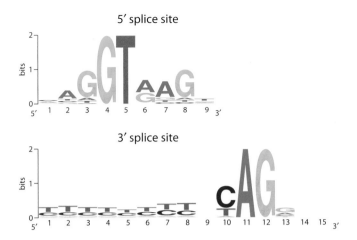

Figure 9.7 Sequence conservation at splice sites. Sequence logos of the 5' and 3' splice sites are shown. Note the strongly conserved GT (or GU at the level of RNA) and AG of the 5' and 3' splice sites, respectively. Reflected in the 3' splice site logo is also the pyrimidine-rich stretch just upstream of the splice site.

THE ENDS OF INTRONS CONTAIN NUCLEOTIDE SEQUENCES CRUCIAL FOR SPLICING

There are critical sequence elements at the 5' and 3' ends of the intron that are recognized by the spliceosome. Particularly essential sequence elements for recognition by the splicing machinery are the dinucleotide sequences GU (GT at the level of DNA sequence) and AG, located at the 5' and 3 ends, respectively, of the intron. However, there are also other components of importance, such as a pyrimidine-rich sequence close to the intron 3' end. The sequence properties of important splice elements are more strictly shown by the **sequence logo** in **Figure 9.7**. This representation is based on a large number of exon/intron boundary region sequences and shows what nucleotide positions are conserved. The height of a letter reflects the extent of conservation (for more mathematical information on sequence logos see Chapter 12). The sequence logo shows that the GU and AG motifs are strongly conserved, whereas the conservation of the pyrimidine-rich sequence is less pronounced.

What are the molecular interactions of the strongly conserved sequences involved in splicing? A lot of the specificity is achieved by base-pairing with RNAs being components of the spliceosomal snRNPs (**Figure 9.8**). An important role in the recognition of the 5' splice site is played by an RNA known as U1 RNA. It has sequence complementarity to the 5' splice site. The 3' end of the intron is recognized by a non-snRNP protein component of the spliceosome, and the branch point adenosine located 20–50 nucleotides upstream of the 3' splice site is recognized by the U2 RNA. One important biological implication of the conserved splicing motifs is that any mutation in these motifs is expected to negatively influence splicing.

Figure 9.8 Spliceosomal RNAs pair with precursor mRNA. Structures of precursor mRNA and spliceosomal RNAs U1 and U2 are shown schematically, highlighting sequences important during splicing. The spliceosomal U1 RNA is part of the U1 snRNP and pairs with the 5' splice site (5' SS) region. The U2 RNA, being part of the U2 snRNP, pairs with a sequence at the adenosine branch site (green A). The 3' splice site (3' SS) sequence AG (blue) does not pair with a spliceosomal RNA. Instead it is recognized by a non-snRNP protein component of the spliceosome (not shown here). Also important for 3' splice site recognition by the spliceosomal machinery is a pyrimidine-rich region (Py) between the branch site A and the 3' splice site. R and Y in the precursor mRNA nucleotide sequences represent purines (A or G) and pyrimidines (C or U), respectively.

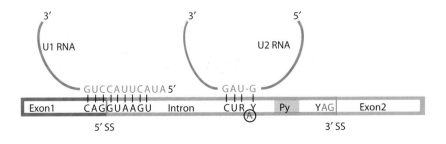

EXONS MAY BE COMBINED IN ALTERNATIVE WAYS

A complexity of splicing is added by the fact that there is no single possible outcome of the splicing process. This is because there are alternative ways that exons may be joined. For instance, under certain conditions, an exon may be skipped by the splicing machinery. **Alternative splicing** is the regulated process where exons are joined in different combinations. In this manner, a number of different RNAs and proteins may be formed from a single gene.

There are different alternative splicing events as shown in **Figure 9.9**. All splice products come about with the same basic machinery of the spliceosome, but additional proteins are responsible for alternative splicing. These proteins determine exactly what exons are to be combined. They do this by binding to specific regions in the RNA. There are RNA sequence elements located in the intron parts of a gene referred to as intronic splicing enhancers (ISEs) and intronic splicing silencers (ISSs) that positively and negatively influence the efficiency of a nearby splicing site. In the same way, there are elements in the exonic parts that affect splicing, exonic splicing enhancers (ESEs) and exonic splicing silencers (ESSs) (**Figure 9.10**). All of these elements have the capacity to attract specific regulatory proteins. The protein binding may involve different molecular mechanisms, but one possible outcome for the splicing silencers is that a splice site is physically blocked, causing the spliceosome to select another site. In the previous chapter, we encountered the MBNL1 protein, one example of the many proteins that regulate splicing.

Alternative splicing is common. It is estimated that about 95% of human genes have multiple splice products. Whereas the number of human genes that code for proteins is approximately 20,000, the number of transcripts of these genes is about 207,000 (numbers of October 2018). In this manner, one single human gene does not code for a single protein. Instead, multiple forms of a protein may be produced. Consider for instance the gene *TPM1*, encoding tropomyosin alpha-1. Tropomyosin is a protein that

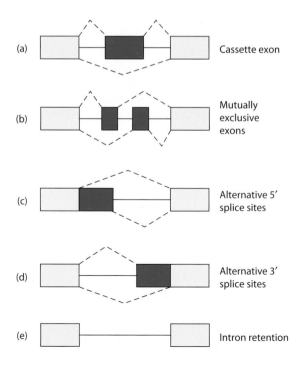

Figure 9.9 Examples of alternative splicing events. Rectangles represent exons. Exons specific to alternative splice products are shown in green. Horizontal lines indicate introns, and connecting dashed lines indicate splicing events. (a) A cassette exon is an exon that may be either included or excluded from the mRNA. (b) In the case of mutually exclusive exons, one exon is selected from an array of two or more exons. (c) Two sites compete as 5′ splice sites. (d) Two sites compete as 3′ splice sites. (e) Splicing between two exons is avoided, and the intron is retained.

(a) Cassette exon

(b) Mutually exclusive exons

(c) Alternative 5′ splice sites

(d) Alternative 3′ splice sites

(e) Intron retention

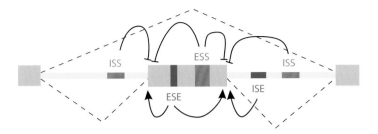

Figure 9.10 The regulation of splicing. Three exons (light green) with introns (yellow) are shown along with possible splicing events (dashed lines). Elements are shown that positively or negatively influence splicing—ISEs, ISSs, ESEs, and ESSs. Enhancers are able to activate nearby splice sites (arrow-headed line) or counteract silencers. Silencers can repress splice sites (bar-headed lines) or enhancers.

is a part of muscle fibers and that regulates muscle contraction. A number of different protein products may be formed from this single gene. These different products have specialized functions and are typically tissue specific. For instance, the tropomyosin in skeletal muscle is different from the tropomyosin found in the brain (**Figure 9.11**).

What is the biological significance of alternative splicing? As we have seen, it is a mechanism to generate diversity. Multiple splice products are critical for a variety of cellular processes, including cell differentiation and development. For instance, alternative splicing is common in brain tissues and contributes to every step of the development of the nervous system.

DNA ANALYSIS SHOWED THAT THE ROYAL BLOOD DISORDER WAS DUE TO A SPLICING ERROR

Now that we know about the process of splicing, we are able to understand the nature of the royal hemophilia mutation. First, the DNA of Alexandra, Alexei's mother, was analyzed. A substitution A > G was found in her gene that encodes the factor IX protein—this gene is referred to as *F9*. The mutation was just three positions upstream of exon 4. Could it be responsible for the disease?

To begin with, we would expect a carrier of the disease gene to have one copy of the healthy gene and one copy of the disease gene. In Alexandra's DNA, there was an equal proportion of A and G at the

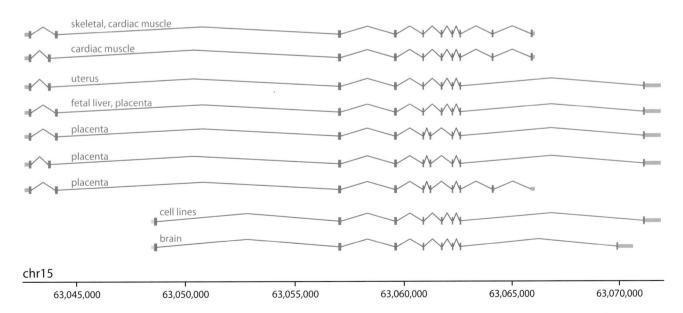

Figure 9.11 Splice variants of the human gene *TPM1*. *TPM1* gene encodes the muscle protein tropomyosin α-1. Introns are shown as red lines, coding exonic regions as green bars, and UTRs as cyan bars. The organs/tissues where the splice form is found are indicated. The bottom line indicates nucleotide positions in chromosome 15 (human genome version GRCh38/hg38).

mutation position. The same was true for one of Alexei's sisters, whose DNA was also analyzed. (This sister is presumed to be Anastasia.) Alexei had only G. This is to be expected if he has one X chromosome with the disease gene together with the Y chromosome that is lacking altogether the *F9* gene.

Splicing was expected to be altered by the mutation, as it occurs in a highly conserved region of a 3′ splice site, a site known to be recognized by the splicing machinery (see Figures 9.7 and 9.8). Scientists verified that splicing was actually altered with an experiment where the mutated sequence was expressed in cultured cells. It was shown that the change A > G introduces a novel 3′ splice site just two nucleotide positions from the normal 3′ splice site (**Figure 9.12**). This is a small change, but it is a dramatic one for the production of factor IX. The effect of the splice mutation is that the reading frame during protein translation is shifted. A new reading frame appears because the difference in size between the normal intron and the "new" intron is not divisible by three (also remember the relationship between coding sequences and intron phase as shown in Figure 4.19). In the new reading frame, a stop codon will appear and give rise to a truncated protein using principles as discussed in Chapter 6.

In conclusion, we now know that the royal hemophilia was caused by a splice mutation in the *F9* gene that gives rise to an erroneous RNA transcript. The transcript is aberrantly spliced, so that production of normal

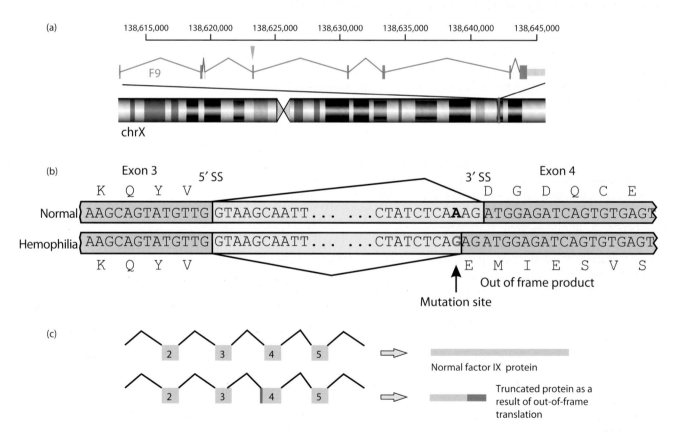

Figure 9.12 Mutation responsible for the royal hemophilia. (a) Overview of the *F9* gene, located on the X chromosome. Numbers are chromosomal positions (version GRCh37/hg19 of the human genome). The exon-intron structure of the gene is shown with coding parts in green, UTRs in cyan, and introns as connecting red lines. The orange triangle indicates the location (close to exon 4) of the hemophilia mutation. (b) Detailed structure of the region with the hemophilia mutation. Exons 3 and 4 are shown in green together with amino acid sequences encoded. Introns are shown in yellow. In the individuals affected by hemophilia, there is a substitution A > G (red). The effect is a novel 3′ splice site, giving rise to a change in reading frame. (c) The effect of the hemophilia mutation is that in the novel reading frame a stop codon will be encountered, giving rise to a truncated protein product.

factor IX protein is prevented. As a result, the efficiency of blood coagulation is severely reduced.

As the disorder resulted from a mutation in the *F9* gene, it is classified as hemophilia B. Both hemophilia A and B are rare disorders—only 1 in 10,000 worldwide have hemophilia A, while hemophilia B is five times less common. Many cases of the B form that we observe today are the result of point mutations, giving rise to amino acid replacements in the factor IX protein. Patients with hemophilia B are typically treated by intravenous injections of the factor IX protein. (For more on therapy of inherited disorders, see Chapter 14.)

A HEARING DISORDER PRESENTS ANOTHER CASE OF SPLICING DEFECT

There are many other examples of inherited disorders that are caused by changes in splicing. We now consider a disorder that is the most common cause of inherited deafblindness.

In the 1860s, the German ophthalmologist Albrecht von Gräfe encountered three brothers with the same symptoms—deafness and vision impairment. The syndrome was studied further by the Scottish ophthalmologist Charles Usher, who examined the pathology and transmission of this illness in 1914. The disease now known as Usher syndrome is, unlike the royal hemophilia, an autosomal disease—a disease caused by mutation in a non-sex chromosome.

The first symptom in Usher disease is deafness or impaired hearing. At the age of 4–5 years, there are signs of vision problems caused by degeneration of the rod photoreceptor cells in the retina. Vision is, however, quite slowly impaired. The disease is relatively rare, affecting approximately 1 in 20,000 individuals. However, it is more common in regions with small, isolated, often inbred populations such as in Israel (Samarathians), Pakistan (Hutterites), Northern Sweden, and Finland.

Usher disease comes in different types, I, II and III. We focus here on type I disease, which is classified into five different subtypes, 1B, 1C, 1D, 1F, and 1G, depending on the gene mutated. In 2000, the molecular basis of Usher type 1C disease was elucidated. This disorder has been described in a Lebanese family as well as in a population of Louisiana Acadians—descendents of seventeenth-century French colonists who developed the Cajun culture. The responsible gene, named *USH1C*, encodes a protein named harmonin, a protein expressed in sensory hair cells in the ear. The exact role of this protein is not understood, but it is known to interact with other proteins in the cell membrane.

All cases of Usher type IC disease among Acadians are caused by a mutation G > A in exon 3 of the *USH1C* gene (**Figure 9.13**). The mutation is in a coding sequence region. If we were to consider the effect of this mutation on the coding sequence only, a valine codon GUG is replaced with another valine codon GUA, an example of a synonymous substitution. Thus, the mutation in the disease leaves the amino acid sequence intact. However, the mutation has an effect on splicing. A new 5′ splice site is being formed and used instead of that of the normal one. The result with respect to the translated protein product is a shift of reading frame (see Figure 9.13). Similar to what we saw in the case of the royal hemophilia, a stop codon will be encountered in the novel reading frame, resulting in a truncated protein. We will return to Usher syndrome in Chapter 14 as we discuss a possible therapy for this disorder.

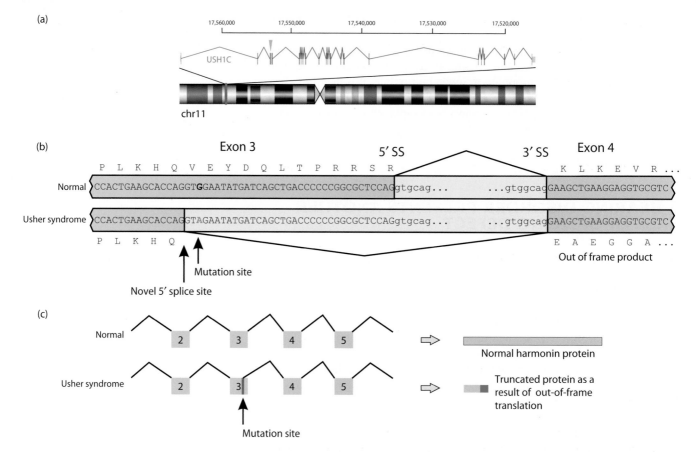

Figure 9.13 A splice mutation gives rise to Usher type IC disorder. (a) Overview of the *USH1C* gene that is located on chromosome 11 and that encodes the protein harmonin. Numbers are chromosomal positions (version GRCh37/hg19 of the human genome). The exon-intron structure of the gene is shown with coding parts in green, UTRs in cyan, and introns as connecting red lines. The orange triangle indicates the location (exon 3) of the Usher mutation. (b) Detailed structure of the region with the mutation. Exons 3 and 4 are shown in green together with amino acid sequences encoded. Introns are shown in yellow. In patients affected by Usher type 1C there is a substitution G > A (red). The effect is a novel 5' splice site, giving rise to a change in reading frame. (c) The effect of the G > A mutation is that in the novel reading frame a stop codon will be encountered, giving rise to a truncated version of the harmonin protein.

A POINT MUTATION CAN HAVE DIFFERENT EFFECTS ON SPLICING

A few more examples of splicing mutations are shown in **Figure 9.14**. As with the cases discussed previously, they illustrate that splicing defects can arise by both intronic and exonic mutations and that a point mutation may result in different types of splicing errors. A mutation in the gene *IKBKAP* (see Figure 9.14a), responsible for the disease of familial dysautonomia, destroys a 5' splice site. As a result, an entire exon is left out from splicing (a phenomenon referred to as **exon skipping**). The mutation is not in the most conserved part of the 5' splice site, but nevertheless, it would seem important for the efficiency of splicing.

A mutation in the *LMNA* gene (see Figure 9.14b) occurs in the coding sequence and results in Hutchinson-Gilford progeria syndrome. The mutation introduces a novel 5' splice site. This is a case similar to that of the royal hemophilia, but the mutation affects the 5' instead of the 3' splice site.

The *SMN1* and *SMN2* genes are associated with spinal muscular atrophy. The *SMN2* gene has a mutation in a coding region, close to the 5' end of the exon (Figure 9.14c). This is presumably a region of regulatory importance and contains either an ESS or an ESE. The mutation leads to either a gain of ESS or a loss of ESE, causing this exon to be skipped.

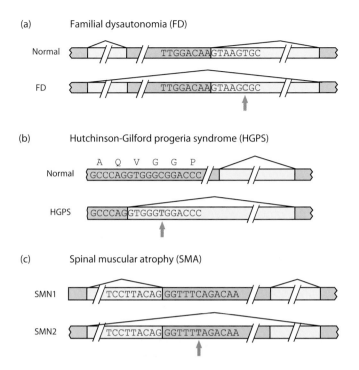

(a)   Familial dysautonomia (FD)

(b)   Hutchinson-Gilford progeria syndrome (HGPS)

(c)   Spinal muscular atrophy (SMA)

Figure 9.14 Three inherited disorders associated with splicing defects. Exons are shown in green, introns in light yellow. Arrows show the sites of substitution mutations, and mutant nucleotides are in bold red. (a) Familial dysautonomia. The mutation T > C in the gene *IKBKAP* destroys the 5′ splice site, resulting in exon skipping and a shift in translational reading frame. (b) Hutchinson-Gilford progeria syndrome. A mutation C > T in a coding sequence of the *LMNA* gene introduces a novel 5′ splice site, in turn causing a translational frameshift. (c) Spinal muscular atrophy. A mutation C > T in the coding sequence of the *SMN2* gene changes an ESS or ESE, resulting in exon skipping and a translational frameshift. The relationship of this mutation to spinal muscular atrophy is further explained in Chapter 14.

PREDICTING SPLICING MUTATIONS

In the search for causative mutations in monogenic disorders, it is essential to take into account any mutations with potential phenotypic consequences. We should therefore include in this analysis splicing mutations—they are indeed relatively common in the disorders analyzed so far, as we have seen previously (Table 6.1). In addition, genomic variants in noncoding regions that affect splicing play a role in complex traits and disorders, as shown by recent analysis of human genomes.

For just any genomic variant within a protein coding gene, we would like to predict the impact on the protein product. Remember that there is the genetic code, a code that reveals what trinucleotide sequences correspond to what amino acids in proteins. This code is distinct in the sense that the outcome in terms of amino acid sequence of a specific mutation, for instance a point mutation, is 100% predictable. Unfortunately, when it comes to many mutations that occur in noncoding regions, such a well-defined code is not available. This is also true in the context of splicing. Thus, in many cases, it is not easy to predict the outcome of a mutation as to splicing based on sequence alone. This is because of a complex biochemical machinery of splicing regulation and because we do not have detailed information about this machinery and the different kinds of RNA-protein interactions that are involved. Mutations in coding regions (see Figures 9.13 and 9.14b,c) present another difficulty as there is a risk that their effect on splicing may be confused with an effect on translation. In the case a novel nonsynonymous mutation is identified in a coding region in the context of disorder, it is often thought to be causative, as many inherited disorders are the result of alterations in the amino acid sequence. However, it cannot be ruled out that it also has an impact on splicing. Once we learn more about the splicing machinery, we may reveal a larger number of inherited disorders that are caused by erroneous splice products.

Is there a splicing code? There are indeed sequences useful in methods to predict the outcome of mutations (see Figures 9.7 and 9.8). Thus, there are the highly conserved sequences GU and AG adjacent to the splice sites, and changes in these elements are highly likely to affect splicing in a negative manner. Furthermore, there are sequences upstream and downstream of these dinucleotide sequences that are of importance as well. Finally, when it comes to alternative splicing, there are the enhancer and silencing elements to consider (see Figure 9.10). Based on all of these signals, a variety of computational tools have been developed for the prediction of splicing defects from a variant genomic sequence.

SUMMARY

- Splicing is the process whereby introns are removed from the primary RNA transcript. It is a two-step process, where exons are combined and the intron is left in a lariat form.
- Splicing is mediated by the spliceosome, a large complex of snRNPs and proteins.
- RNAs of the spliceosome have essential functions—specific RNAs are responsible for the catalytic activity, and other RNAs base-pair with the mRNA precursor and interact with the 5′ splice site and the branch site.
- The most important signals for splicing are dinucleotide sequences GU and AG at the 5′ and 3′ ends of the intron.
- Alternative splicing means that exons of an mRNA precursor may be combined differently, resulting in multiple transcripts for each gene.
- Important in alternative splicing are regulatory sequences present in both introns and exons that determine what splice site is selected.
- Point mutations may alter splicing in different ways. These mutations may occur in introns as well as exons. One common outcome of a splicing mutation is a shift of reading frame, causing premature translation termination.
- A variant gene that gives rise to hemophilia B was passed on from Queen Victoria to many of her descendents. Analysis of remains of the Romanov family showed that the causative mutation is a point mutation that introduces a novel 3′ splice site. The effect of the splice mutation is that the reading frame during protein translation is shifted, and in the new reading frame a premature stop codon appears.
- A mutation may also generate a novel 5′ splice site. An example of a disorder resulting from this type of mutation is the Usher type IC disease among Acadians.
- Other possible effects of splice mutations include (1) the elimination of a 5′ splice site, in turn causing exon skipping and (2) a change in a regulatory splicing enhancer.

QUESTIONS

1. What does it mean when we say that hemophilia is a *sex-linked recessive* disease?
2. What are the proteins affected in hemophilia, and what role do they play in blood coagulation?
3. What are the two important chemical reactions that lead to the joining of two exons during splicing?
4. What are the biochemical components of the *spliceosome*? What functions do the different components have?
5. What are the conserved sequences characteristic of the 5′ and 3′ splice sites, respectively?

6. Consider a gene with the following exon structure:
   E1-E2-E3-E4-E5-E6
   Which of the following are possible splice products, considering alternative splicing?
   a. E1-E2-E3-E5-E6
   b. E1-E5-E6
   c. E1-E3-E2-E6

7. Where are sequence elements responsible for alternative splicing located?

8. How many human genes are subject to alternative splicing? How many human protein coding transcripts have been identified?

9. What is the point mutation shown to be responsible for the royal hemophilia? What is the effect of this mutation on splicing? What is the effect on the protein produced?

10. Discuss the different examples of substitution mutations accounted for in this chapter that result in disease. How are they similar and different in terms of the functional outcome of the mutation?

URLS

Hemophilia: The Royal Disease (http://sciencecases.lib.buffalo.edu/cs/files/hemo.pdf)

FURTHER READING

Royal hemophilia

Coble MD, Loreille OM, Wadhams MJ et al. 2009. Mystery solved: The identification of the two missing Romanov children using DNA analysis. *PLOS ONE* 4(3):e4838.

McKusick VA. 1965. The Royal Hemophilia. *Scientific American* 213:88–95.

Rogaev EI, Grigorenko AP, Faskhutdinova G et al. 2009. Genotype analysis identifies the cause of the "royal disease." *Science* 326(5954):817.

Rogaev EI, Grigorenko AP, Moliaka YK et al. 2009. Genomic identification in the historical case of the Nicholas II royal family. *Proc Natl Acad Sci USA* 106(13):5258–5263.

Steinberg MD, Khrustalëv VM. 1995. *The fall of the Romanovs: Political dreams and personal struggles in a time of revolution.* Yale University Press, New Haven, CT.

Discovery of splicing

Berget SM, Moore C, Sharp PA. 1977. Spliced segments at the 5′ terminus of adenovirus 2 late mRNA. *Proc Natl Acad Sci USA* 74(8):3171–3175.

Alternative splicing

Chen M, Manley JL. 2009. Mechanisms of alternative splicing regulation: Insights from molecular and genomics approaches. *Nat Rev Mol Cell Biol* 10(11):741–754.

Lee Y, Rio DC. 2015. Mechanisms and regulation of alternative Pre-mRNA splicing. *Ann Rev Biochem* 84:291–323.

Wang ET, Sandberg R, Luo S et al. 2008. Alternative isoform regulation in human tissue transcriptomes. *Nature* 456(7221):470–476.

Splicing code

Barash Y, Calarco JA, Gao W et al. 2010. Deciphering the splicing code. *Nature* 465(7294):53–59.

Rosenberg AB, Patwardhan RP, Shendure J, Seelig G. 2015. Learning the sequence determinants of alternative splicing from millions of random sequences. *Cell* 163(3):698–711.

Wang Z, Burge CB. 2008. Splicing regulation: From a parts list of regulatory elements to an integrated splicing code. *RNA* 14(5):802–813.

Xiong HY, Alipanahi B, Lee LJ et al. 2015. RNA splicing. The human splicing code reveals new insights into the genetic determinants of disease. *Science* 347(6218):1254806.

RNA splicing: Genetic variation and disease

Li YI, van de Geijn B, Raj A et al. 2016. RNA splicing is a primary link between genetic variation and disease. *Science* 352(6285):600–604.

Diseases and splicing

Ward AJ, Cooper TA. 2010. The pathobiology of splicing. *J Pathol* 220(2):152–163.

Usher type IC disease

Bitner-Glindzicz M, Lindley KJ, Rutland P et al. 2000. A recessive contiguous gene deletion causing infantile hyperinsulinism, enteropathy and deafness identifies the Usher type 1C gene. *Nat Genet* 26(1):56–60.

Ebermann I, Lopez I, Bitner-Glindzicz M et al. 2007. Deafblindness in French Canadians from Quebec: A predominant founder mutation in the *USH1C* gene provides the first genetic link with the Acadian population. *Genome Biol* 8(4):R47.

Lentz J, Savas S, Ng SS et al. 2005. The USH1C 216G—A splice-site mutation results in a 35-base-pair deletion. *Hum Genet* 116(3):225–227.

Mathur P, Yang J. 2015. Usher syndrome: Hearing loss, retinal degeneration and associated abnormalities. *Biochim Biophys Acta* 1852(3):406–420.

Ouyang XM, Hejtmancik JF, Jacobson SG et al. 2003. USH1C: A rare cause of USH1 in a non-Acadian population and a founder effect of the Acadian allele. *Clin Genet* 63(2):150–153.

Ouyang XM, Yan D, Du LL et al. 2005. Characterization of Usher syndrome type I gene mutations in an Usher syndrome patient population. *Hum Genet* 116(4):292–299.

Verpy E, Leibovici M, Zwaenepoel I et al. 2000. A defect in harmonin, a PDZ domain-containing protein expressed in the inner ear sensory hair cells, underlies Usher syndrome type 1C. *Nat Genet* 26(1):51–55.

Other splicing disorders

Eriksson M, Brown WT, Gordon LB et al. 2003. Recurrent *de novo* point mutations in lamin A cause Hutchinson-Gilford progeria syndrome. *Nature* 423(6937):293–298.

Lorson CL, Hahnen E, Androphy EJ, Wirth B. 1999. A single nucleotide in the SMN gene regulates splicing and is responsible for spinal muscular atrophy. *Proc Natl Acad Sci USA* 96(11):6307–6311.

Monani UR, Lorson CL, Parsons DW et al. 1999. A single nucleotide difference that alters splicing patterns distinguishes the SMA gene *SMN1* from the copy gene *SMN2*. *Hum Mol Genet* 8(7):1177–1183.

Slaugenhaupt SA, Blumenfeld A, Gill SP et al. 2001. Tissue-specific expression of a splicing mutation in the *IKBKAP* gene causes familial dysautonomia. *Am J Hum Genet* 68(3):598–605.

Computational prediction of the effect of mutations on splicing and the roles of these mutations in disease

Xiong HY, Alipanahi B, Lee LJ et al. 2015. RNA splicing. The human splicing code reveals new insights into the genetic determinants of disease. *Science* 347(6218):1254806.

The Regulation of Transcription

10

We have considered sequence elements in the regions of the human genome that are part of mRNAs or mRNA precursors subject to splicing. Now we turn to another group of important elements—those that are involved in the regulation of gene transcription. DNA sequences that control the transcription of a gene may be located upstream of, within, or downstream of the gene under control. Such elements are quantitatively important in the human genome. It has been estimated that they make up as much as 20% of the genome, although more studies are required to arrive at a reliable percentage. We look more closely into some of the elements of transcription regulation. We also introduce epigenetic mechanisms—transcription regulatory mechanisms that involve either DNA or histone covalent modifications.

DNA SEQUENCES AND GENERAL TRANSCRIPTION FACTORS

What is the molecular machinery behind transcription? First, the production of the RNA, using a DNA strand as template, is carried out by the enzyme RNA polymerase. In eukaryotic cells, there are three different such polymerases, named I, II, and III. We restrict the discussion in this chapter to RNA polymerase II, the enzyme responsible for transcription of all protein genes as well as a selection of noncoding RNA genes[1].

Remember that a **gene** is a portion of DNA in the genome responsible for production of a protein or RNA. Transcription is initiated at a **promoter** region of the gene. A large complex with DNA and different proteins is formed at this site as a prerequisite for transcription. We discuss the DNA sequences typically found within promoters and that are critical for the initiation of transcription, and we see what proteins—**transcription factors**—are attracted by these sequences.

---

[1]Ribosomal RNA genes 5.8S, 18S, and 28S are transcribed by polymerase I and tRNA and ribosomal 5S RNA genes by polymerase III.

Figure 10.1 Promoter and transcription initiation with RNA polymerase II. (a) A number of DNA regions are important for the initiation of transcription by RNA polymerase II. Enhancer, proximal promoter, core promoter, transcription start site, and downstream promoter elements are shown. (b) During transcription initiation, a complex is formed with transcription factors and other proteins. Several basal transcription factors (light gray, proteins TFIIB, TFIIF, TFIIE, and TFIIH) and RNA polymerase II are bound to the core promoter and interact with transcription factors bound to the proximal promoter region. Specific transcription factors are shown bound to the proximal promoter region and to the enhancer. The mediator (darker gray) acts as a bridge between the RNA polymerase complex and transcription factors. The interactions of the enhancer cause a looped structure in DNA to form.

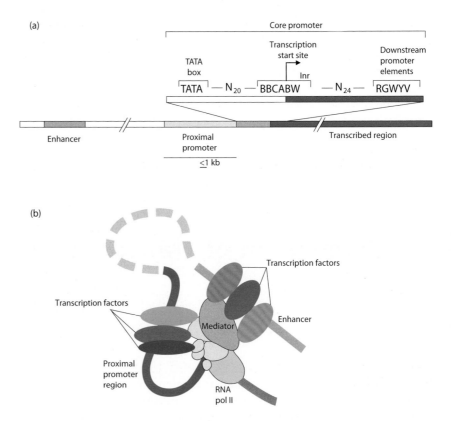

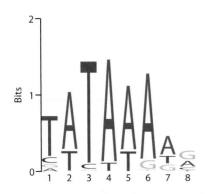

Figure 10.2 Sequence conservation among TATA promoters. Sequence logo of TATA box motif shows the extent of sequence conservation of the TATA box motif and was derived from 1,000 human TATA-containing promoter sequences.

Different sequence elements of importance in transcription initiation are shown in **Figure 10.1a**. First, there are elements of the **core promoter**, the minimal region of the promoter required for transcription initiation. The **proximal promoter** is a region upstream of the core promoter that may be up to 1,000 nucleotides in size. Important in transcriptional regulation are also **enhancer** elements that often are located distant to the transcription start site.

Each of the transcription factors typically recognizes DNA in a sequence-specific manner. However, efficient recognition does not depend on one specific DNA sequence—rather there is typically room for some sequence variation. The DNA site recognized by a transcription factor may be represented by either a **sequence logo** (see also previous chapter), or by a **consensus sequence** showing for every sequence position the most common nucleotide. In a consensus sequence, ambiguity symbols are often used to allow for variation. For instance, the sequence "RWY" is interpreted in the following way: In the first position, we find R which is either A or G. The next two positions are W (A or T) and Y (T or C). For more on nucleotide ambiguity symbols, see Appendix 2.

We first consider the DNA elements and proteins of the core promoter. There are DNA elements characteristic of the RNA polymerase II promoters, although none of them occur in all genes (see Figure 10.1a). A **TATA box** is a DNA motif that includes the consensus sequence TATA, hence its name. The sequence conservation of the TATA box is illustrated as a sequence logo in **Figure 10.2**. (In comparison, a possible *consensus sequence* derived from this sequence logo is TATAAADR.) The TATA box motif may be present in a region 30–100 nucleotides upstream of the transcription start site. But although this motif has been extensively studied, it is present only in a minority of genes. The most common core promoter sequence is the **initiator element** (**Inr**) (consensus sequence BBCABW, where B is C, G, or T), which is located at positions

−3 and +3 relative to the transcription start site. The initiator element is present in about 46% of all core promoters. Furthermore, a **downstream promoter element** (DPE) with consensus sequence RGWYV may be located in between positions +28 and +32 relative to the transcription start site.

A number of proteins, **basal** or **general transcription factors**, are attracted by the different DNA sequence elements of the core promoter. The binding of these proteins is required for the recruitment of RNA polymerase. The TATA-binding protein (TBP) binds to the TATA box motif. The binding of this protein causes a sharp bend in the DNA structure, thereby assisting in the separation of DNA strands necessary for transcription to occur (**Figure 10.3**). TBP is one of many different subunits in a larger complex named *transcription factor II D* (TFIID). Additional transcription factors are then recruited as well as the RNA polymerase. The initiator element and DPE are recognized by the transcription factors TAF1/2 and TAF6/9, respectively, which are all part of the complex TFIID. In genes that are lacking a TATA element, the basal transcription machinery will instead recognize the initiator element and downstream elements. TBP will in this situation still bind to DNA in a region corresponding to the TATA box, but not in a sequence-specific manner.

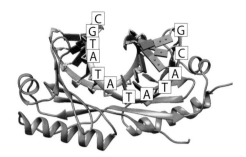

Figure 10.3 TATA-binding protein binds in a sequence-specific manner to DNA. The binding of TATA-binding protein causes a sharp bend in the DNA structure, thereby assisting in the separation of DNA strands necessary in transcription. DNA sequence of one of the strands is 5'-CGTATATATACG-3'. (Adapted from Juo ZS et al. 1996. *J Mol Biol* 261:239–254. PBD ID: 1TGH.)

β-THALASSEMIA AND CORE PROMOTER MUTATIONS

β-Thalassemia is an inherited disorder where the production of β-globin is defect. As with sickle cell anemia, the affected gene is the β-globin gene *HBB*. We have previously seen a broad spectrum of mutations affecting the coding sequence of this gene (Figure 6.2).

As β-thalassemia is a common disorder and affected patients have been screened for mutations, we have access to a whole catalog of functionally important elements in this gene. Thus, mutations in the promoter regions have also been observed. The sequence of the gene *HBB* that corresponds to the TATA box is CATAAA. Patients with a disorder known as β-**plus thalassemia**[2] have reduced amounts of β-globin. In some of these patients, mutations in the CATAAA element have been identified. These mutations are all suspected to decrease the affinity of transcription factors to DNA, thereby reducing transcriptional efficiency. Mutations have been identified that affect not only the TATA box, but also the initiator and downstream elements in the *HBB* promoter. In **Figure 10.4** are shown known thalassemia mutations of the TATA box and DPE in the *HBB* gene. They are all expected to negatively affect transcription initiation.

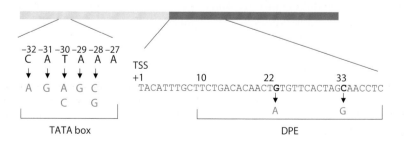

Figure 10.4 Mutations in β-globin promoter associated with disease. The mutations in the TATA box as well as the mutations in the region of downstream promoter elements (DPEs) are observed in patients with β-plus thalassemia. Transcribed region is shown in green. The transcription start site (TSS) is indicated and the first nucleotide of the transcript is numbered +1. In dbSNP the TATA box mutations are represented by the entries rs34500389 (−32), rs33981098 (−31), rs33980857 (−30), rs34598529 (−29), and rs33931746 (−28).

---

[2]Patients affected by β-plus thalassemia have one allele with the sickle variant and one allele with another mutation in the β-globin gene causing a reduced amount of normal β-globin protein.

DNA SEQUENCES AND SPECIFIC TRANSCRIPTION FACTORS

The promoter elements discussed so far are those that interact with the basal transcription factors, factors that are part of a preinitiation complex and are necessary for the transcription of all genes transcribed by RNA polymerase II. But not all genes are expressed with the same efficiency. Rather, each gene is transcribed at an appropriate level in a specific location in the body and at a suitable point in time, and expression depends on changing requirements. For instance, some genes may be efficiently turned on, while others are completely shut down. This means that each cell type under a certain condition has a characteristic gene expression profile.

Critical molecular players in the regulation of transcription are the **specific transcription factors**, proteins that bind to a subset of genes, rather than all of them. These transcription factors may act either as repressors or activators of transcription. Each of the proteins will bind to specific regions in DNA. The DNA binding sites may be located within 1,000 nucleotides from the transcription start site, the promoter proximal region. They interact with the general transcription factors to influence the formation of the initiation complex.

Other essential control elements bound by specific transcription factors are the enhancers. These elements are able to stimulate transcription, often at a large distance—several thousand nucleotides—from the actual transcription start site. An enhancer is in the size range of 50–1,500 nucleotides and may be located upstream, downstream, or within the actual gene under control. Enhancers are abundant—it has been estimated that the human genome contains in the order of 400,000 of these regulatory regions. Transcription factors that bind to enhancers are also able to interact with the transcription initiation complex at the core promoter, thereby causing a loop structure in chromatin (**Figure 10.1b**). Both the promoter proximal region of the promoter and enhancers have binding sites for several different transcription factors. The mediator, a complex of more than 20 different protein subunits, plays an important role in transcription initiation, transducing signals from the enhancer bound transcription factors to the initiation complex that includes RNA polymerase II. The contribution of transcription factors to the formation of the transcription initiation complex is summarized in Figure 10.1b.

TRANSCRIPTION FACTOR DNA BINDING SPECIFICITY

It is interesting to note that a considerable portion of all human genes encodes transcription factors. Thus, it is estimated that there are about 1,600 different human proteins in this category—or about 8% of all human proteins. This is one illustration of the importance of transcriptional control in the human genome. For about three-quarters of the transcription factors, we know what DNA sequences they target.

The action of a transcription factor is dependent on its recognition of a site in DNA. The transcription factor typically identifies structural features of the double-stranded DNA, which is to say that the helical structure is not opened. When it comes to specific interactions between protein and DNA, we distinguish two categories. One is specificity for the actual base sequence in DNA (direct or base readout), and the other is specificity for DNA shape (indirect or shape readout). Direct readout interactions are mediated either by van der Waals interactions or by hydrogen bonds, and examples of interactions are shown in **Table 10.1**.

Table 10.1 Interaction preferences between DNA bases and amino acids

DNA	Amino acids	Type of interaction
Adenine	Asn, Gln	Hydrogen bonding
	Asn, Gln, Phe, Pro	van der Waals
Cytosine	Glu, Phe	van der Waals
Guanine	Arg, Lys	Hydrogen bonding
	Arg, His, Ser, Gln	van der Waals
Thymine	Thr, Phe, Pro	van der Waals

We know the structure of a number of different complexes of DNA and transcription factors through x-ray crystallography or nuclear magnetic resonance (NMR). Selected structures are shown in **Figure 10.5**. These are all examples of how shorter DNA sequences are recognized by these proteins. For instance, P63 binds to a region with two or more repeats with consensus sequence RRRCWWGYYY. OCT1 and SOX2 bind to the consensus sequences ATGCTAAT and CTTTGTT, respectively. The binding of OCT1 and SOX2 is an example of synergistic action of two different transcription factors. Finally, the estrogen receptor (ER) DNA-binding domain binds as a dimer to DNA sequence elements known as estrogen response elements. These elements are such that a sequence on one strand of DNA (CAGGTCA in Figure 10.5) is repeated on the other strand—an example of an **inverted repeat** or **palindrome**.

Can we predict on the basis of the DNA sequence what transcription factor is binding where on DNA? There is a sequence specificity in the recognition between a transcription factor and DNA, as we see from Figure 10.5. However, transcription factors recognize only a short sequence (typically between 6 and 12 nucleotides) that is often degenerate. For this reason, a specific transcription factor is able to bind to

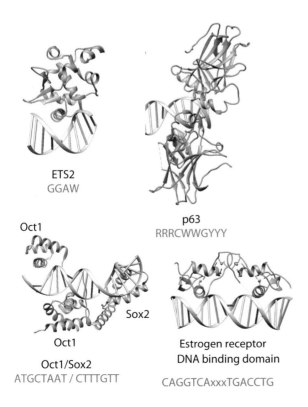

ETS2
GGAW

Oct1

p63
RRRCWWGYYY

Sox2

Oct1

Oct1/Sox2
ATGCTAAT / CTTTGTT

Estrogen receptor
DNA binding domain

CAGGTCAxxxTGACCTG

Figure 10.5 Structures of complexes between DNA and transcription factors. Structures of four different DNA-protein complexes are shown. The proteins involved are (1) the ETS domain of human ETS2, (2) P63, (3) a complex of SOX2 and OCT1, and (4) the estrogen receptor (ER) DNA-binding domain. Text in red is the consensus DNA sequence (R = A or G—Y = C or T and W = A or T) recognized by the respective transcription factors.

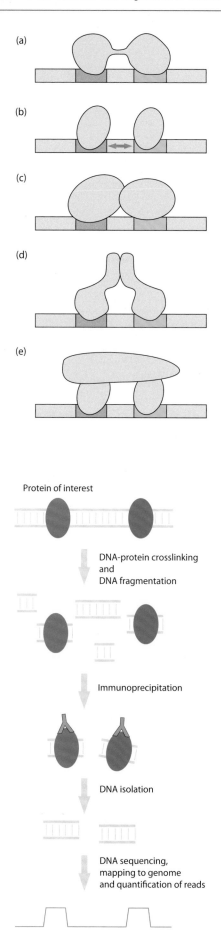

(a)

(b)

(c)

(d)

(e)

Protein of interest

DNA-protein crosslinking and
DNA fragmentation

Immunoprecipitation

DNA isolation

DNA sequencing,
mapping to genome
and quantification of reads

Figure 10.6 Protein-DNA and protein-protein interactions involving transcription factors. A multitude of mechanisms add to the complexity of interaction between DNA and transcription factors. Two protein binding sites in DNA (orange and yellow) are shown. (a) One transcription factor may have multiple DNA binding sites. (b–e) Different mechanisms where the binding of one transcription factor influences the binding of another factor. (b) Transcription factors may cause changes in DNA structure (red arrow) that affect binding of subsequent transcription factors. (c) DNA binding domains of transcription factors may interact directly. (d) Non-DNA binding domains of transcription factors may interact. (e) Another protein may bind two transcription factors and influence their structures and DNA affinities.

a large number of sites in a human genome, but only a subset of these binding interactions is functionally significant. Therefore, we are not able to predict gene regulation on the basis of the sequences that attracts a specific transcription factor.

Then, how is specificity achieved in transcriptional control? It turns out that it often comes about by the formation of complexes involving multiple transcription factors and/or multiple DNA binding sites. The binding of multiple proteins to both the promoter proximal region and to enhancers is illustrated in Figure 10.1b. It must also be noted that individual transcription factors are built from multiple domains, including domains that do not bind to DNA but have other functions. Overall, there are many different protein-protein and protein-DNA interactions that add to the complexity of the recognition of transcription factors of DNA sequences as highlighted in **Figure 10.6**. In addition, it has been shown that the efficiency of transcription factor binding depends on whether the DNA binding site is part of a nucleosomal structure or not. These circumstances make it difficult to predict transcription factor binding sites based on DNA sequence alone. Rather, we must rely on experimental methods that monitor the binding of these proteins to sites in the human genome. One such method is **ChIP-sequencing**, also known as **ChIP-seq**, a method designed to identify the DNA sites where a specific protein binds (**Figure 10.7**).

THE ENCODE PROJECT MAPPED FUNCTIONAL SITES IN THE HUMAN GENOME

ChIP-sequencing to map transcription factor binding sites is one of many experimental methods used in the Encyclopedia of DNA Elements (ENCODE) project. This project was launched in 2003 by the US National Human Genome Research Institute (NHGRI) with the aim to map all kinds of functional elements in the human genome, mainly elements of regulatory importance. Important results from the project include (1) mapping of transcribed regions, (2) identification of many noncoding transcripts, (3) analysis of chromatin accessibility and histone modification patterns (see section "Epigenetic mechanisms of transcriptional control"), and (4) the mapping of disease-associated single-nucleotide polymorphisms (SNPs) to noncoding regions of the genome, in regions outside of protein coding genes. One significant finding from the ENCODE project is that a much larger fraction of the genome is functionally significant as compared

Figure 10.7 Outline of ChIP-sequencing. In the first step, DNA and protein are covalently cross-linked. Then DNA is fragmented, so that to each protein molecule is attached only a smaller piece of DNA. Then, an antibody against a protein of interest, such as a specific transcription factor, is used to immunoprecipitate the protein of interest (antibody is shown in orange). From this precipitate, DNA is isolated and subjected to DNA sequencing. All sequence reads are finally aligned to the human genome and quantitation of reads along the genome is shown as a graph. ChIP-sequencing will thus show where a protein of interest is binding in the genome sequence.

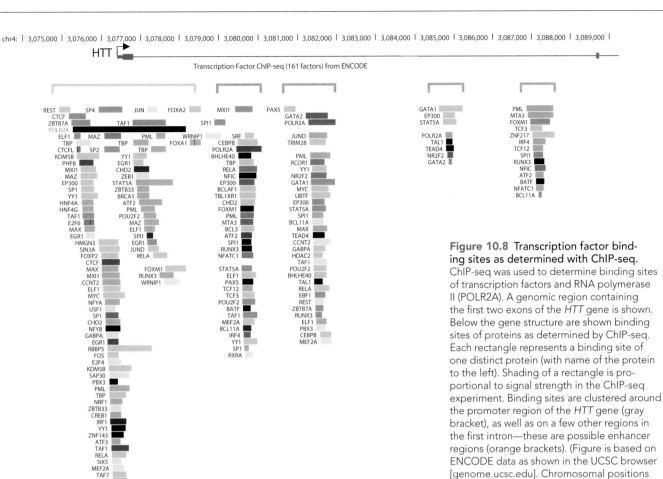

Figure 10.8 Transcription factor binding sites as determined with ChIP-seq. ChIP-seq was used to determine binding sites of transcription factors and RNA polymerase II (POLR2A). A genomic region containing the first two exons of the *HTT* gene is shown. Below the gene structure are shown binding sites of proteins as determined by ChIP-seq. Each rectangle represents a binding site of one distinct protein (with name of the protein to the left). Shading of a rectangle is proportional to signal strength in the ChIP-seq experiment. Binding sites are clustered around the promoter region of the *HTT* gene (gray bracket), as well as on a few other regions in the first intron—these are possible enhancer regions (orange brackets). (Figure is based on ENCODE data as shown in the UCSC browser [genome.ucsc.edu]. Chromosomal positions refer to chromosome 4 and genome version is GRCh37/hg19.)

to what was previously thought. Thus, it was reported that biochemical activity occurs in about 80% of the genome.[3] Many of these regions are involved in transcriptional control. An example from the transcription factor binding analysis by the ENCODE project is shown in **Figure 10.8**, in which the binding of 89 different proteins is demonstrated in promoter and enhancer regions of the *HTT* gene.

PROXIMAL PROMOTER MUTATIONS AND DISEASE: GATA-1 AND δ-THALASSEMIA

We turn to one of the transcription factors known to be associated with disease. GATA-1 is a member of the GATA family of transcription factors. The name "GATA" is derived from the DNA recognition sequence—the consensus sequence is 5′-WGATAR-3′. GATA-1 belongs to a category of transcription factors that contain one or more DNA binding domains known as **zinc fingers**. The interaction between GATA-1 and DNA, the basic structure of a zinc finger, as well as a sequence logo describing the conservation of the GATA motif are shown in **Figure 10.9**.

GATA-1 regulates a large number of genes and is instrumental in the development of specific types of blood cells from precursor cells. DNA recognition elements of GATA-1 can occur in both promoter-proximal regions

---

[3]This number has been subject to debate, and more studies are required before we are able to more fully understand how different specific biological functions distribute in the human genome.

Figure 10.9 Interaction of transcription factor GATA-1 with DNA. (a) GATA-1 may recognize either a single GATA motif or, as in the structure here, in addition a pseudopalindromic sequence GATG on the other strand (a "double GATA site"). In the sequence shown on one of the DNA strands, 5'-CCATCTGATAAG-3, the two motifs are shown in blue. They are recognized by N- and C-terminal zinc finger domains (ZFs), respectively. Spheres in the protein structure are zinc atoms. A region between the ZFs is unordered in the structure ("disordered linker"). (b) Structure of zinc finger domain where a zinc atom is coordinated to cysteine and histidine residues. (c) Sequence conservation of GATA-1 binding sites is shown with a sequence logo.

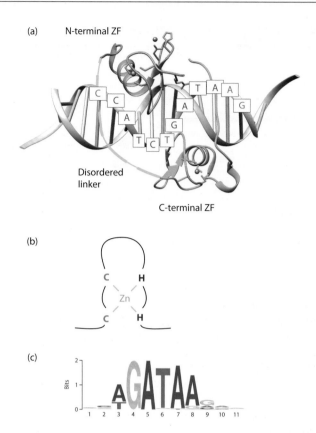

as well as in enhancers. For instance, there is one GATA-1-binding site 77 nucleotides upstream of the transcription start site in the gene *HBD*, a gene encoding δ-globin. The GATA-1 binding site is CTTATCT (AGATAAG on the complementary strand). As with many other transcription factors, the orientation of the GATA sequence element is not critical for the action of GATA-1. Thus, the GATA sequence may be placed on either DNA strand.

δ-Globin is part of a hemoglobin named HbA-2. This protein has the subunit structure $\alpha_2\delta_2$ and makes up only 3% of all adult hemoglobin. The effect of GATA-1 binding to the promoter element is to increase transcription of *HBD*. The importance of the GATA element is illustrated by a case of δ-thalassemia, a form of thalassemia where the production of δ-globin is defective. A mutation T > C disrupts the GATA-1 binding site, giving rise to the sequence TCATCT (complementary strand AGATGA). The mutation affects binding of GATA-1, in turn leading to lowered expression of δ-globin (**Figure 10.10**).

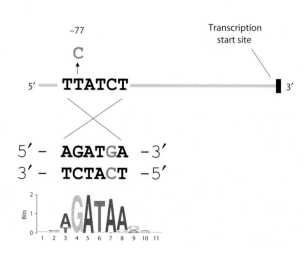

Figure 10.10 δ-Thalassemia as caused by mutation in a GATA-1 binding site. The gene *HBD* has a GATA-1-binding site 77 nucleotides upstream of the transcription start site. A sequence logo is shown aligned to this region. In one case of δ-thalassemia, a mutation T > C disrupts the GATA-1 binding site.

PROXIMAL PROMOTER MUTATIONS AND DISEASE: *TERT* PROMOTER AND CANCER

The importance of transcription factor binding sites is also illustrated by studies of cancer. In a large number of cancers, there are mutations in the promoter of the *TERT* gene. This gene encodes a protein that is part of a complex referred to as telomerase. The telomerase is responsible for regulating the length of **telomeres** (repetitive region consisting of TTAGGG repeats) that occur at the ends of chromosomes.

In a majority of tissues, expression of the *TERT* gene is turned off. In the absence of telomerase activity, telomeres will shorten with every cell division. When a certain critical length of the telomere is reached, cells enter into a nonproliferate state. Cells that require many cell divisions, such as blood stem cells, are characterized by high telomerase activity. In most cancers, telomerase is active. This activation is the result of mutations in the *TERT* gene, most commonly in its promoter. There are two particularly frequent mutations, in positions 124 and 146 upstream of the ATG translation start site of the gene. These are both C > T (complementary strand G > A) mutations. The two mutations form the same sequence CCCCTTCCGGG (complement CCCGGAAGGGG). Thereby, a consensus sequence is formed bound by a family of transcription factors that includes GABP ("GA-binding protein"). This family of proteins recognizes a GGAW motif that is part of the complement (underlined portion in CCC<u>GGAA</u>GGGG).

The GABP protein is a dimer or tetramer built from the proteins GABPA and GABPB, where GABPA is the chain that binds to DNA. The result of the above mutations in the *TERT* promoter is an activation of the *TERT* gene. A model has been put forward for this activation where GABP is recruited as a tetramer, and where one subunit of GABPA binds to a region with the promoter mutation and the other binds to a native binding site farther downstream (**Figure 10.11**). The specific binding of GABPA to DNA is illustrated with the structure of the corresponding protein-DNA complex in mice in **Figure 10.12**, in which the interaction with the sequence GGAA is mediated through hydrogen bonds.

It should be noted that the mutations of the *TERT* promoter discussed here are creating a novel binding site. This is contrary to our case of δ-thalassemia, where the mutations reduce transcription factor binding.

MUTATIONS IN ENHANCERS MAY GIVE RISE TO DISEASE

In the sections to follow, we focus on enhancer elements and their role in disease. Remember that enhancers are regions, often distant to the transcription start site, that have important regulatory functions. As an example, consider enhancer regions that regulate transcription of the gene *ALAS2*. This gene produces an enzyme critical for the synthesis of heme, a nonprotein component of hemoglobin. *ALAS2* is specifically expressed in developing red blood cells. Once again, we encounter GATA-1 as an important transcription factor as it regulates the expression of *ALAS2*.

A disorder named X-linked sideroblastic anemia (XLSA) is often caused by missense mutations in the *ALAS2* gene coding sequence. However, there is also a report of five families with mutations in an enhancer in the first intron of the *ALAS2* gene (**Figure 10.13**). The mutations are predicted to disrupt the binding of GATA-1.

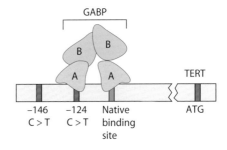

Figure 10.11 Mutations in cancer cells that activate the *TERT* gene. In cancer there are two particularly frequent mutations C > T, in positions −124 and −146 relative to ATG translation start site of the *TERT* gene. These mutations affect the binding of GABP, a dimer or tetramer of GABPA and GABPB subunits. A model is shown here where one subunit of GABPA binds to a region with one of the promoter mutations and the other binds to a native binding site farther downstream. The effect is an activation of transcription.

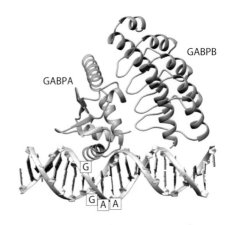

Figure 10.12 Specific recognition of GGAA sequence by protein GABPA. The A subunit of the GABP proteins binds in a sequence-specific manner to DNA. The interaction with the sequence GGAA is mediated through hydrogen bonds. (Based on Batchelor AH et al. 1998. *Science* 279:1037–1041. PBD ID: 1AWC.)

Figure 10.13 Mutations in enhancer that are associated with X-linked sideroblastic anemia (XLSA). On top is shown the genomic structure of the gene *ALAS2*. This gene is located on the reverse strand of the reference human genome. Exons are shown in green. In families with XLSA (f1-3), there are mutations in an enhancer located in the first intron of the gene.

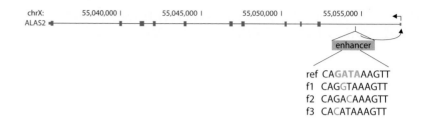

SICKLE CELL DISEASE AND ENHANCER GENETIC VARIANTS

We encounter another example of GATA-1 binding to an enhancer in the context of sickle cell disease (SCD), a group of disorders where the individual carries two abnormal copies of the β-globin gene and where at least one allele is the sickle β-globin. In Chapter 2 we discussed a special case of this disorder, sickle cell anemia in which there are two copies of the sickle cell gene. The result is a hemoglobin where the β-chain has been replaced with the sickle variant.

In the human fetus, there is a specific hemoglobin (HbF) that is different from the adult form as it has a higher affinity for oxygen. In this way it is able to "steal" oxygen from the blood of the mother. Fetal hemoglobin has the structure $\alpha_2\gamma_2$. The γ subunits, encoded by the genes *HBG1* or *HBG2*, are the fetal counterparts of the adult hemoglobin β-globin chains.

Interestingly, there is a distinct relationship between HbF and SCD. Some patients with SCD have high levels of HbF, although the expression of this protein normally is shut down in the adult. These patients have milder symptoms because HbF tends to prevent the harmful aggregation characteristic of sickle cell hemoglobin. This observation is of interest, because it suggests that a possible therapy for SCD is to increase the expression of HbF.

To identify any genetic variants (SNPs) responsible for the higher levels of HbF in SCD patients, **genome-wide association studies** (GWAS) (Chapter 5) have been carried out. One particularly interesting locus was identified in this search, that of the gene *BCL11A*. The product of this gene is a zinc finger protein, and one of its functions is to repress the transcription of the genes *HBG1* and *HBG2*. These genes are part of a locus containing five different genes being members of the β-chain family (**Figure 10.14**).

Figure 10.14 The human β-globin locus and regulation of fetal globin chains. The β-globin locus contains five different genes encoding β-chains of hemoglobin. Expression of these genes is controlled by a locus control region (LCR). The gene *HBE1* is embryonically expressed, *HBG2* and *HBG1* expressed at fetal stage and proteins of *HBD* and *HBB* are produced in the adult. The protein BCL11A negatively regulates the fetal genes *HBG1* and *HBG2* by interacting with an enhancer located in between *HBG2* and *HBD*.

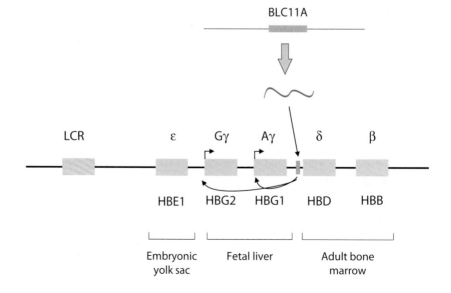

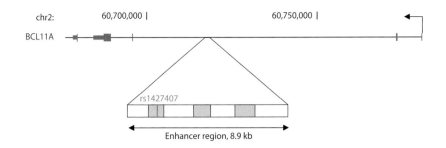

Figure 10.15 Gene *BCL11A* and location of enhancer region. The genomic structure of the gene *BCL11A* is shown on top where exons are shown in green. An enhancer region is located in the second intron. The shaded parts of the enhancer region correspond to three different regions sensitive to cleavage by the enzyme DNase I. Sensitivity to the enzyme DNase I is used to identify regions such as promoters and enhancers as explained later in the context of epigenetic mechanisms. The shaded regions also have binding sites for the transcription factors GATA-1 and TAL1. The SNP rs1427407 is located in one of these regions. The enhancer region is located at chr2:60,717,007–60,725,926 in genome version GRCh37/hg19.

Many of the SNPs identified that are related to high levels HbF are in regions that regulate the expression of *BCL11A* (**Figure 10.15**). Further studies of the *BCL11A* gene revealed that one SNP associated with elevated HbF values is within an enhancer, and it disrupts a binding site for the transcription factor TAL1. This transcription factor recognizes the consensus sequence TG (a motif known as a "half E-box"), which is changed to TT as a result of the mutation. The TAL1 binding site is close to a binding site for GATA-1, and these two transcription factors presumably bind in a cooperative manner to their respective DNA motifs (**Figure 10.16**). The disruption of the half E-box is therefore likely to disturb also the binding of GATA-1. Thereby, the expression of *BCL11A* is reduced, in turn leading to an increase in fetal hemoglobin expression.

AN ENHANCER GENETIC VARIANT IS ASSOCIATED WITH PARKINSON'S DISEASE

Another disorder that has been associated with enhancers is Parkinson's disease (PD). A gene of interest in PD is the gene *SNCA*, a gene encoding

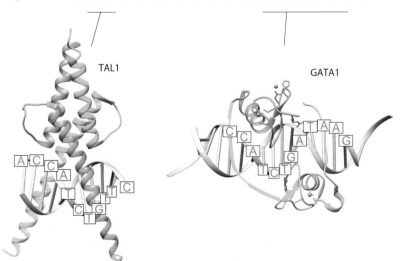

Figure 10.16 A SNP in a *BCL11A* enhancer may affect the binding of transcription factors TAL1 and GATA-1. A mutation G > T is associated with an elevated expression of fetal hemoglobin (HbF). This mutation may disrupt a binding site for TAL1 and GATA-1, two cooperating transcription factors that bind to a partial E-box located seven or eight base-pairs upstream of a WGATAA motif. GATA and the structure of the GATA-1/DNA complex are the same as in Figure 10.9. (TAL1 structure is based on El Omari K et al. 2013. *Cell Rep* 4:135. PDB ID: 2YPB.)

Figure 10.17 An enhancer variant associated with Parkinson's disease. (a and b) The genomic structure of the gene *SNCA* where exons are shown in green. The SNP rs356168 is located in an enhancer element contained within intron 4 of the *SNCA* gene. Nucleotide positions refer to the human genome version GCRh37/hg19. (a) When the variant position is occupied by A, the transcription factors EMX2 and NKX6-1 bind effectively, the enhancer is repressed, and *SNCA* expression is inefficient. (b) When the variant position is changed to G, the consensus sequence AATTA is disrupted, in turn affecting the binding of the two transcription factors. Thereby the *SNCA* gene is activated. (c) Sequence logos show consensus properties of EMX2 and NKX6-1 binding sites.

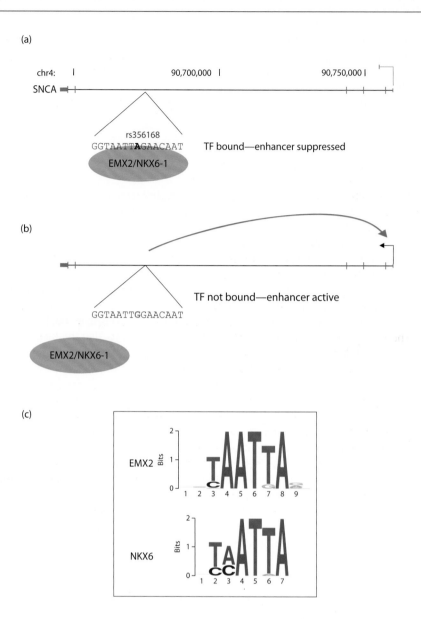

α-synuclein. While familial PD may be caused by coding mutations in *SNCA*, there are variants in the same gene identified by GWAS studies that increase the risk of developing the sporadic form of the disease. For instance, GWAS studies identified the SNP rs356168, which is located in an enhancer element contained within intron 4. The SNP affects the binding of the transcription factors EMX2 and NKX6-1. When the variant position is occupied by A, these transcription factors bind effectively. As a result, the enhancer is suppressed. But when the position instead has a G, the binding of these transcription factors is disrupted, thereby activating the *SNCA* gene (**Figure 10.17**).

TRANSCRIPTIONAL CONTROL IS MEDIATED BY HIGHER-ORDER ORGANIZATION OF THE GENOME

We have seen that enhancers are important elements in transcriptional control. Substitution or indel mutations in these elements may result in an altered affinity for a transcription factor. We have encountered such mutations that are associated with disorders. We turn now to a group of disorders that also are related to the function of enhancers. But rather than being caused by mutations within an enhancer element, the genetic change is a larger deletion or rearrangement that causes the enhancer to be incorrectly placed with respect to the gene it is normally controlling. Relevant to this discussion is a higher organization of DNA. We previously discussed the three-dimensional structure of chromatin (Chapter 4) and that DNA needs to be extensively packed to fit within the nucleus. The first level of packing is achieved with nucleosomes. We now consider a higher-order structure of DNA that was elucidated with methods known as chromosome conformation capture (3C). These methods identify interactions between locations in the genome that are close in three-dimensional space, but that are widely separated in the linear genome.

Such methods have revealed regions in the genome known as **topologically associating domains** (**TADs**). These are regions that show high levels of interaction within the regions but little or no interaction with neighboring regions. TADs may be viewed as extensive loops in chromatin, loops that are often nested. Each TAD may be large, encompassing up to millions of nucleotides. There are boundary regions that separate TADs. They range in size from a few to hundreds of kilobases. Furthermore, in most boundaries are found specific proteins. CCCTC-binding factor (CTCF) and cohesin are two such particularly well-studied proteins. It is thought that they contribute to the formation of TADs. CTCF is a zinc finger protein that binds to DNA in a sequence-specific manner. Important principles of TADs are shown in **Figure 10.18**. It should be noted that the domains are formed by interactions of boundary regions and that enhancers are restricted to action within their own TAD.

The significance of TADs is further emphasized by their conservation among animals. For instance, homologous regions of the genomes in human and mouse have the same TAD boundaries and the same number of TADs. In addition, TADs are also similar when comparing different tissues in an animal.

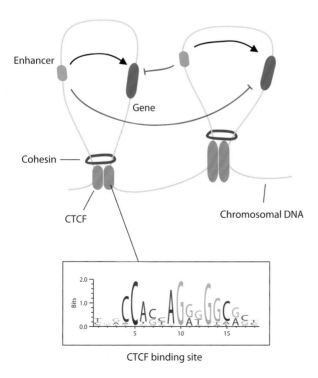

Figure 10.18 Looped chromatin structure and topologically associating domains. TADs are regions whose DNA sequences preferentially contact each other. Based on TADs, chromatin may be represented as a looped structure as shown here. Proteins CTCF and cohesin are shown here as part of a boundary region separating two TADs. The sequence logo (bottom) shows the conserved portions of the CTCF DNA binding site. One consequence of the TAD organization of chromatin is that enhancer elements may only affect gene promoters that are within the same TAD.

ENHANCERS THAT GET OUT OF CONTEXT

There are specific developmental disorders with a genetic background that nicely illustrate the biological significance of TADs and boundary elements, and how enhancers can only act on genes within the same TAD. One set of these disorders is associated with a chromosomal locus containing a group of genes important for correct limb development. One of the genes in that locus is *EPHA4* that specifies ephrin type-A receptor 4, a protein mediating developmental events. In one scientific study, different patients were examined who have disorders resulting from genetic changes in this genomic region. The patients were all characterized by limb deformation, but depending on the genomic change, the phenotypes were different.

First, in one group of patients, there were deletions that included the *EPHA4* gene and the TAD boundary located to the telomeric side of *EPHA4*. These deletions give rise to a phenotype where fingers and toes are shorter than normal, a phenotype known as brachydactyly. It could be shown that the molecular basis of this malformation is that enhancers get out of context. Thus, there is an enhancer region in the *EPHA4* TAD that normally regulates the *EPHA4* gene. But in brachydactyly, this region is instead activating a gene *PAX3* (**Figure 10.19a**). *PAX3* encodes a transcription factor critical during development.

Second, in another group of patients, an inversion of a region including a TAD boundary upstream of *EPHA4* results in a disorder where fingers are fused (syndactyly). Again, the enhancer is now out of its normal TAD context and is activating the gene *WNT6*, which encodes a signaling protein essential for a variety of patterning events during development (**Figure 10.19b**).

Finally, a family was examined that had a duplication of a region where one breakpoint is just centromeric to a gene named *IHH*. The gene *IHH* is encoding a protein with a role in a signaling pathway. The affected family members were characterized by polydactyly—that is, an increase in the number of fingers (at least seven) (**Figure 10.19c**). The molecular background to the polydactyly is that the *IHH* gene is erroneously activated.

CANCER AND DNA ORGANIZATION

Mutations in the gene *IDH1* are common in many cancers. This gene encodes the enzyme isocitrate dehydrogenase 1, an important enzyme in the citric acid cycle, a metabolic process essential for conserving energy in the form of ATP molecules. The mutations found in cancer causes *IDH1* to produce abnormal metabolites. These metabolites are, in turn, able to inhibit enzymes that remove methyl groups from DNA. As a result, specific regions of chromatin are methylated to a larger extent than what is normal. An effect at TAD boundary regions is that elevated methylation causes a reduction in the binding of the CTCF protein. This, in turn, leads to the disruption of TAD boundaries and breakdown of TAD structure. Enhancers may now activate the transcription of many genes they are not normally controlling, including oncogenes (**Figure 10.20**). These observations of the consequences of *IDH1* mutations demonstrate that the higher structural organization of DNA may be important in the development of cancer.

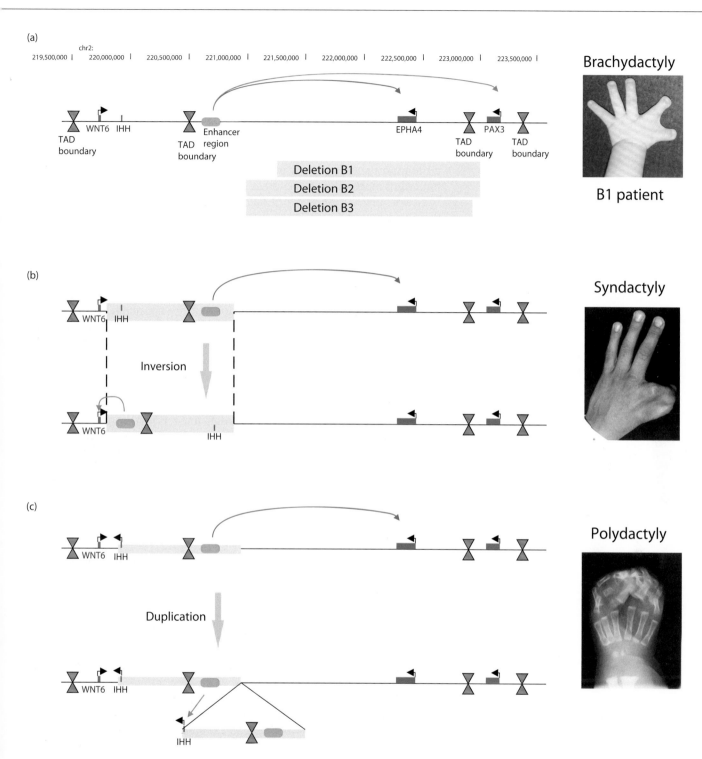

Figure 10.19 Locus of *EPHA4* and genomic rearrangements that put enhancers out of context. Three examples of structural changes in a locus containing the gene *EPHA4* that give rise to limb deformation are shown. All examples involve the same enhancer, normally controlling the gene *EPHA4*. TAD boundaries are shown with red triangles, enhancers in orange, and genes in green. Arrows show the transcription direction. Chromosomal position numbers refer to human genome version GRCh37/hg19. (a) Three cases of brachydactyly (patients B1, B2, and B3) are caused by deletions that remove a TAD boundary. (b) A case of syndactyly is the result of inversion of a region that includes a TAD boundary. (c) A duplication of a region causes a gene *IHH* to be regulated by the enhancer, giving rise to polydactyly. (Adapted from *Cell*, 161, Lupiáñez DG et al., 1012–1025, Copyright 2015, with permission from Elsevier.)

(a)

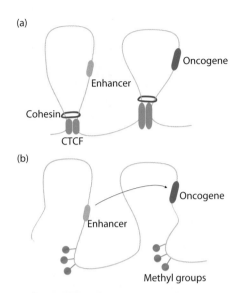

Figure 10.20 Cancer mutations may give rise to the disruption of TAD boundaries. (a) The TAD structure of a genomic region is shown with an enhancer in one TAD and an oncogene in another. (b) A mutation has led to methylation of a region, which causes CTCF protein to bind less efficiently. This leads, in turn, to breakdown of the TAD structure, and the enhancer is now able to activate the oncogene.

ONE GENE: MULTIPLE TRANSCRIPTION START AND TERMINATION SITES

Adding to the complexity of transcription initiation and regulation is the fact that many genes have more than one transcription start site, and the selection of the start site is subject to control. An example of how different transcripts may be formed from the gene *TP73* using alternative start site selection is shown in **Figure 10.21**. It is estimated that more than 50% of all human genes have alternative transcription start sites, and this is an important mechanism to generate tissue-specific transcripts.

Not only do many genes have multiple transcription initiation sites, they also have alternative sites where transcription is terminated. Thus, about 70% of all genes have multiple termination sites. The termination of transcription uses a specific mechanism that involves addition of a **poly(A) tail** at the 3′ end of the mRNA precursor (**Figure 10.22**). The primary event in transcription termination is the recognition of a sequence element in the synthesized RNA with the consensus sequence AAUAAA. This element is recognized by a protein cleavage factor. Further specificity is provided by a G/U rich region downstream of the cleavage site. The RNA is cleaved downstream of the AAUAAA element, and to the 3′ end thus formed is then added a sequence of A's with the help of a specific enzyme. Transcription then terminates farther downstream in a manner not fully understood.

For the genes with multiple poly(A) sites, each of them generates a different 3′ end of the transcript and as a consequence, a different exon structure (**Figure 10.23**). Further studies are required to understand the DNA and RNA elements that control alternative polyadenylation.

There are inherited disorders where the molecular defect is a mutation in a poly(A) signal. For instance, in patients with thalassemia, mutations have been identified, such as T > C (U > C in RNA) in position 3 and A > G in position 6 of the AAUAAA motif. In the case of the T > C mutation, it was shown that it leads to the formation of an elongated β-globin mRNA isoform using another AAUAAA polyadenylation signal 900 nucleotides downstream of the mutation site.

In summary, we have in this section seen that a single gene may generate multiple transcripts and proteins by way of alternative transcription start sites as well as alternative termination sites. Remember from the previous chapter that different variants of transcripts and proteins may also be produced using alternative usages of exons. Thus, we see that three different mechanisms contribute to the significant principle of gene product diversity, "one gene—multiple transcripts."

EPIGENETIC MECHANISMS OF TRANSCRIPTIONAL CONTROL

So far in this text, we regarded all inherited changes as a result of modifications to the sequence of nucleotides in DNA. However, there is another

Figure 10.21 Two alternative transcription start sites in gene *TP73*. The genomic structure and two alternative transcripts of gene *TP73* are shown. Coding regions are in green, and UTRs in cyan. Arrows indicate transcription direction. Chromosomal position numbers refer to human genome version GRCh37/hg19.

important category of changes that may be inherited, **epigenetic** changes, and as the mechanisms of this category are highly important for transcriptional control, it is time to present an overview of them and how they are related to the actual DNA sequence.

Epigenetics may be defined as inheritance that does not result from changes in the sequence of nucleotides in DNA (genetic changes), but rather other modifications that affect chromatin structure, modifications that are associated with transcriptional control. There is another difference between genetic and epigenetic changes. Genetic variants are present already in the germline and are propagated faithfully to all daughter cells during development. Epigenetic changes also occur during development, where they for instance can aid in stabilizing the gene expression pattern and identity of a specific cell lineage. This means that different cells in a human body have different epigenetic profiles. This is unlike the sequence of nucleotides—although there are a number of somatic mutations, all cells have virtually the same DNA sequence.

Two important epigenetic mechanisms to be discussed in the following sections are (1) covalent modification of histone proteins and (2) methylation of DNA.

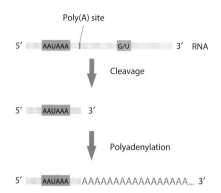

Figure 10.22 Mechanism of nuclear polyadenylation. A biochemical machinery recognizes the signal AAUAAA and a G/U rich region in the newly formed RNA strand during transcription. Cleavage then occurs and to the 3′ end thus generated is added a poly(A) tail of about 250 As. After cleavage and polyadenylation, transcription is terminated (not shown here).

HISTONE MODIFICATION

In Chapter 4, the basic structure of chromatin was discussed. To achieve packing of DNA, it is wrapped around histone proteins (H2A, H2B, H3, and H4) to form a nucleosome structure (Figure 4.5). However, histones not only have a passive role in packing DNA, they are also important players in transcriptional control. Histone proteins are highly basic proteins with unstructured N-terminal and C-terminal "tails" that protrude from the nucleosome. These tails may be subject to several different covalent modifications, including phosphorylation, ubiquitination, methylation, and acetylation.

Specific enzymes are responsible for introducing ("writing") these modifications—for instance, histone acetyltransferase (HAT) acetylate specific residues in histones. At the same time, there are enzymes that remove modifications ("erasers"), such as histone deacetylase that reverses the modification introduced by HAT.

There are many different states of chromatin, but historically we distinguish between **euchromatin**, which is a lightly packed form, and **heterochromatin**, which is condensed and compact. This is originally a classification based on staining of nuclear DNA with microscopy techniques. Typically, euchromatin is associated with active genes, and heterochromatin is transcriptionally silent. Some of the histone modifications achieve an "open" conformation of chromatin, correlated with active transcription (euchromatin), whereas others have an opposite role to produce a more condensed and compact state of chromatin, associated with a low level of transcription (heterochromatin).

Some of the modifications in the N-terminal tail of histones are shown in **Figure 10.24**. The histone marks are typically represented with an abbreviation to show the identity of the histone and the position of

Figure 10.23 Two alternative transcription polyadenylation sites in gene *CSTF3*. The genomic structure and two alternative transcripts of gene *CSTF3* are shown. Coding regions are in green and UTRs in cyan. Arrow indicates transcription direction. Chromosomal position numbers refer to human genome version GRCh37/hg19.

Figure 10.24 A selection of histone modifications. The N-terminal ends of histones are shown with the rest of the amino acid sequences represented by shaded boxes. Histones also have modifications in other parts of the protein, but only the N-terminal modifications are shown here. Modifications that are typically associated with transcriptionally active chromatin are acetylation, arginine methylation (orange Me hexagon), and some lysine methylation (H3K4, green Me hexagon and H3K36, yellow Me hexagon). Repressive modifications include methylation at H3K9, H3K27, and H4K20 (red hexagon). (From Allis CD et al. 2015. *Epigenetics*, 2nd ed. Cold Spring Harbor, NY: Cold Spring Harbor Laboratory Press. With permission from Cold Spring Harbor Laboratory Press.)

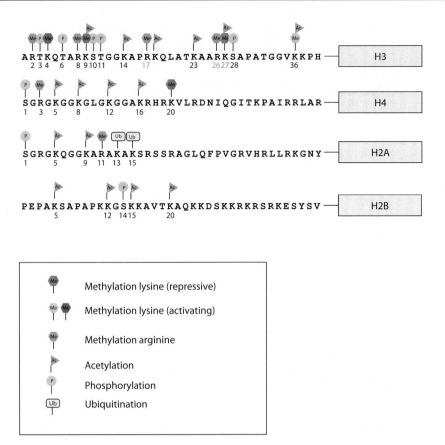

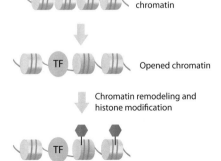

Condensed chromatin

Opened chromatin

Chromatin remodeling and histone modification

Figure 10.25 Transcription factor binding may give rise to epigenetic modifications of chromatin. Binding of transcription factor(s) may lead to an open configuration of chromatin. This is followed by chromatin remodeling and histone modification to produce transcriptionally active chromatin.

the modification. For instance, H3K4 means that histone 3 is modified at lysine (K) residue at position 4. Histone modifications may alter the chromatin state. Modifications that are often associated with transcriptionally active chromatin are acetylation, arginine methylation, and some lysine methylation (H3K4 and H3K36). Repressive modifications include methylation at H3K9, H3K27, and H4K20. Histone marks constitute epigenetic information, as they may be inherited across many cell generations.

What signals in DNA lead to the transformation of chromatin into an active state? An important primary event is the binding of a transcription factor to a regulatory element. In **Figure 10.25** is shown how transcription factor binding is followed by chromatin remodeling[4] and histone modification to produce transcriptionally active chromatin. The epigenetic status may be inherited to the next generation of cells. In such a case, the transcriptionally active status of chromatin is maintained through the epigenetic modifications even though the transcription factors are lost. Considering the important role of transcription factors in initiating epigenetic changes, we may expect that mutations in transcription factor binding sites cause aberrant epigenetic regulation, but further studies are required to establish details of the relationship between DNA sequence and epigenetic regulation.

DNA METHYLATION AND CpG ISLANDS

Another important epigenetic mechanism is a reaction where DNA is methylated. Typically, a methyl group is added to cytosine, and in mammals this

---

[4]Chromatin remodeling is a change in chromatin structure that comes about through histone modification and chromatin remodeling complexes.

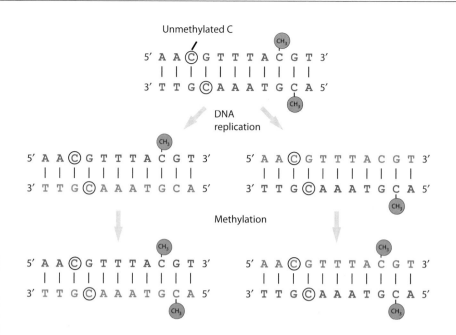

Figure 10.26 An epigenetic methylation pattern may be inherited. At a CpG site in DNA, both strands are methylated. A newly produced strand is not methylated, but DNA methyltransferase can methylate at a CpG site which is paired with a methylated CpG. The same enzyme cannot methylate an unmethylated CpG that is paired to another unmethylated CpG. With these principles, the methylation pattern of a cell may be copied to a daughter cell.

reaction primarily occurs at **CpG sites**. (We previously encountered these sites and saw that they have a tendency to mutate [Chapter 6].) The methylation at CpG sites is an epigenetic modification, as the methylation pattern can be transmitted to daughter cells by way of DNA replication that takes place during cell division (**Figure 10.26**).

The methylated C residues in CpG dinucleotides are easily deaminated to T's (a spontaneous reaction). This has the effect that CpG tends to mutate to TpG. For this reason, CpG dinucleotides are markedly underrepresented in the human genome. However, there are regions in the human genome that are unusually rich in CpG content, regions referred to as CpG islands. They may be defined as regions that are (1) longer than 200 nucleotides, (2) have a G + C content greater than 50%, and (3) have a ratio of observed to expected CpG greater than 0.6 (**Figure 10.27**).

About 60%–70% of all human genes have a CpG island in their promoter region, and they are believed to be important in transcriptional

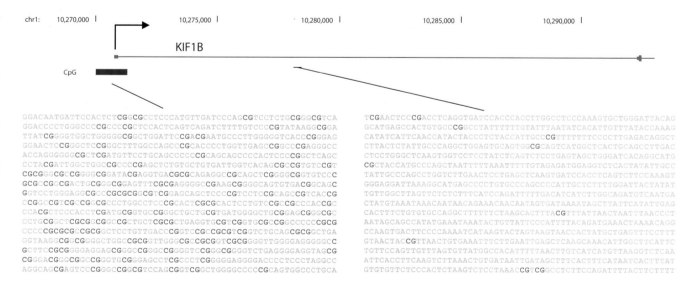

Figure 10.27 CpG islands are present in the promoter region of many genes. CpG islands contain a much higher frequency of CpG dinucleotide sequences as compared to the rest of the human genome. Shown here is a CpG island of the *KIF1B* gene, as well as a portion of the first intron of the same gene with a CpG frequency characteristic of a major fraction of the genome.

regulation. Methylation of CpG islands tends to silence genes. But CpG islands are mostly unmethylated and have histone modifications that render them permissive for transcription. Because CpG islands are so common close to the transcription start sites of genes, and because they may easily be identified computationally from the genome sequence, they are useful in the prediction of transcription start sites.

An important question is how the methylation and demethylation of CpG sites are regulated. For instance, how is a nonmethylated status preserved in CpG islands? Research is still in progress to address these questions. As with histone modifications, transcription factors are likely to play a role in the context of CpG methylation. In this regard, it is interesting to note that for a subset of transcription factors, the affinity for the CpG sequence is increased as a result of cytosine methylation, whereas for other factors the affinity is decreased. With this in mind, it is likely that transcription factors are used to monitor methylation status as well as to regulate methylation and demethylation.

Methylation of CpG islands has been shown to be an important component in the development of cancer. Thus, in cancer cells, CpG islands are often methylated to a much larger extent than normal. This gives rise to a transcriptional silencing that may be inherited by daughter cells following cell division. Conversely, CpG islands may also be undermethylated in cancer, giving rise to an activation of transcription. So, although a number of genes may be affected by point and indel mutations in the DNA sequence (Chapter 5), the progression of cancer may also be caused by aberrant DNA methylation, in which case the expression of numerous genes may be affected. For instance, whereas a point mutation may result in an altered expression of a single gene, the expression of hundreds of genes may be altered by DNA methylation.

THE EPIGENOME

We have seen that covalent modifications of histones and DNA methylation play an important role in transcriptional control, and these covalent changes may be inherited. With this knowledge, we realize that to understand genetics of human individuals and specific human cells, we need not only consider the actual DNA sequence, but also the **epigenome**. Therefore, epigenetic changes are being mapped in different types of cells through laboratory work. For instance, the ENCODE project referred to previously has mapped a whole range of epigenetically significant elements.

Different methods are used to identify transcriptionally significant regions in the genome. Sensitivity to the enzyme DNase I identifies regions such as promoters and enhancers. This is because chromatin regions that are active in transcription control are depleted of nucleosomes. For this reason, they have a more open structure and are more sensitive to cleavage with DNase I. There are also methods available to map sites of DNA methylation. Histone modifications are typically identified with ChIP-seq. For an example of ChIP-seq data showing histone marks and transcription factor binding sites in the *ALAS2* gene, see **Figure 10.28**.

FRAGILE X SYNDROME: AN EPIGENETIC DISORDER?

We encountered Fragile X syndrome as an example of a trinucleotide repeat expansion (Chapter 7). The disorder is characterized by different degrees of

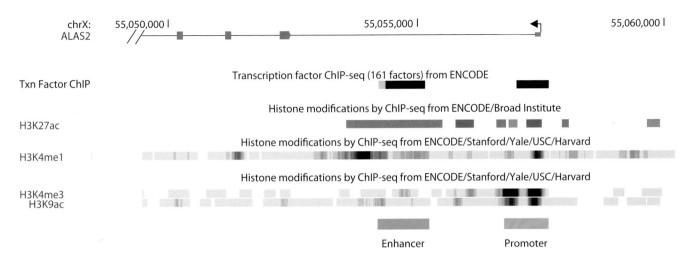

Figure 10.28 Histone marks reveal promoters and enhancers. On top is shown the genomic structure of a 5′ portion of the gene *ALAS2*, including four exons (green). Below the genomic region is shown the results of different ChIP-seq experiments. One experiment shows the binding sites for transcription factors (track "Txn Factor ChIP"), where the signals for each of the transcription factors have been merged. The other tracks, H3K27ac, H3K4me1, H3K4me3, and H3K9ac, reveal histone marks in this region of the *ALAS2* gene. The shading in each of the five tracks reflects the intensity of signal from the ChIP-seq experiment. The H3K27ac and H3K4me1 modifications are associated with enhancers, while H3K4me3 and H3K9ac are suggestive of promoter regions. The transcription factor binding data and the histone marks predict promoter (cyan) and enhancer (orange) regions as indicated. The chromosomal positions refer to the human genome version GRCh37/h19.

mental retardation, from mild to severe. The symptoms typically include autism, social anxiety, lack of ability to concentrate, and hyperactivity. There are often physical characteristics such as an elongated face and protruding ears. The syndrome is the most common inherited cause of mental retardation. Fragile X syndrome is a dominant X-linked disorder that affects 15–25 of 100,000 boys and 8–12 of 100,000 girls.

The responsible gene is *FMR1* (fragile X mental retardation protein 1), a gene localized to the X chromosome. This gene was identified as a cause of the disease in 1991. The *FMR1* gene encodes a protein FMRP which is needed for proper development of the brain and its signaling system. Fragile X syndrome is in a majority of cases caused by an expansion of a repeat of the triplet CGG. Previously, we have seen examples of triplet repeats in the context of disorders such as Huntington's disease and spinocerebellar ataxias (Chapter 7). The fragile X repeat is located in a 5′ UTR of the first exon of the gene and is believed to influence gene transcription. Early diagnosis of the disease involved a staining of DNA and then monitoring for discontinuity of staining in the region of the trinucleotide repeat. This discontinuity was referred to as a "fragile" site, hence, the name "Fragile X."

According to one hypothesis, the repeat region of the disease *FMR1* mRNA forms a RNA-DNA hybrid with the complementary strand in the DNA, which in turn results in epigenetic silencing, as seen by the specific introduction of repressive histone marks in Fragile X syndrome (**Figure 10.29**) The histone mark is H3K9me2—remember from Figure 10.24 that H3K9 is correlated with transcriptional repression. The molecular details of this epigenetic mechanism are yet to be identified. Furthermore, recent studies suggest an alternative model for altered *FMR1* expression in Fragile X syndrome, in which the CGG repeat disrupts a TAD boundary located in this region.

We encounter a few more examples of epigenetic control and its relation to disease in Chapter 11, which discusses noncoding RNAs.

Figure 10.29 Repeat expansion in fragile X syndrome. The 5′ end of the *FMR1* gene is shown, with a CGG repeat close to its promoter region. In the fragile X syndrome, this repeat is expanded, and as a result there is repression of gene transcription. A putative mechanism is shown here where the repeat region of the disease *FMR1* mRNA forms a RNA-DNA hybrid with the complementary strand in the DNA, which in turn results in epigenetic silencing.

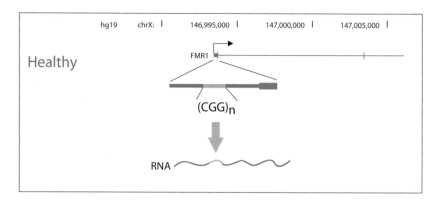

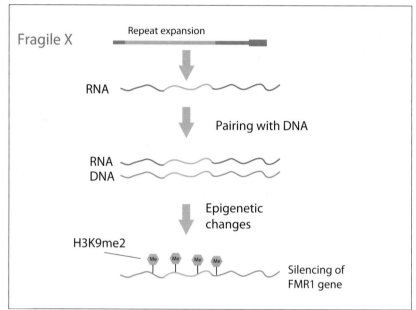

SUMMARY

- A large fraction of the human genome is allocated to transcriptional control. Regions of transcriptional control are the most common functional elements of the human genome. For instance, transcriptional control elements make up a larger fraction of the human genome than protein coding sequences.

- A transcription initiation complex is formed close to the transcription start site (at the core promoter). RNA polymerase and a number of basal transcription factors are part of this complex.

- Whereas *basal* transcription factors are involved in the transcription of all genes, *specific* transcription factors regulate the transcription of a subset of genes. Transcription factors interact with the core promoter, the proximal promoter, or *enhancers*, regions distant from the transcription start site.

- Transcription factors typically recognize a short (6–12) nucleotide sequence in DNA.

- Transcription factor binding sites on DNA may be predicted using computational tools, but more reliable assessments of these sites are determined experimentally using, for instance, ChIP-seq methodology.

- The ENCODE project has mapped numerous sites of regulatory importance in the human genome.

- Mutations in the promoter regions of the *TERT* gene are common in cancer. They lead to activation of *TERT* transcription.

- Examples of disorders associated with enhancer mutations are XLSA and Parkinson's disease.
- The human genome is organized into TADs. These are extensive loops that may encompass up to millions of nucleotides. It is believed that enhancers are restricted to action within their own TAD.
- In cases where large deletions or inversions disrupt a TAD structure, this may lead to erroneous transcription regulation and disease. One such example is in a locus containing the gene *EPHA4*, where mutations cause developmental disorders.
- One gene may typically give rise to multiple transcripts. The mechanisms responsible for this diversity are (1) alternative transcription start sites, (2) alternative splicing (Chapter 9), and (3) alternative polyadenylation.
- Two important epigenetic mechanisms of transcriptional control are *histone modification* and *DNA methylation*. Both of these modifications may be inherited to daughter cells.
- The tails of histones may be subject to several different covalent modifications, including phosphorylation, ubiquitination, methylation, and acetylation. These modifications may either activate or repress transcription.
- Methylation of DNA mainly takes place at cytosine in CpG sites. CpG islands are regions that are rich in CpG sites and that are involved in transcriptional regulation.
- In cancer cells, CpG islands may either be methylated to a larger extent than normal, giving rise to transcriptional silencing, or undermethylated, causing transcriptional activation.

QUESTIONS

1. What different regions of a human gene are important for transcriptional control? What is characteristic of the sequence elements located in these regions?
2. How are *basal or general* transcription factors different from those that are *specific* transcription factors?
3. How are *promoters* different from *enhancers*?
4. What is characteristic of the DNA sequences recognized by human (and other eukaryotic) transcription factors?
5. What experimental methods may be used to infer what transcription factors bind to what DNA sequences?
6. In this chapter, a number of diseases are accounted for that are the result of point mutations in regulatory regions. But there are a limited number of such diseases as compared to monogenic disorders that result from mutations in protein coding sequences. Why could it be that a coding region is more sensitive to point mutation than a region of transcriptional control?
7. What is a sequence logo?
8. What is the relationship between cancer and the *TERT* gene?
9. How does the protein *BCL11A* control the expression of β-globin?
10. How could fetal hemoglobin (HbF) be exploited in treating sickle cell anemia?
11. What are *topologically associating domains*? What are the experimental methods used to reveal these structures?
12. Explain why different structural alterations in the *EPHA4* locus give rise to different developmental disorders.
13. There are at least three different mechanisms that give rise to the effect that multiple transcripts may be formed from the same gene. One of these mechanisms is alternative splicing. What are the other two?

14. Give an example of a disease mutation affecting *polyadenylation*.
15. Explain the words *epigenetics* and *epigenetic*.
16. Give examples of histone modifications that are associated with transcriptionally active chromatin. Also provide examples of repressive modifications.
17. What is a *CpG island*?
18. What are the experimental methods used to reveal epigenetic modification of chromatin?
19. In what respect could epigenetic control be related to Fragile X syndrome?

URLs

Roadmap Epigenomics Project (www.roadmapepigenomics.org/)

FURTHER READING

General on transcriptional regulatory elements and transcription factors

Lambert SA, Jolma A, Campitelli LF et al. 2018. The human transcription factors. *Cell* 172(4):650–665.

Maston GA, Evans SK, Green MR. 2006. Transcriptional regulatory elements in the human genome. *Annu Rev Genomics Hum Genet* 7:29–59.

Spitz F, Furlong EE. 2012. Transcription factors: From enhancer binding to developmental control. *Nat Rev Genet* 13(9):613–626.

The TATA box and other elements of promoters

Carninci P, Sandelin A, Lenhard B et al. 2006. Genome-wide analysis of mammalian promoter architecture and evolution. *Nat Genet* 38(6):626–635.

Suzuki Y, Tsunoda T, Sese J et al. 2001. Identification and characterization of the potential promoter regions of 1031 kinds of human genes. *Genome Res* 11(5):677–684.

Yang C, Bolotin E, Jiang T et al. 2007. Prevalence of the initiator over the TATA box in human and yeast genes and identification of DNA motifs enriched in human TATA-less core promoters. *Gene* 389(1):52–65.

The TATA box and disease

DPE mutations in HBB: Lewis BA, Kim TK, Orkin SH. 2000. A downstream element in the human beta-globin promoter: Evidence of extended sequence-specific transcription factor IID contacts. *Proc Natl Acad Sci USA* 97(13):7172–7177.

Drachkova I, Savinkova L, Arshinova T et al. 2014. The mechanism by which TATA-box polymorphisms associated with human hereditary diseases influence interactions with the TATA-binding protein. *Hum Mutat* 35(5):601–608.

Structures of transcription factors in complex with DNA

Chen C, Gorlatova N, Herzberg O. 2012. Pliable DNA conformation of response elements bound to transcription factor p63. *J Biol Chem* 287(10):7477–7486.

Newman JA, Cooper CD, Aitkenhead H, Gileadi O. 2015. Structural insights into the autoregulation and cooperativity of the human transcription factor Ets-2. *J Biol Chem* 290(13):8539–8549.

Schwabe JW, Chapman L, Finch JT, Rhodes D. 1993. The crystal structure of the estrogen receptor DNA-binding domain bound to DNA: How receptors discriminate between their response elements. *Cell* 75(3):567–578.

Wilkinson-White L, Lester KL, Ripin N et al. 2015. GATA1 directly mediates interactions with closely spaced pseudopalindromic but not distantly spaced double GATA sites on DNA. *Protein Sci* 24(10):1649–1659.

Williams DC Jr., Cai M, Clore GM. 2004. Molecular basis for synergistic transcriptional activation by Oct1 and Sox2 revealed from the solution structure of the 42-kDa Oct1.Sox2. Hoxb1-DNA ternary transcription factor complex. *J Biol Chem* 279(2):1449–1457.

Specificity in the interaction between transcription factors and DNA

Rohs R, Jin X, West SM et al. 2010. Origins of specificity in protein-DNA recognition. *Annu Rev Biochem* 79:233–269.

Smith NC, Matthews JM. 2016. Mechanisms of DNA-binding specificity and functional gene regulation by transcription factors. *Curr Opin Struct Biol* 38:68–74.

Wunderlich Z, Mirny LA. 2009. Different gene regulation strategies revealed by analysis of binding motifs. *Trends Genet* 25(10):434–440.

The ENCODE project

Djebali S, Davis CA, Merkel A et al. 2012. Landscape of transcription in human cells. *Nature* 489(7414):101–108.

ENCODE Project Consortium. 2012. An integrated encyclopedia of DNA elements in the human genome. *Nature* 489(7414):57–74.

Gerstein MB, Bruce C, Rozowsky JS et al. 2007. What is a gene, post-ENCODE? History and updated definition. *Genome Res* 17(6):669–681.

Gerstein MB, Kundaje A, Hariharan M et al. 2012. Architecture of the human regulatory network derived from ENCODE data. *Nature* 489(7414):91–100.

Kellis M, Wold B, Snyder MP et al. 2014. Defining functional DNA elements in the human genome. *Proc Natl Acad Sci USA* 111(17):6131–6138.

Neph S, Vierstra J, Stergachis AB et al. 2012. An expansive human regulatory lexicon encoded in transcription factor footprints. *Nature* 489(7414):83–90.

Pennisi E. 2012. Genomics. ENCODE project writes eulogy for junk DNA. *Science* 337(6099):1159, 1161.

Sanyal A, Lajoie BR, Jain G, Dekker J. 2012. The long-range interaction landscape of gene promoters. *Nature* 489(7414):109–113.

Sloan CA, Chan ET, Davidson JM et al. 2016. ENCODE data at the ENCODE portal. *Nucleic Acids Res* 44(D1):D726–D732.

Thurman RE, Rynes E, Humbert R et al. 2012. The accessible chromatin landscape of the human genome. *Nature* 489(7414): 75–82.

Transcriptional regulation and disease

Lee TI, Young RA. 2013. Transcriptional regulation and its misregulation in disease. *Cell* 152(6):1237–1251.

Human genetic variation and transcription factor binding

Deplancke B, Alpern D, Gardeux V. 2016. The genetics of transcription factor DNA binding variation. *Cell* 166(3):538–554.

δ-Thalassemia and GATA-1

Matsuda M, Sakamoto N, Fukumaki Y. 1992. δ-Thalassemia caused by disruption of the site for an erythroid-specific transcription factor, GATA-1, in the δ-globin gene promoter. *Blood* 80(5):1347–1351.

Transcriptional regulation and cancer

Bradner JE, Hnisz D, Young RA. 2017. Transcriptional addiction in cancer. *Cell* 168(4):629–643.

The *TERT* gene and promoter mutations in cancer

Batchelor AH, Piper DE, de la Brousse FC et al. 1998. The structure of GABPα/β: An ETS domain-ankyrin repeat heterodimer bound to DNA. *Science* 279(5353):1037–1041.

Bell RJ, Rube HT, Xavier-Magalhaes A et al. 2016. Understanding *TERT* promoter mutations: A common path to immortality. *Mol Cancer Res* 14(4):315–323.

Chiba K, Lorbeer FK, Shain AH et al. 2017. Mutations in the promoter of the telomerase gene *TERT* contribute to tumorigenesis by a two-step mechanism. *Science* 357(6358):1416–1420.

Horn S, Figl A, Rachakonda PS et al. 2013. *TERT* promoter mutations in familial and sporadic melanoma. *Science* 339(6122):959–961.

Enhancers

Visel A, Rubin EM, Pennacchio LA. 2009. Genomic views of distant-acting enhancers. *Nature* 461(7261):199–205.

XLSA and enhancer GATA site mutations

Campagna DR, de Bie CI, Schmitz-Abe K et al. 2014. X-linked sideroblastic anemia due to ALAS2 intron 1 enhancer element GATA-binding site mutations. *Am J Hematol* 89(3):315–319.

BCL11A enhancer and genetic variation

Bauer DE, Kamran SC, Lessard S et al. 2013. An erythroid enhancer of *BCL11A* subject to genetic variation determines fetal hemoglobin level. *Science* 342(6155):253–257.

Canver MC, Smith EC, Sher F et al. 2015. *BCL11A* enhancer dissection by Cas9-mediated *in situ* saturating mutagenesis. *Nature* 527(7577):192–197.

Hardison RC, Blobel GA. 2013. Genetics. GWAS to therapy by genome edits? *Science* 342(6155):206–207.

TAL1/GATA-1 sequence logo

Soler E, Andrieu-Soler C, de Boer E et al. 2010. The genome-wide dynamics of the binding of Ldb1 complexes during erythroid differentiation. *Genes Dev* 24(3):277–289.

An oncogenic super-enhancer

Mansour MR, Abraham BJ, Anders L et al. 2014. Oncogene regulation. An oncogenic super-enhancer formed through somatic mutation of a noncoding intergenic element. *Science* 346(6215):1373–1377.

CRISPR modification of GATA1 binding sites associated with disorders

Wakabayashi A, Ulirsch JC, Ludwig LS et al. 2016. Insight into GATA1 transcriptional activity through interrogation of cis elements disrupted in human erythroid disorders. *Proc Natl Acad Sci USA* 113(16):4434–4439.

The role of an enhancer in Parkinson's disease

Abeliovich A, Rhinn H. 2016. Parkinson's disease: Guilt by genetic association. *Nature* 533(7601):40–41.

Soldner F, Stelzer Y, Shivalila CS et al. 2016. Parkinson-associated risk variant in distal enhancer of alpha-synuclein modulates target gene expression. *Nature* 533(7601):95–99.

Hirschsprung disease and enhancer variants

Chatterjee S, Kapoor A, Akiyama JA et al. 2016. Enhancer variants synergistically drive dysfunction of a gene regulatory network in Hirschsprung disease. *Cell* 167(2):355–368.e310.

Emison ES, Garcia-Barcelo M, Grice EA et al. 2010. Differential contributions of rare and common, coding and noncoding ret mutations to multifactorial Hirschsprung disease liability. *Am J Hum Genet* 87(1):60–74.

Higher-order structure of DNA and topologically associating domains

Beagrie RA, Scialdone A, Schueler M et al. 2017. Complex multi-enhancer contacts captured by genome architecture mapping. *Nature* 543:519.

Belton JM, McCord RP, Gibcus JH et al. 2012. Hi-C: A comprehensive technique to capture the conformation of genomes. *Methods* 58(3):268–276.

Bickmore WA, van Steensel B. 2013. Genome architecture: Domain organization of interphase chromosomes. *Cell* 152(6):1270–1284.

Bouwman BA, de Laat W. 2015. Getting the genome in shape: The formation of loops, domains and compartments. *Genome Biol* 16:154.

Dixon JR, Selvaraj S, Yue F et al. 2012. Topological domains in mammalian genomes identified by analysis of chromatin interactions. *Nature* 485(7398):376–380.

Kleckner N, Zickler D, Witz G. 2013. Molecular biology. Chromosome capture brings it all together. *Science* 342(6161):940–941.

Larson DR, Misteli T. 2017. The genome—Seeing it clearly now. *Science* 357(6349):354–355.

Lieberman-Aiden E, van Berkum NL, Williams L et al. 2009. Comprehensive mapping of long-range interactions reveals folding principles of the human genome. *Science* 326(5950): 289–293.

Nasmyth K. 2017. How are DNAs woven into chromosomes? *Science* 358(6363):589–590.

Ou HD, Phan S, Deerinck TJ et al. 2017. ChromEMT: Visualizing 3D chromatin structure and compaction in interphase and mitotic cells. *Science* 357(6349):eaag0025.

Rao SS, Huntley MH, Durand NC et al. 2014. A 3D map of the human genome at kilobase resolution reveals principles of chromatin looping. *Cell* 159(7):1665–1680.

Terakawa T, Bisht S, Eeftens JM et al. 2017. The condensin complex is a mechanochemical motor that translocates along DNA. *Science* 358(6363):672–676.

Topologically associating domains and disease

Flavahan WA, Drier Y, Liau BB et al. 2016. Insulator dysfunction and oncogene activation in IDH mutant gliomas. *Nature* 529(7584): 110–114.

Franke M, Ibrahim DM, Andrey G et al. 2016. Formation of new chromatin domains determines pathogenicity of genomic duplications. *Nature* 538(7624):265–269.

Grimmer MR, Costello JF. 2016. Cancer: Oncogene brought into the loop. *Nature* 529(7584):34–35.

Lupianez DG, Kraft K, Heinrich V et al. 2015. Disruptions of topological chromatin domains cause pathogenic rewiring of gene-enhancer interactions. *Cell* 161(5):1012–1025.

Lupianez DG, Spielmann M, Mundlos S. 2016. Breaking TADs: How alterations of chromatin domains result in disease. *Trends Genet* 32(4):225–237.

Spielmann M, Mundlos S. 2016. Looking beyond the genes: The role of non-coding variants in human disease. *Hum Mol Genet* 25(R2):R157–R165.

Spielmann M, Brancati F, Krawitz PM et al. 2012. Homeotic arm-to-leg transformation associated with genomic rearrangements at the PITX1 locus. *Am J Hum Genet* 91(4):629–635.

CTCF protein and topologically associating domains

Guo Y, Xu Q, Canzio D et al. 2015. CRISPR inversion of CTCF sites alters genome topology and enhancer/promoter function. *Cell* 162(4):900–910.

Polyadenylation in health and disease

Curinha A, Oliveira Braz S, Pereira-Castro I et al. 2014. Implications of polyadenylation in health and disease. *Nucleus* 5(6):508–519.

Orkin SH, Cheng TC, Antonarakis SE, Kazazian HH Jr. 1985. Thalassemia due to a mutation in the cleavage-polyadenylation signal of the human beta-globin gene. *EMBO J* 4(2):453–456.

Epigenetics text book

Allis CD. 2015. *Epigenetics.* Cold Spring Harbor Laboratory Press, New York.

Epigenetic core concepts

Ptashne M. 2013. Epigenetics: Core misconcept. *Proc Natl Acad Sci USA* 110(18):7101–7103.

Williams SC. 2013. Epigenetics. *Proc Natl Acad Sci USA* 110(9):3209.

Epigenetics and inheritance

Heard E, Martienssen RA. 2014. Transgenerational epigenetic inheritance: Myths and mechanisms. *Cell* 157(1):95–109.

Hughes V. 2014. Epigenetics: The sins of the father. *Nature* 507(7490):22–24.

Hughes V. 2014. Sperm RNA carries marks of trauma. *Nature* 508(7496):296–297.

Susiarjo M, Bartolomei MS. 2014. Epigenetics. You are what you eat, but what about your DNA? *Science* 345(6198):733–734.

Roadmap epigenomics project

Roadmap Epigenomics Consortium, Kundaje A, Meuleman W et al. 2015. Integrative analysis of 111 reference human epigenomes. *Nature* 518(7539):317–330.

Romanoski CE, Glass CK, Stunnenberg HG et al. 2015. Epigenomics: Roadmap for regulation. *Nature* 518(7539):314–316.

Relationship between genetic variants and epigenetics

Furey TS, Sethupathy P. 2013. Genetics. Genetics driving epigenetics. *Science* 342(6159):705–706.

Mapping of epigenomes

Rivera CM, Ren B. 2013. Mapping human epigenomes. *Cell* 155(1):39–55.

Functional output of the human genome is a result of sequence, epigenetics, three-dimensional organization, and other factors

Misteli T. 2013. The cell biology of genomes: Bringing the double helix to life. *Cell* 152(6):1209–1212.

Cancer epigenetics

Dawson MA, Kouzarides T. 2012. Cancer epigenetics: From mechanism to therapy. *Cell* 150(1):12–27.

Polycomb and Trithorax proteins

Bauer M, Trupke J, Ringrose L. 2016. The quest for mammalian Polycomb response elements: Are we there yet? *Chromosoma* 125:471–496.

Li H, Liefke R, Jiang J et al. 2017. Polycomb-like proteins link the PRC2 complex to CpG islands. *Nature* 549:287.

Schuettengruber B, Bourbon HM, Di Croce L, Cavalli G. 2017. Genome regulation by Polycomb and Trithorax: 70 years and counting. *Cell* 171(1):34–57.

DNA methylation

Hughes TR, Lambert SA. 2017. Transcription factors read epigenetics. *Science* 356(6337):489–490.

Schubeler D. 2015. Function and information content of DNA methylation. *Nature* 517(7534):321–326.

Yin Y, Morgunova E, Jolma A et al. 2017. Impact of cytosine methylation on DNA binding specificities of human transcription factors. *Science* 356(6337).

Ziller MJ, Gu H, Muller F et al. 2013. Charting a dynamic DNA methylation landscape of the human genome. *Nature* 500(7463):477–481.

FMR1 and epigenetic silencing

Colak D, Zaninovic N, Cohen MS et al. 2014. Promoter-bound trinucleotide repeat mRNA drives epigenetic silencing in fragile X syndrome. *Science* 343(6174):1002–1005.

The Noncoding RNAs

11

There are studies that suggest that more than 80% of the human genome is transcribed. A subset of these transcipts gives rise to protein coding mRNA. As we saw earlier, the human genome encodes some 20,000 different proteins. Other transcripts give rise to RNAs that do not code for proteins, the **noncoding RNAs** (ncRNAs). Estimates of the number of ncRNAs vary between >20,000 and 100,000. The current Ensembl annotation of the human genome lists about 22,000 different ncRNAs (Table 4.5), but these numbers may increase over time as human genome annotation is improved with respect to ncRNAs. The biological functions of many ncRNAs are unknown. Some of the noncoding transcripts from the human genome may not even have a distinct function, but instead represent "noise" of the transcription system. We here focus on some of the more well-studied ncRNAs. They may be classified into two major groups. In one of them, we find the **small ncRNAs**. These are typically less than 200 nucleotides, and many of them have a characteristic secondary structure based on internal base-pairing. Included in this group of RNAs are molecules that were discovered early and that have been well studied. We first consider this group and later return to the other category of RNAs referred to as **long ncRNAs**.

SMALL NONCODING RNAs

The classic central dogma in molecular biology states that "DNA makes RNA makes protein." However, already in the late 1950s it was realized that certain genes do not code for protein. Rather, they were responsible for producing an RNA with some other function, a function distinct from the protein-coding function of messenger RNA. Transfer RNAs (tRNAs) and ribosomal RNAs (rRNAs), both important components of the translational apparatus, are examples of such RNAs that were discovered early. The very first ncRNA to be characterized with respect to its nucleotide sequence was a tRNA in 1965. This was before the era of DNA sequencing and was the first nucleic acid to be sequenced.

Additional ncRNAs were identified in the following decades. Splicing was discovered in the late 1970s, and as we have seen this process requires specific RNAs (Chapter 9). Furthermore, snoRNAs (small nucleolar RNAs) are essential components in an enzymatic reaction carrying out specific base modifications in rRNA. The RNA component of the enzyme RNaseP is the catalytic subunit responsible for the processing of tRNA precursors. It is an example of an RNA molecule with catalytic activity. The structurally and evolutionarily related MRP enzyme is involved in the processing of rRNA precursors. The signal recognition particle (SRP), involved in targeting specific proteins to membranes, has an essential RNA component. The enzyme telomerase has an RNA subunit guiding the synthesis of the tandem repeat characteristic of telomeres. Finally, another class of small ncRNA are the microRNAs (miRNAs) (Chapter 8).

In each family of small RNAs, there are typically certain characteristic primary sequence elements that are specific to that family. But in addition, there are **RNA secondary structure** characteristics, produced by internal base-pairing. We earlier presented examples of this in the case of tRNA (Chapter 3). Studies of the three-dimensional structure of many small ncRNAs reveal that not only does standard base-pairing play an important role in the folding of the RNA, but also do additional interactions (as an example for tRNA, see Figures 3.10 and 3.11).

THE RNA WORLD

Quite a few ncRNAs have critical functions in cellular life. For instance, a critical step during protein synthesis when peptide bonds are formed is catalyzed by ribosomal RNA. Furthermore, some of the RNAs involved in splicing are responsible for the actual catalytic step in splicing. The **RNA world hypothesis** was put forward to explain the presence of RNA molecules in central biochemical processes. According to this hypothesis, RNA had an essential role at an early stage in the development of life on our planet, and actually predated proteins and DNA. Two properties of RNA made such a prominent role possible. First, because it is a nucleic acid, it is able to carry genetic information in much the same way as DNA. Second, RNA is structurally and functionally versatile and is able to catalyze different chemical reactions. Catalytic RNAs were first discovered by Sidney Altman and Thomas Cech—in 1989 they received the Nobel Prize in Chemistry for their important pioneering work. In the RNA world, there could therefore have existed RNA molecules that on the one hand carried genetic information, and on the other hand were able to replicate themselves. At a later stage during the evolution of life, proteins enhanced or replaced many of the reactions mediated by RNA molecules, because proteins are more versatile than RNA in terms of catalysis. In addition, DNA replaced RNA as a carrier of genetic information because it is chemically more stable. But today, we still see remnants of the RNA world in the form of several RNAs with critical functions in cellular life.

BIOSYNTHESIS OF SMALL ncRNAs

In all eukaryotic cells, there are three different RNA polymerases with different roles, as discussed in Chapter 10. RNA polymerase I synthesizes most of the ribosomal RNAs, and RNA polymerase II produces mRNA, most spliceosomal RNA, miRNA, and long noncoding RNA. Finally, RNA polymerase III produces tRNA and a class of ribosomal RNA, 5S rRNA.

Many of the genes encoding ncRNA form an independent transcriptional unit with a specific promoter and regulatory elements similar to a protein gene. But some ncRNA genes do not form a transcriptional unit on

Figure 11.1 Some ncRNAs are located in introns of protein coding genes. The three snoRNAs, SNORA48, SNORD10, and SNORA67, are located within introns of the protein gene SENP3-EIF4A1. SENP3-EIF4A1 is at the chromosomal location chr17:7,476,160–7,481,804 (human genome version GRCh37/hg19).

their own but are located within introns of protein coding genes and lack their own promoter. These RNAs are mainly miRNAs, snoRNAs, and sca-RNAs (small Cajal body-specific RNAs, these RNAs take part in modification of spliceosomal RNAs). The expression of such an RNA is coupled to the transcription of a host gene, and the mature form of the RNA is produced by processing of the intron after splicing. An example of intronic localization of ncRNA genes is shown in **Figure 11.1**, in which three different snoRNAs are within introns of a protein gene.

Presented in the following sections are a few examples of diseases resulting from mutations in three different small RNAs—tRNA, MRP RNA, and snoRNA.

A NUMBER OF MITOCHONDRIAL DISEASES ARE THE RESULT OF MUTATIONS IN tRNA GENES

While there are numerous examples of genetic disorders that affect protein coding genes, the number of disorders with mutations in ncRNA genes is much lower (Table 6.1). One possible explanation for this low frequency is that many important ncRNA genes are present in multiple copies in the genome; therefore, a mutation in one gene is not expected to have a strong phenotypic effect. Other possibilities are that many ncRNAs are not as essential as most proteins are or that they have fewer nucleotide positions that are functionally important. Finally, many experimental studies are focusing on protein genes and avoid the ncRNAs. Nevertheless, considered here are a number of examples of inherited disorders that result from mutations in ncRNA genes.

The first example of an ncRNA disorder is concerned with **tRNA**. Whereas many tRNA genes of the nuclear genome are present in more than one copy, the tRNA genes of the mitochondrial genome are each represented by only one gene. Therefore, many of the mutations affecting these tRNA genes seriously affect protein synthesis in the mitochondrion.

There are several mitochondrial disorders where the molecular background is a tRNA mutation. One such example is the disorder MELAS (acronym based on "mitochondrial encephalomyopathy with lactic acidosis and stroke-like episodes"). Characteristic of this disease are repeated episodes that resemble stroke and that result in brain damage with seizures and dementia. A buildup of lactate in the body and muscle weakness are other symptoms.

The exact incidence of MELAS is not known, but together, mitochondrial diseases occur in about 1 in 4,000 individuals. About 80% of MELAS cases are the result of mutations in a leucine tRNA gene with an anticodon normally able to read the codons UUA and UUG (**wobble** reading, Chapter 3). The most common mutation in the tRNA gene is a substitution A > G in position 14 of the RNA (also known as A3243G, where the number refers to the mitochondrial genome position) (**Figure 11.2**). This mutation has the effect that the three-dimensional structure is disturbed, as an interaction between A14 and U8 in the normal tRNA stabilizes the structure.

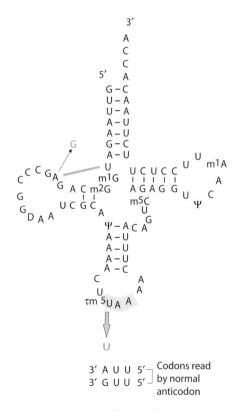

Figure 11.2 A mitochondrial tRNA associated with disease. The leucine tRNA depicted has an anticodon able to read the codons UUA and UUG. The mutation A > G in position 14 of the RNA disturbs the structure because of the tertiary base-pair involving A14 and U8 (orange line). The structural changes will also hinder the modification of the wobble nucleotide ($\tau m^5 U$). The unmodified U (red) is not able to read UUG codons properly.

The structural perturbation has a number of effects, such as defects in the process where mature tRNAs are formed from larger precursor RNAs. In the normal leucine tRNA, there is in the first (wobble) position of the anticodon a modified nucleotide, 5-taurino-methyl-2-thio-uridine (τm^{5}U). This position is modified in many tRNAs with the purpose to achieve correct reading of codons. However, structural changes in the leucine tRNA resulting from the A > G mutation hinder this modified nucleotide from being formed. With a normal U in this position, UUG codons are not read efficiently, resulting in termination of polypeptide chain elongation.

CARTILAGE-HAIR HYPOPLASIA IS CAUSED BY MALFUNCTIONING MRP RNA

The ribonucleoprotein complex MRP is an enzyme that processes ribosomal RNA precursors, generating individual RNAs being building blocks of ribosomes. (The acronym MRP is derived from "mitochondrial RNA processing," which is somewhat misleading as the most important function of MRP appears to be in the nucleus.) The RNA component of MRP is structurally and functionally related to another RNA enzyme, ribonuclease P, which is responsible for processing tRNA precursors. Like many other small ncRNAs, the RNase P and MRP RNAs have sequences that may be folded into a secondary structure characteristic of that RNA family.

Cartilage-hair hypoplasia (CHH) is a rare inherited disorder first identified in 1965 in the Amish population. It was later found in other ethnic groups, and is common also among the Finns. Carrier frequencies among the Amish and Finns are 1:19 and 1:76, respectively. CHH is characterized by insufficient development of bone and cartilage. It gives rise to dwarfism and thin, sparse hair. Abnormal immune system function is also characteristic of the disease.

The RNase MRP RNA gene (also known as *RMRP*) was identified as being responsible for CHH in 2001. This was actually the first time a nuclear encoded noncoding RNA was shown to be responsible for a genetic disease.

The most common mutation in CHH is the substitution A > G at position 70 in the RNA sequence. It is present in 92% of Finnish and 48% of non-Finnish patients with CHH. The A70 together with many other mutation sites causing CHH are indicated in the context of the MRP RNA secondary structure in **Figure 11.3a**. MRP RNA contains two regions CR-I and CR-V that base-pair in the three-dimensional structure of the RNA to form an essential catalytic core of the enzyme. The A70 nucleotide is part of a single-stranded region of a highly conserved part of the CR-I region (see **Figure 11.3b**). The A > G mutation is therefore expected to interfere with the catalytic activity of MRP.

More than 80 different mutations have been identified within the *RMRP* gene, and in all of these cases, each allele has only one mutation. Most of the mutations are associated with disease. They affect evolutionarily conserved nucleotides or are likely to destabilize RNA secondary structure. In addition, CHH-associated mutations may reduce the binding of specific protein components of MRP. There are also mutations that affect the gene promoter (see Figure 11.3a).

The large majority of mutations described in the *RMRP* gene are associated with CHH, but other diseases may result from *RMRP* mutations. A wide phenotypic spectrum is observed. Some patients have predominantly skeletal symptoms, whereas others are immune deficient with only limited skeletal manifestations. Some *RMRP* mutations result in severe immunodeficiency but without any skeletal involvement.

Although a large number of mutations in *RMRP* have now been described, and we may interpret these in the context of known properties

(a)

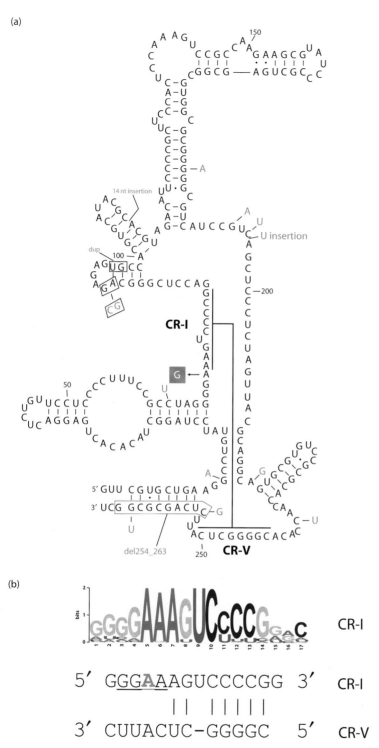

Figure 11.3 Disease mutations in MRP RNA. (a) Secondary structure model of MRP RNA and location of disease mutations. Mutations are either substitutions or indels (shown in red). The substitution 70A > G, common in CHH, is highlighted with a red background. Conserved regions I and V (CR-I and CR-V) are shown. (b) A sequence logo displays the sequence conservation of the conserved region 1 (CR-I) in animals. Below the sequence logo is shown the pairing of the CR-I and CR-V regions. A single-stranded region that contains A70 (red) is underlined.

(b)

of RNase MRP, it is not at all clear how these mutations give rise to the particular phenotypes observed in diseases such as CHH.

PRADER–WILLI SYNDROME AND THE ABSENCE OF SPECIFIC snoRNAs

The function of snoRNAs is to guide chemical modification of certain RNAs. The best-characterized modifications are those in ribosomal RNAs. For

Figure 11.4 The snoRNAs guide chemical modification of specific RNAs. (a) The base-pairing of a C/D box snoRNA to its target RNA (green) is shown schematically. The C/D box snoRNAs take part in base methylation and contain two conserved sequence motifs, C (RUGAUGA) and D (CUGA), located near the 5′ and 3′ ends of the snoRNA, respectively. The RNAs also have copies of the C- and D-box motifs that are less well conserved (C′ and D′ box). The positions of methyl groups are shown as green circles. (b) The H/ACA box RNAs are involved in the formation of pseudouridine, an isomer of uridine. The pairing of this RNA to its target (green) and positions of pseudouridines (ψ) are indicated.

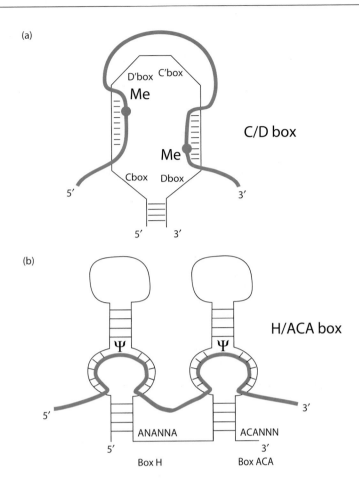

instance, in the large ribosomal subunit RNA (28S RNA), there are approximately 120 nucleotide modifications that are all guided by snoRNAs.

There are two different classes of snoRNA. The C/D box snoRNAs take part in methylation, and the H/ACA box RNAs are responsible for the formation of pseudouridine, an isomer of uridine. All snoRNAs act as guide RNAs with sequences complementary to that of the target RNA (**Figure 11.4**). The human genome encodes more than 500 different C/D box and H/ACA snoRNAs. Out of these, about 76 have no known target gene and are referred to as orphan snoRNAs.

Prader–Willi syndrome (PWS) is an inherited disorder that is probably caused by a deficiency in snoRNAs. It is typically caused by extensive deletions in chromosome 15. The syndrome was first described in 1956 by the Swiss pediatricians Andrea Prader and Heinrich Willi. It is characterized by muscle weakness and poor feeding in the newborn, and in childhood, short stature and excessive appetite leading to obesity and type II diabetes. There is also a varying degree of intellectual impairment. About 3–10 of 100,000 children are born with PWS.

In about 70% of the cases, PWS is caused by deletions in a specific region of the paternal copy of chromosome 15 (15q11.2-q13). This region includes **imprinted** genes. For each of our autosomal genes, we inherit one copy from our father and one from our mother. Genomic imprinting refers to the situation where only one of the two genes is expressed (**Figure 11.5**). Imprinting is typically caused by an **epigenetic** mechanism. There are about 80 human genes that are imprinted—these genes are particularly important during embryonic development.

The inherited diseases PWS and Angelman syndrome (AS) are the result of imprinting combined with a mutation and were the first imprinting

(a)

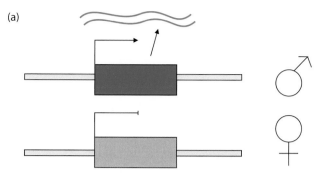

Maternal imprinting, only paternal gene copy active

(b)

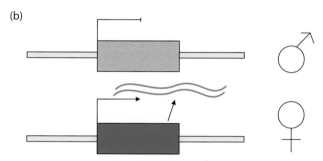

Paternal imprinting, only maternal gene copy active

(c)

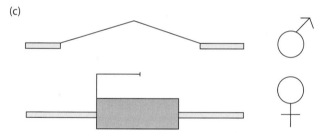

Maternal imprinting, paternal copy deleted (like in PWS)

Figure 11.5 Imprinting and relationship to disease. (a) Maternal imprinting. Paternal copy of gene is transcriptionally active (green), while the maternal copy is epigenetically silenced. (b) Paternal imprinting, opposite situation as compared to (a). (c) Maternal imprinting combined with deletion of the corresponding gene on the paternal chromosome. No gene product is formed.

disorders to be examined. They may both be caused by the loss of the chromosomal region 15q11-13. Many of the genes in this region are expressed only from the paternal chromosome, and two genes are expressed only from the maternal chromosome. The region includes the protein genes SNRPN (small nuclear ribonucleoprotein polypeptide N) and NDN (necdin) as well as a large number of snoRNA genes—SNORD64, SNORD107, SNORD108, and two copies of SNORD109, 29 copies of SNORD116 (also known as HBII-85), and 48 copies of SNORD115 (also known as HBII-52) (**Figure 11.6a**).

In PWS, the paternally inherited genes are deleted, and at the same time the maternal genes are silenced. In AS, it is the opposite situation, as discussed under diseases related to long noncoding RNAs later in this chapter.

How do deletions in these disorders come about? Large genomic deletions are typically the result of **homologous recombination** events. In most cases, the deletion has arisen by chance during the formation of a sperm cell, and in such a case the probability that the parents will give birth to more children with the disorder is low.

The most common deletions in PWS are large deletions (5.7 and 4.8 MB, respectively) that occur through sites known as BP1 (breakpoint 1), BP2, and BP3 (**Figure 11.6b**, the breakpoints are shown in orange vertical zigzag lines). A large number of imprinted genes are located within this

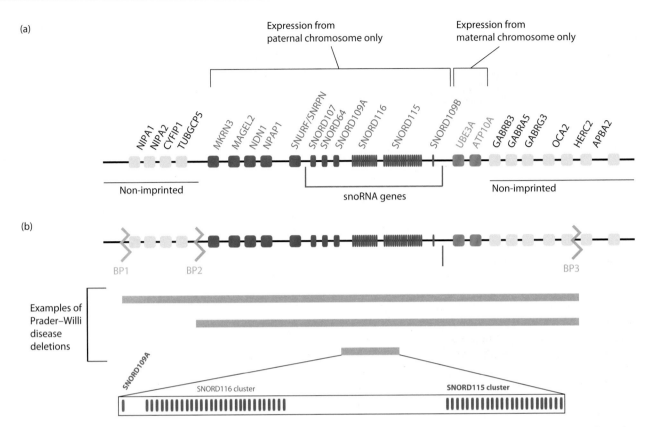

Figure 11.6 Genomic locus related to Prader–Willi and Angelman syndromes. The positions of genes are shown schematically and are not to scale. (a) Genes that are expressed from the paternal chromosome only (and that are maternally imprinted) are shown in blue. Conversely, paternally imprinted genes are shown in red. Nonimprinted genes are in gray. (b) Genes are shown as in (a), but three different deletions affecting the genomic region are shown. Two of these deletions are associated with the breakpoints BP1, BP2, and BP3. The third deleted region shown at the bottom contains only *SNORD116* and *SNORD115* snoRNA genes. The locations of the breakpoints BP1, BP2, and BP3 are at chr15:22876632, 23758390 and 28557186, respectively (human genome version GRCh37/hg19). The breakpoints of the shorter deletions are chr15:25284501 and 25459086.

large region. What are the important genes missing in PWS? A clue is provided by genetic analyses of a range of patients with PWS or related disorders. For instance, in one child with PWS, a much smaller deletion has been identified, affecting a region limited to about 175,000 nucleotides. The only known genes contained within this region are snoRNAs. Thus, there are clusters of the two C/D box snoRNAs: SNORD115 and SNORD116. These findings, in combination with other studies, indicate that the absence of the snoRNA *SNORD116* cluster plays a crucial role in PWS. Unfortunately, the molecular pathways leading to the symptoms in PWS are completely unknown.

LONG NONCODING RNAs, A UNIVERSE OF RNAs YET TO BE EXPLORED

The noncoding RNAs that we have sampled so far in this chapter are all relatively short RNAs, each of them with a characteristic secondary structure as elucidated by structural studies or through the comparison of evolutionary related sequences. More recently in the history of ncRNA, a class of longer RNAs has been identified whose function in many cases is unknown. Such **long noncoding RNAs (lncRNAs)** do not have a completely adequate definition, but they are often defined as RNAs longer than 200 nucleotides. (This is a somewhat arbitrary size limit and not entirely suitable as some traditional ncRNAs, like certain ribosomal RNAs, are above this limit.) Furthermore, the lncRNAs are often mRNA-like in the sense that they are transcribed by RNA polymerase II and are polyadenylated and spliced. LncRNA genes are typically shorter than protein-coding genes and have fewer exons. Two exons are the most common architecture of the long ncRNA transcript. It should be noted that unlike mRNAs, they do not have a recognizable open reading frame and protein product. Finally, many lncRNAs remain physically associated with the DNA locus at which they are encoded, and they exert their function at that site.

Ensembl listed in 2018 about human 22,000 ncRNAs, of which about 15,000 are lncRNAs. However, some studies suggest an even larger number of lncRNAs. They are typically not associated with a conserved secondary structure and with conserved primary sequence elements like the small ncRNAs. Furthermore, the lncRNAs are less conserved in evolution than protein genes. For instance, about 30% of human lncRNAs are primate specific. The lncRNAs are in general characterized by lower expression as compared to the protein genes, and they show more tissue-specific expression. The brain is the tissue with the largest fraction of tissue-specific lncRNAs.

In this book, we focus on the functional impact of genomic sequence elements. In this regard, it is interesting to note that very few lncRNAs seem to have sequence-specific functions. In fact, in many cases it is not the lncRNA itself that has a specific transcription regulatory effect, but rather the very act of transcription and/or splicing at the lncRNA locus.

What is the genomic organization of lncRNAs? With respect to protein coding genes, they can either be in a position in between two different protein genes (intergenic) or overlap with a protein-coding gene. In case of an overlap, a lncRNA can be encoded on the same strand as the protein or, which is more common, on the complementary strand (an "antisense" lncRNA) (**Figure 11.7**).

LncRNAs have been attributed to different molecular functions, most of them related to transcriptional control. Thus, they may have a signaling function to regulate transcription of a downstream gene. They may influence transcription by acting as decoys and sequester chromatin proteins or miRNAs. They may act as guides to recruit chromatin-modifying enzymes. Finally, they may act as scaffolds to bring together multiple proteins.

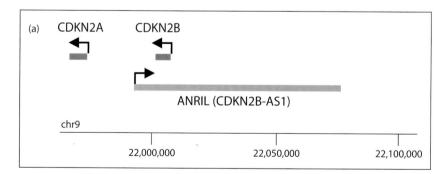

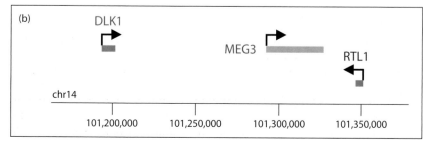

Figure 11.7 Genomic organization of lncRNAs. Protein genes are shown in green and lncRNA genes in cyan. Arrows indicate the direction of transcription. (a) The lncRNA ANRIL (also known as CDKN2B-AS1) is antisense to and overlapping with protein gene CDKN2B. (b) The lncRNA MEG3, an example of a lncRNA with an intergenic location, in between the protein genes DLK1 and RTL1. At the bottom in both panels are shown genomic positions (human genome version GRCh37/hg19).

LncRNAs REGULATE EXPRESSION USING MANY DIFFERENT MECHANISMS

The function of most lncRNAs is yet unknown. For those that we do have functional information, a large number of different mechanisms of transcriptional control have been revealed. Here, we briefly examine a selection of these. Many of them affect the expression of one or more nearby genes (they act in **cis**). However, our first example, the lncRNA HOTAIR, is an RNA working in **trans**, as it can exert its effect on a gene located on another chromosome.

HOTAIR

The lncRNA HOTAIR, encoded on chromosome 12, is involved in an epigenetic control mechanism (**Figure 11.8a**). It is necessary for a histone modification, trimethylation of histone 3 on lysine 27 (H3K27me3), a

Figure 11.8 Examples of lncRNA mechanisms. (a) The lncRNA HOTAIR, encoded on chromosome 12, is part of a mechanism promoting trimethylation of histone 3 on lysine 27 (red pins), leading to silencing of the HOXD locus genes on chromosome 2. (b) The Xist RNA is specifically expressed from the inactive X chromosome. It is able to coat the entire chromosome, thereby repressing all genes (left side of panel). In the active X chromosome, Xist is silenced by the action of the antisense lncRNA Tsix (orange, right side of panel). The transcription of the Tsix gene across the Xist promoter is required for repression of Xist. (c) The lncRNA Kcnq1ot1 is an example of an RNA essential for imprinting (epigenetically controlled allele-specific expression). It is antisense to the protein gene Kcnq1 (green). An imprinting control region (ICR, orange) contains the promoter of the Kcnq1ot1 ncRNA. On the paternal allele, the ICR is unmethylated (gray pins), allowing Kcnq1ot1 expression. Expression of this lncRNA silences nearby protein-coding genes (red rectangles) on the paternal allele. On the maternal allele, the ICR is methylated (red pins). Thereby, Kcnq1ot1 is not transcribed and the nearby protein coding imprinted genes are activated. Nonimprinted genes are shown as gray rectangles. (d) An enhancer RNA (eRNA) is produced by transcription of DNA and regulates transcription.

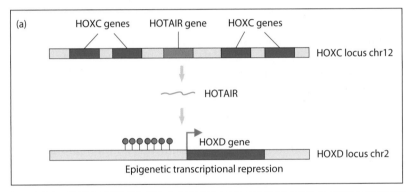

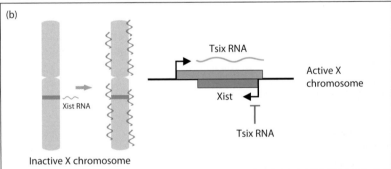

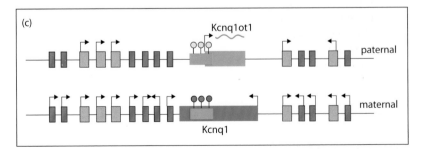

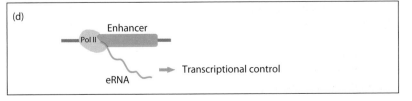

modification promoting transcriptional repression of chromatin. This chromatin modification occurs at the HOXD locus on chromosome 2, thereby lowering gene expression at this locus.

X chromosome inactivation and Xist RNA

Females have two X chromosomes, whereas males have only one. In mammals there must be a mechanism to ensure that the gene expression from an X chromosome in females is reduced so that it matches that of males. This mechanism involves epigenetic silencing of one of the X chromosomes. A lncRNA, Xist RNA, is a major player in this silencing (**Figure 11.8b**). It was one of the first lncRNAs to be discovered and is part of a locus containing additional lncRNAs. The Xist RNA is specifically expressed from the inactive X chromosome. The Xist transcripts are able to coat the entire X chromosome from which they are transcribed, thereby silencing all genes on that chromosome. Other lncRNAs take part in the regulation of Xist expression. For instance, in the active X chromosome, Xist is silenced by the action of the antisense lncRNA Tsix.

Imprinting

LncRNAs are often essential for epigenetically controlled allele-specific expression at imprinted loci. One example is the RNA Kcnq1ot1, which is antisense to the protein gene *KCNQ1* and regulates the expression of nearby protein genes (**Figure 11.8c**).

Enhancer RNAs

We have seen that **enhancers** are important parts of the machinery of transcriptional regulation. It was recently realized that many enhancer regions are transcribed. The RNA products, **enhancer RNAs** (**eRNAs**), are in a size range of 50–2,000 nucleotides (**Figure 11.8d**). Like DNA enhancer elements, eRNAs are involved in transcriptional control. However, further studies are required to elucidate the functional significance of these RNAs.

A REPEAT EXPANSION IN A lncRNA CAUSES A SPINOCEREBELLAR ATAXIA

We now examine a few cases where lncRNAs are implicated in disease. We discussed the spinocerebellar ataxias (SCAs) in Chapter 7. They are in many cases caused by repeat expansion in a protein gene, either in the coding sequence or in a noncoding part of the protein gene. In this regard, it is interesting to note that SCA8 is caused by expansion of a repeat in a lncRNA, ATXN8OS (ataxin 8 opposite strand). The repeated triplet is CTG (complementary to CAG) reminiscent of many other SCAs. This lncRNA overlaps with and is antisense to the protein gene KLHL1.

A lncRNA AND CELIAC DISEASE

Celiac disease is a common disorder affecting about 1 in 100 individuals. When affected individuals eat gluten—a protein present in wheat, rye, and barley—an immune response is elicited that causes inflammation in the small intestine.

A lncRNA has been associated with susceptibility to the disease. Genome-wide association studies have identified a limited number of single-nucleotide polymorphisms (SNPs) associated with the risk of being

Figure 11.9 A genetic variant associated with celiac disease and a model for the action of the lncRNA lnc13. (a) Genomic organization of *IL18RAP* and lnc13. The SNP rs917997, associated with celiac disease, is located 1.5 kb downstream of the gene *IL18RAP* and within the lnc13 gene. (b) When the lnc13 RNA is bound in a complex including the proteins hnRNPD (ribonucleoprotein D) and HDAC1 (histone deacetylase), transcription of inflammatory genes is silenced. This is characteristic of normal macrophages. The C shown in green is the normal variant of SNP rs917997. (c) The interaction between lnc13 and hnRNPD is reduced when the genomic variant is T (red) instead of C. As a result of inflammatory response or celiac disease, the protein DC2P is activated, leading to degradation of RNAs including lnc13. In this case, inflammatory genes are erroneously activated. The lnc13 RNA is at the genomic location chr2:103,068,510–103,071,054 and rs917997 is at position 103,070,568 (genome version GRCh37/hg19).

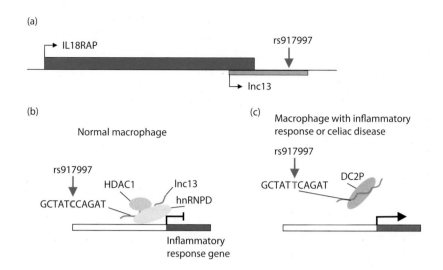

diagnosed with celiac disease. One of the SNPs (rs917997) is located 1.5 kb downstream of the gene *IL18RAP* (interleukin-18 receptor accessory protein), a gene involved in inflammatory response and associated with susceptibility of celiac disease. In the region of the SNP is located a lncRNA named lnc13. It is transcribed from the same strand as *IL18RAP*, but the two genes are transcribed independently. A model for the action of lnc13 is shown in **Figure 11.9**. When the lnc13 RNA is bound in a complex including the protein hnRNPD, transcription of inflammatory genes is silenced. The interaction between lnc13 and hnRNPD is reduced when there is T instead of the normal C in the position of the SNP rs917997. In this case, inflammatory genes are erroneously activated.

A lncRNA PLAYS AN IMPORTANT ROLE IN A MUSCLE DISORDER

Facioscapulohumeral muscular dystropy (FSHD) is an autosomal dominant disease with a prevalence in the range of 4–12 in 100,000. FSHD is one of the most common muscle disorders and is characterized by wasting of facial, upper arm, and shoulder girdle muscles. In about 95% of these patients, the genetic defects are deletions that reduce the copy number of a nonidentical repeat referred to as D4Z4. The repeat is about 3,300 nucleotides in size. Included in the repeat is the open reading frame (ORF) for the gene *DUX4*, encoding a transcription factor known as a homeobox protein. The repeats are located close to the telomere end of the chromosome 4. A healthy individual carries 11–150 copies, but in FSHD patients the number of copies is less than 11.

Why does the decrease in the number of repeats cause disease? A lower number of repeat copies causes deregulation of genes that are close to or part of the repeat. The molecular background to this aberrant regulation involves epigenetic mechanisms as well as a lncRNA named DBE-T (**Figure 11.10**). In healthy individuals, the D4Z4 repeat is bound by Polycomb-group (PcG) proteins. There are sequence elements that may act to recruit these proteins. One of them is a sequence CNGCCATNDNND (where D is A, G, or T), conserved among animals. It should also be noted that PcG protein recruitment may be dependent on CpG-rich regions—the region occupied by D4Z4 repeats in healthy subjects is one of the largest CpG islands in the human genome. The PcG proteins bring about histone

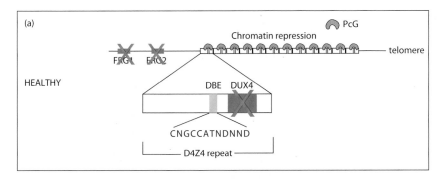

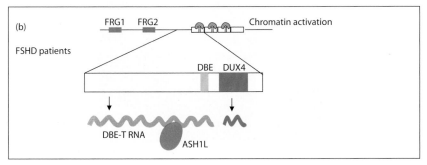

Figure 11.10 Model for gene de-repression in FSHD as mediated by the lncRNA DBE-T. (a) Multiple copies of the D4Z4 repeats (each about 3,300 nucleotides) are extensively bound by PcG proteins, leading to epigenetic repression of chromatin in this chromosomal region. Thereby, protein genes *FRG1*, *FRG2*, and *DUX4* are silenced. Located within the repeat is the sequence element DBE with the consensus sequence as indicated. (b) In FSHD patients, D4Z4 repeats are deleted. PcG proteins are insufficiently bound, and transcriptional silencing is less efficient. Therefore, the *DBET* gene is activated. The *DBET* recruits the protein ASH1L, leading to activation of all nearby genes. One of these genes is *DUX4*.

H3-lysine-27 trimethylation (H3K27me3). This methylation acts to repress transcription of neighboring genes, including *DUX4*. In FSHD patients, however, there is a limited number of D4Z4 repeats. As a result, transcriptional silencing is less efficient, and the genes nearby are instead activated. One of the activated genes is *DBET*, encoding a lncRNA with its transcription start site just upstream of the D4Z4 repeat array. The DBE-T RNA interacts with a protein named ASH1L, a component of a chromatin activator complex known as Trithorax. This interaction leads to H3 lysine-36 dimethylation (H3K36me2) at the locus, a modification that is activating the chromatin. As a result, nearby protein coding genes such as *DUX4* are activated. Increased expression of *DUX4* is believed to be a crucial step for the development of the FSHD phenotype.

A lncRNA AND ANGELMAN SYNDROME

We have previously seen that PWS is typically caused by deletions in a region of the paternal chromosome 15, a region involving a large number of snoRNAs as well as protein genes. We now turn to AS, which is in most cases caused by the same type of deletions, but in the maternal chromosome.

AS got its name from the British pediatrician Harry Angelman who described three children with the disorder. Harry Angelman first referred to children with the syndrome as "Puppet Children." He described his choice of this title to describe the cases as being related to an oil painting he had seen during a vacation in Italy (**Figure 11.11**):

> When on holiday in Italy I happened to see an oil painting in the Castelvecchio Museum in Verona called ... *a Boy with a Puppet*. The boy's laughing face and the fact that my patients exhibited jerky movements gave me the idea of writing an article about the three children with a title of Puppet Children. It was not a name that pleased all parents but it served as a means of combining the three little patients into a single group. Later

Figure 11.11 "Boy with a Puppet" or "A child with a drawing" by Giovanni Francesco Caroto. Castelvecchio Museum in Verona.

the name was changed to Angelman syndrome. This article was published in 1965 and after some initial interest lay almost forgotten until the early eighties.[1]

The syndrome was renamed "Angelman syndrome" in 1982. It affects the nervous system, and symptoms include intellectual disability and problems with movement and balance. Seizures (epilepsy) are also common, and most children with AS are characterized by frequent laughter or smiling. The incidence of AS is about 1 in 10,000 live births.

Although large deletions in the chromosomal region 15q11.2-q13 are associated with AS (see Figure 11.6a), there is one gene in particular that is crucial in developing the disorder. This is *UBE3A* that encodes an enzyme, an E3 ubiquitin ligase. This enzyme is part of an important pathway for the degradation of proteins. The paternal copy of *UBE3A* is silenced (imprinted), and for adequate expression of *UBE3A*, the maternal copy of the gene must be present and functional. In most AS patients, the cause of the disorder is a deletion in the region 15q11.2-q13 of the maternal chromosome that includes the *UBE3A* gene, but there are also cases that result from coding sequence mutations in the *UBE3A* gene. For instance, there is one mutation that disrupts a phosphorylation site in the protein, and this leads to increased activity of the protein.

On the paternal chromosome, the *UBE3A* gene is completely repressed. What is the mechanism behind this repression? It turns out it involves a lncRNA, named UBE3A-ATS, which is produced antisense to *UBE3A* (**Figure 11.12a**) and overlaps that gene. The UBE3A-ATS RNA is a product obtained by the processing of a large transcript initiating upstream of the protein gene SNRPN (small nuclear ribonucleoprotein polypeptide N). It has been suggested that UBE3A-ATS RNA controls the production of the UBE3A mRNA through a transcriptional interference mechanism (**Figure 11.12b**). This mechanism would lead to repression of the paternal *UBE3A* gene.

---

[1]Seventh edition, Facts about Angelman Syndrome, by Charles A. Williams, M.D., Sarika U. Peters, Ph.D., Stephen N. Calculator, Ph.D. in 2009

(a)

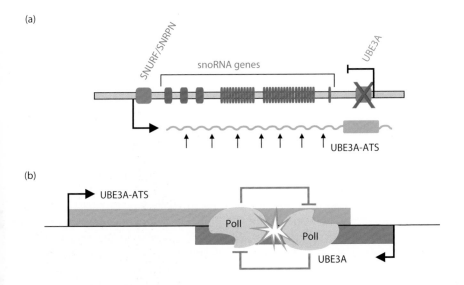

(b)

Figure 11.12 The role of a lncRNA in Angelman syndrome. The protein gene *UBE3A* plays a critical role in AS. Illustrated is how UBE3A is silenced on the paternal chromosome. (a) A chromosomal region containing the gene *UBE3A* (green) and the antisense lncRNA transcript UBE3A-ATS is shown. A long transcript initiating upstream of the protein gene *SNRPN* (green) is processed (multiple small arrows) to form the mature UBE3A-ATS RNA. The UBE3A-ATS RNA represses transcription of UBE3A. (b) According to one model, expression of the *UBE3A* gene product is hindered by a transcriptional interference mechanism in which elongating RNA polymerases physically collide.

Although disorders such as AS and FSHD are not caused by mutations in lncRNAs, there may be important therapeutic implications of these RNAs. For instance, in AS the disorder might be corrected by inactivating the UBE3A-ATS transcript and thereby activating the silenced copy of the UBE3A gene. (For more on gene therapy, see Chapter 14.)

SUMMARY

- Noncoding RNAs (ncRNAs) are RNAs that unlike mRNAs do not code for protein but have other functions.
- Each family of small ncRNAs is typically characterized by specific primary sequence elements as well as specific secondary structure elements that are based on base-pairing within the molecule.
- According to the RNA world hypothesis, RNAs had an important role early in the history of life. This was made possible as RNA molecules may carry genetic information as well as catalyze chemical reactions, thereby taking on many of the functions today carried by proteins and DNA.
- Some small ncRNAs have their own promoter, whereas others are encoded within a protein or ncRNA gene without their own promoter.
- A number of mitochondrial disorders result from mutations in mitochondrial tRNA genes.
- Mutations in the MRP RNA gene may cause cartilage-hair hypoplasia. One common mutation is $A > G$ in position 70 of the mature RNA, expected to affect its catalytic activity.
- PWS is an inherited disorder associated with a deficiency in snoRNAs.
- The 15q11.2-q13 region implicated in both PWS and AS contains imprinted genes. Imprinting occurs when only one of two copies of a gene is expressed. In maternal imprinting, the paternal copy of the gene is transcriptionally active, while the maternal copy is epigenetically silenced. Conversely, in paternal imprinting, the paternal copy of the gene is silenced.
- The lncRNAs are typically mRNA-like in the sense that they are transcribed by RNA polymerase II, are polyadenylated, and are formed by the removal of introns.
- The lncRNAs may exert their function both in cis and in trans.

- Unlike the small ncRNAs, lncRNAs are poorly conserved during evolution, and they are not associated with specific RNA secondary structure features.
- The Xist RNA plays an important role in X chromosome inactivation, a mechanism to reduce expression from one of the two X chromosomes in females.
- Many enhancer regions are transcribed to form enhancer RNAs, important in transcriptional control.
- A repeat expansion in a lncRNA causes the spinocerebellar ataxia SCA8.
- The lncRNA lnc13 is associated with celiac disease.
- The cause of FSHD is deletions that reduce the copy number of a nonidentical 3.3 kb repeat. In healthy individuals, the repeat is bound by PcG proteins, leading to repressive histone modification. In FSHD patients with fewer repeats, transcriptional silencing is less efficient, leading to activation of the lncRNA gene *DBET*. This activation leads in turn to the activation of *DUX4*, presumably an essential step for development of the FSHD phenotype.
- AS is intimately linked to *UBE3A*, a paternally imprinted gene. Responsible for AS are mutations that affect the maternal copy of the *UBE3A* gene.
- Responsible for repression of the paternal *UBE3A* gene is a lncRNA, UBE3A-ATS. Therefore, a potential therapy of AS includes inactivation of the UBE3A-ATS gene, leading to activation of the silenced copy of the *UBE3A* gene.

QUESTIONS

1. What is meant by the *RNA world*?
2. Name at least seven types of *small* noncoding RNAs (tRNA is one example) and their functions.
3. Why does the mutation A > G in position 14 of a human mitochondrial leucine tRNA have such dramatic consequences on mitochondrial function?
4. What was the first disease identified where a nuclear noncoding RNA is defective? What is the most common mutation in this disease, and why is it harmful?
5. How are the C/D box and H/ACA box snoRNAs different in terms of structure and function?
6. What noncoding RNAs are implicated in PWS? What mutations give rise to this disorder?
7. What is imprinting?
8. What is characteristic of *long noncoding RNAs*?
9. What does it mean when we say that lncRNAs may be operating in *cis* or *trans*?
10. What is an *antisense RNA*? Give an example.
11. What is the mutation responsible for *spinocerebellar ataxia 8*?
12. How is a lncRNA related to gluten intolerance?
13. What cases in the chapter are discussed that are examples of the role of lncRNAs in epigenetic silencing?
14. What is the most common mutation in patients with FSHD? How is the transcript DBE-T related to this disorder?
15. What kinds of mutations may give rise to AS?
16. How could the transcript UBE3A-ATS be exploited in AS therapy?

FURTHER READING

Extensive transcription of the human genome

Djebali S, Davis CA, Merkel A et al. 2012. Landscape of transcription in human cells. *Nature* 489(7414):101–108.

Hangauer MJ, Vaughn IW, McManus MT. 2013. Pervasive transcription of the human genome produces thousands of previously unidentified long intergenic noncoding RNAs. *PLOS Genet* 9(6):e1003569.

Small ncRNAs

Cech TR, Steitz JA. 2014. The noncoding RNA revolution-trashing old rules to forge new ones. *Cell* 157(1):77–94.

Davila Lopez M, Rosenblad MA, Samuelsson T. 2009. Conserved and variable domains of RNase MRP RNA. *RNA Biol* 6(3):208–220.

Esakova O, Krasilnikov AS. 2010. Of proteins and RNA: The RNase P/MRP family. *RNA* 16(9):1725–1747.

Goldfarb KC, Cech TR. 2017. Targeted CRISPR disruption reveals a role for RNase MRP RNA in human preribosomal RNA processing. *Genes Dev* 31(1):59–71.

Holley CL, Topkara VK. 2011. An introduction to small non-coding RNAs: miRNA and snoRNA. *Cardiovasc Drugs Ther* 25(2):151–159.

Podlevsky JD, Chen JJL. 2016. Evolutionary perspectives of telomerase RNA structure and function. *RNA Biol* 13(8):720–732.

Rosenblad MA, Larsen N, Samuelsson T, Zwieb C. 2009. Kinship in the SRP RNA family. *RNA Biol* 6(5):508–516.

The RNA world hypothesis

Gesteland RF, Cech T, Atkins JF. 1999. *The RNA world: The nature of modern RNA suggests a prebiotic RNA*. Cold Spring Harbor Laboratory Press, Cold Spring Harbor, NY.

Gilbert W. 1986. Origin of life: The RNA world. *Nature* 319(6055):618–618.

Woese CR. 1967. *The genetic code: The molecular basis for genetic expression*. Harper and Row, New York.

Yarus M. 2010. *Life from an RNA world: The ancestor within*. Harvard University Press, Cambridge, MA.

Catalytic RNA

Guerrier-Takada C, Altman S. 1984. Catalytic activity of an RNA molecule prepared by transcription *in vitro*. *Science* 223(4633):285–286.

Guerrier-Takada C, Gardiner K, Marsh T et al. 1983. The RNA moiety of ribonuclease P is the catalytic subunit of the enzyme. *Cell* 35(3 Pt 2):849–857.

Kruger K, Grabowski PJ, Zaug AJ et al. 1982. Self-splicing RNA: Autoexcision and autocyclization of the ribosomal RNA intervening sequence of tetrahymena. *Cell* 31(1):147–157.

Noncoding RNAs and disease

Esteller M. 2011. Non-coding RNAs in human disease. *Nat Rev Genet* 12(12):861–874.

Mitochondrial leucine tRNA disorder

Finsterer J. 2007. Genetic, pathogenetic, and phenotypic implications of the mitochondrial A3243G tRNALeu(UUR) mutation. *Acta Neurol Scand* 116(1):1–14.

Yasukawa T, Suzuki T, Ueda T et al. 2000. Modification defect at anticodon wobble nucleotide of mitochondrial tRNAs(Leu)(UUR) with pathogenic mutations of mitochondrial myopathy, encephalopathy, lactic acidosis, and stroke-like episodes. *J Biol Chem* 275(6):4251–4257.

MRP RNA and disease

Bonafe L, Dermitzakis ET, Unger S et al. 2005. Evolutionary comparison provides evidence for pathogenicity of RMRP mutations. *PLOS Genet* 1(4):e47.

Martin AN, Li Y. 2007. RNase MRP RNA and human genetic diseases. *Cell Res* 17(3):219–226.

Ridanpaa M, van Eenennaam H, Pelin K et al. 2001. Mutations in the RNA component of RNase MRP cause a pleiotropic human disease, cartilage-hair hypoplasia. *Cell* 104(2):195–203.

Thiel CT, Mortier G, Kaitila I et al. 2007. Type and level of RMRP functional impairment predicts phenotype in the cartilage hair hypoplasia-anauxetic dysplasia spectrum. *Am J Hum Genet* 81(3):519–529.

The annotation of human snoRNAs

Jorjani H, Kehr S, Jedlinski DJ et al. 2016. An updated human snoRNAome. *Nucleic Acids Res*.

Disease and snoRNAs

Anderlid BM, Lundin J, Malmgren H et al. 2014. Small mosaic deletion encompassing the snoRNAs and SNURF-SNRPN results in an atypical Prader-Willi syndrome phenotype. *Am J Med Genet Part A* 164A(2):425–431.

de Smith AJ, Purmann C, Walters RG et al. 2009. A deletion of the HBII-85 class of small nucleolar RNAs (snoRNAs) is associated with hyperphagia, obesity and hypogonadism. *Hum Mol Genet* 18(17):3257–3265.

McMahon M, Contreras A, Ruggero D. 2015. Small RNAs with big implications: New insights into H/ACA snoRNA function and their role in human disease. *Wiley Interdiscip Rev RNA* 6(2):173–189.

Sahoo T, del Gaudio D, German JR et al. 2008. Prader-Willi phenotype caused by paternal deficiency for the HBII-85 C/D box small nucleolar RNA cluster. *Nat Genet* 40(6):719–721.

Long ncRNAs

Adelman K, Egan E. 2017. More uses for genomic junk. *Nature* 543:183.

Augui S, Nora EP, Heard E. 2011. Regulation of X-chromosome inactivation by the X-inactivation centre. *Nat Rev Genet* 12(6):429–442.

Batista PJ, Chang HY. 2013. Long noncoding RNAs: Cellular address codes in development and disease. *Cell* 152(6):1298–1307.

Cerase A, Pintacuda G, Tattermusch A, Avner P. 2015. Xist localization and function: New insights from multiple levels. *Genome Biol* 16:166.

Derrien T, Johnson R, Bussotti G et al. 2012. The GENCODE v7 catalog of human long noncoding RNAs: Analysis of their gene structure, evolution, and expression. *Genome Res* 22(9):1775–1789.

Engreitz JM, Haines JE, Perez EM et al. 2016. Local regulation of gene expression by lncRNA promoters, transcription and splicing. *Nature* 539(7629):452–455.

Hon C-C, Ramilowski JA, Harshbarger J et al. 2017. An atlas of human long non-coding RNAs with accurate 5′ ends. *Nature* 543:199.

Kopp F, Mendell JT. 2018. Functional classification and experimental dissection of long noncoding RNAs. *Cell* 172(3):393–407.

Liu SJ, Horlbeck MA, Cho SW et al. 2017. CRISPRi-based genome-scale identification of functional long noncoding RNA loci in human cells. *Science* 355(6320).

Meseure D, Drak Alsibai K, Nicolas A et al. 2015. Long noncoding RNAs as new architects in cancer epigenetics, prognostic biomarkers, and potential therapeutic targets. *Bio Med Res Int* 2015:320214.

Miao YR, Liu W, Zhang Q, Guo AY. 2017. lncRNASNP2: An updated database of functional SNPs and mutations in human and mouse lncRNAs. *Nucleic Acids Res* 46:D276–D280.

Quinn JJ, Chang HY. 2016. Unique features of long non-coding RNA biogenesis and function. *Nat Rev Genet* 17(1):47–62.

St Laurent G, Wahlestedt C, Kapranov P. 2015. The landscape of long noncoding RNA classification. *Trends Genet* 31(5):239–251.

Ulitsky I, Bartel DP. 2013. lincRNAs: Genomics, evolution, and mechanisms. *Cell* 154(1):26–46.

Zhao Y, Li H, Fang S et al. 2016. NONCODE 2016: An informative and valuable data source of long non-coding RNAs. *Nucleic Acids Res* 44(D1):D203–D208.

Enhancer RNAs and protein interactions

Bose DA, Donahue G, Reinberg D et al. 2017. RNA binding to CBP stimulates histone acetylation and transcription. *Cell* 168(1):135–149.e122.

Long ncRNAs and disease

Aung T, Ozaki M, Lee MC et al. 2017. Genetic association study of exfoliation syndrome identifies a protective rare variant at LOXL1 and five new susceptibility loci. *Nat Genet* 49:993.

Cabianca DS, Casa V, Bodega B et al. 2012. A long ncRNA links copy number variation to a polycomb/trithorax epigenetic switch in FSHD muscular dystrophy. *Cell* 149(4):819–831.

Castellanos-Rubio A, Fernandez-Jimenez N, Kratchmarov R et al. 2016. A long noncoding RNA associated with susceptibility to celiac disease. *Science* 352(6281):91–95.

Hauser MA, Aboobakar IF, Liu Y et al. 2015. Genetic variants and cellular stressors associated with exfoliation syndrome modulate promoter activity of a lncRNA within the LOXL1 locus. *Hum Mol Genet* 24(22):6552–6563.

Huarte M. 2016. RNA. A lncRNA links genomic variation with celiac disease. *Science* 352(6281):43–44.

Lalevee S, Feil R. 2015. Long noncoding RNAs in human disease: Emerging mechanisms and therapeutic strategies. *Epigenomics* 7(6):877–879.

Lee JT, Bartolomei MS. 2013. X-inactivation, imprinting, and long noncoding RNAs in health and disease. *Cell* 152(6):1308–1323.

Wapinski O, Chang HY. 2011. Long noncoding RNAs and human disease. *Trends Cell Biol* 21(6):354–361.

Angelman syndrome

Meng L, Person RE, Beaudet AL. 2012. Ube3a-ATS is an atypical RNA polymerase II transcript that represses the paternal expression of Ube3a. *Hum Mol Genet* 21(13):3001–3012.

Meng L, Person RE, Huang W et al. 2013. Truncation of Ube3a-ATS unsilences paternal Ube3a and ameliorates behavioral defects in the Angelman syndrome mouse model. *PLOS Genet* 9(12):e1004039.

Meng L, Ward AJ, Chun S et al. 2015. Towards a therapy for Angelman syndrome by targeting a long non-coding RNA. *Nature* 518(7539):409–412.

Williams C. 1990. *Facts about Angelman syndrome*. University of Florida Health Science Center, Gainesville, FL.

Computational Methods Are Critical in the Analysis of Molecular Sequences

12

I n recent years we have seen a tremendous increase in the amount of information collected within the field of biomedicine. For instance, the remarkable development of DNA sequencing technology has given rise to a vast amount of genomic information. Every human haploid genome is 3 billion nucleotides, and more than a million people have been subject to sequencing. In addition, there are tens of thousands of human **cancer genomes** that have been sequenced. Outside of the world of humans, there are more than half a million other animals on our planet—thousands of these have been subject to genome sequencing (Tables 4.1 and 4.2). In addition, **epigenomes** are being mapped, and **genome-wide association studies** (GWASs) identify numerous genetic variants associated with traits and disorders. Furthermore, not only DNA sequences and variants are accumulating—there are large-scale projects to sequence all RNA molecules that are expressed in different types of cells. Traditionally, DNA and RNA sequencing have been carried out using whole animals or tissue samples where the DNA to be analyzed originates from a large number of individual cells. But now sensitive methods have been developed that allow DNA and RNA sequencing to be carried out even at a single-cell level, allowing us to examine the evolution of cell populations in great detail. Finally, a vast number of proteins are being characterized with respect to their amino acid sequences, and the three-dimensional structures of proteins are being determined using x-ray crystallography, nuclear magnetic resonance (NMR), and cryo-electron microscopy.

BIOINFORMATICS METHODS ARE WIDELY USED

How are we to keep track of and analyze such large volumes of data? Obviously, computers are necessary, as well as knowing how to tell the computers what to do. The discipline **bioinformatics** has emerged, a discipline that may be defined as the handling and analysis of information originating from current biomedical research. It is an interdisciplinary science with contributions from computer science, mathematics, and biomedicine.

Table 12.1 Selected highly cited scientific publications presenting bioinformatics methods

Publication title	Number of citations	Publication
Basic local alignment search tool	50,967	Altschul, S.F. et al. *J Mol Biol*, 1990. 215(3): p. 403–10.
CLUSTAL-W—improving the sensitivity of progressive multiple sequence alignment through sequence weighting, position-specific gap penalties, and weight matrix choice	48,445	Thompson, J.D., D.G. Higgins, and T.J. Gibson, *Nucleic Acids Res*, 1994. 22(22): p. 4673–80.
Gapped BLAST and PSI-BLAST: A new generation of protein database search programs	46,175	Altschul, S.F. et al., *Nucleic Acids Res*, 1997. 25(17): p. 3389–402.
The neighbor-joining method—A new method for reconstructing phylogenetic trees	39,108	Saitou, N. and M. Nei, *Mol Biol Evol*, 1987. 4(4): p. 406–25.
The CLUSTAL_X windows interface: Flexible strategies for multiple sequence alignment aided by quality analysis tools	29,823	Thompson, J.D. et al., *Nucleic Acids Res*, 1997. 25(24): p. 4876–82.
MEGA5: Molecular evolutionary genetics analysis using maximum likelihood, evolutionary distance, and maximum parsimony methods	28,742	Tamura, K. et al., *Mol Biol Evol*, 2011. 28(10): p. 2731–9.
MrBayes 3: Bayesian phylogenetic inference under mixed models	18,736	Ronquist, F. and J.P. Huelsenbeck, *Bioinformatics*, 2003. 19(12): p. 1572–4.
The Protein Data Bank	17,994	Berman, H.M. et al., *Nucleic Acids Res*, 2000. 28(1): p. 235–42.
MUSCLE: Multiple sequence alignment with high accuracy and high throughput	16,828	Edgar, R.C., *Nucleic Acids Res*, 2004. 32(5): p. 1792–7.
Gene Ontology: Tool for the unification of biology	15,982	Ashburner, M. et al., *Nat Genet*, 2000. 25(1): p. 25–9.
The Sequence Alignment/Map format and SAMtools	12,868	Li, H. et al., *Bioinformatics*, 2009. 25(16): p. 2078–9.
A simple, fast, and accurate algorithm to estimate large phylogenies by maximum likelihood	11,311	Guindon, S. and O. Gascuel, *Syst Biol*, 2003. 52(5): p. 696–704.
Fast and accurate short read alignment with Burrows-Wheeler transform	9211	Li, H. and R. Durbin, *Bioinformatics*, 2009. 25(14): p. 1754–60.
Ultrafast and memory-efficient alignment of short DNA sequences to the human genome	8645	Langmead, B. et al., *Genome Biol*, 2009. 10(3): p. R25.
Predicting transmembrane protein topology with a hidden Markov model: Application to complete genomes	5918	Krogh, A. et al., *J Mol Biol*, 2001. 305(3): p. 567–80.
Improved prediction of signal peptides: SignalP 3.0	4942	Bendtsen, J.D. et al., *J Mol Biol*, 2004. 340(4): p. 783–95.

Note: Publications are ordered according to the number of times they have been cited in other scientific papers. The list is an extract from a search in "Web of Science," a scientific indexing service maintained by Clarivate Analytics.

Important for the development of bioinformatics have been researchers who are competent in computer science and/or mathematics and at the same time are appreciating and fully understanding the biological problems addressed. Many of them remain unsung heroes in the history of the human genome as well as in molecular biology in general. But a large number of computational tools have been developed, and many of them have been cited extensively in the scientific literature (**Table 12.1**).

THE HUMAN GENOME AND THE REST OF BIOLOGY CAN ONLY BE UNDERSTOOD IN THE LIGHT OF EVOLUTION

In this chapter, we discuss some of the most important computational tools related to the human genome and the proteins encoded by that genome. Many of these tools are concerned with molecular evolution. The human genome was shaped by evolution, and its organization and functions can only be fully understood if we understand how it developed from other species. There is a famous quote of Theodosius Dobzhansky, "Nothing in biology makes sense except in the light of evolution," where he argues against creationism and intelligent design. This statement is highly relevant in the context of the human genome. We begin this chapter by introducing some of the fundamental principles of evolution and later discuss the computational methods used to address evolutionary problems.

The first forms of life arose on our planet some 4 billion years ago. It is hard to imagine this time perspective. A human generation is about 22–32 years, and with some luck you can deduce your family ancestry some 10 generations back in time. The first modern humans presumably came about 300,000 years ago, corresponding to about 10,000 human generations. However, the first life forms arose at a time that corresponds to some 100,000,000 human generations ago. A brief timeline of evolution is shown in **Figure 12.1**.

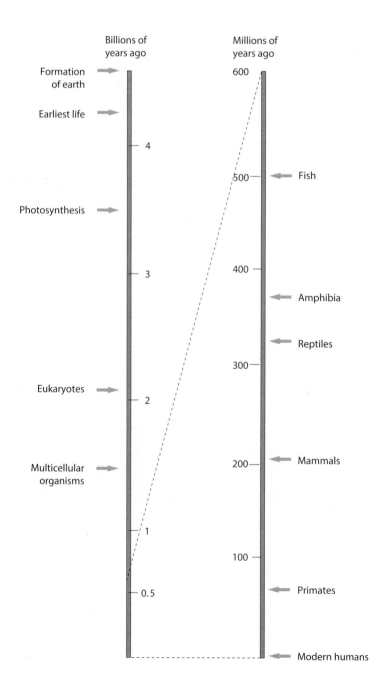

Figure 12.1 Evolution of life on earth. Earth was formed around 4.5 billion years ago. First forms of life appeared about 4.3 billion years ago, whereas anatomically modern humans arose around 300,000 years ago.

Figure 12.2 Simplified tree of life. The three domains of life, Bacteria, Archaea, and Eukaryota, are shown with example phyla. (Based on Wikimedia figure.)

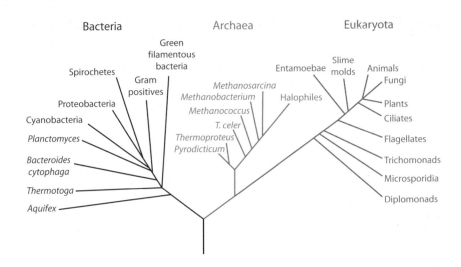

We have seen that there are three major domains of life: **eukaryotes, bacteria**, and **archaea** (Chapter 3). A schematic tree to show the relationship between these three major life forms is shown in **Figure 12.2**.

Important mechanisms of evolution include genetic changes and selective forces. The genetic changes are the mutations that we encountered in Chapter 5, and they largely come about by chance. Selective forces determine what genetic changes stay and are propagated to future generations. Given the time perspective of evolution of species on our planet, with chance playing an important role, it is not surprising that a complexity has arisen, a complexity in terms of the number of biological species and in terms of a large number of proteins and nucleic acids within a species, not to mention a complexity of the human genome.

An important consequence of evolution is that all species on our planet are related. This is also reflected in DNA and protein sequences. Consider, for instance, the sequences of a specific class of proteins as shown in **Figure 12.3**. This protein family has members from all three major domains of life. The individual proteins are strikingly similar in sequence, and the probability that these sequences have arisen by independent events is infinitesimal. The sequence similarity is convincing evidence that the proteins have a common origin during the evolution of life.

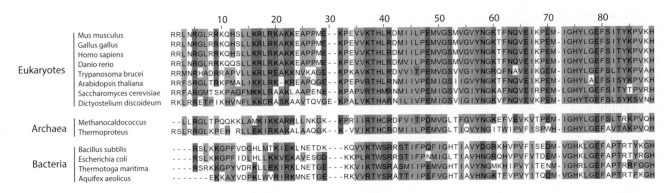

Figure 12.3 Evolution is reflected in molecular sequences. A family of ribosomal proteins has members from all three domains of life. Selected members of this family are compared here. The remarkable sequence similarity between protein sequences from many different species demonstrates an evolutionary relationship, as there is a negligible probability that the sequence relationship has arisen by chance. Amino acids are shown in a one-letter code, and the background colors reflect physical and chemical properties of the amino acids.

HISTORY OF SEQUENCE COMPUTING STARTED WITH PROTEINS

The sequence comparison in Figure 12.3 is one example of how sequences may be used to examine evolution. Protein sequences rather than DNA were first used for this purpose, because methods for determining the sequence of nucleotides in DNA were developed long after protein sequencing methods. The first protein to be sequenced was insulin, whose complete amino acid sequence was reported in 1953 by Frederick Sanger. The sequence was determined through a laborious protocol analyzing overlapping fragments of insulin. For this work Sanger was awarded the 1958 Nobel Prize in Chemistry. Using the same type of technology, researchers sequenced additional proteins during the 1950s and 1960s. The first collection of available protein sequences—around 70 at the time—was published in 1965. One of the co-authors of the book, *Atlas of Protein Sequence and Structure*, was Margaret Dayhoff, a pioneer in bioinformatics. Having access to these protein sequences, Dayhoff realized that they could be used to address significant biological problems. For instance, molecular sequences could be exploited to construct phylogenetic trees to show how animals are related to one another by evolution. The 1965 edition of *Atlas of Protein Sequence and Structure* contains a few of the very first phylogenetic trees based on molecular sequences.

HOW SEQUENCES ARE COMPARED

Sequence comparison is critical for evolutionary studies. A simple way to compare two sequences is by way of a **dotplot** as shown in **Figure 12.4**. Any diagonal pattern observed in such a plot reflects an identity between the two sequences. The **alignment** between the two sequences that is most likely from an evolutionary perspective may be found by investigating all possible paths through the graph, where different scores are applied for matches, mismatches, and gaps (see Figure 12.4). Clever algorithms were developed by researchers in the field of bioinformatics to solve the problem of efficient sequence alignment. For instance, a method was devised that allows the identification of the optimal path in the graph, without having to explore every possible path.

The example in Figure 12.4 is concerned with nucleotide sequences where the mismatch scores are identical for every possible nucleotide substitution. Thus, we assume that all nucleotide substitutions occur with the same frequency. The scoring of protein sequence alignments is different as we take advantage of the fact that not all amino acid changes take place with the same frequency. These frequencies may be observed in alignments of protein sequences—alignments that are reliable because they are based on similar sequences with few mismatches. For instance, a change from glutamic acid to aspartic acid is much more common than a change from glutamic acid to cysteine. These differences in the frequencies of amino acid substitutions may be exploited in the scoring of alignments. One of the first **substitution matrices** was constructed by Margaret Dayhoff. Later, other matrices were more widely used, such as BLOSUM62 (**Figure 12.5**). An example of protein sequence alignment is shown in **Figure 12.6**, where human α- and β-globins have been computationally aligned.

If two DNA or protein sequences are related by evolution, they are said to be **homologous**. Homology may be revealed by sequence alignments. Whenever two sequences have an extent of similarity that is unlikely to have arisen by chance only, we may conclude that they are homologous,

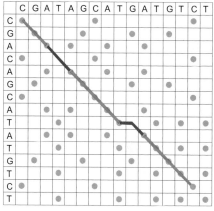

CGATAGCATGATGTCT
*** ***** ******
CGACAGCAT-ATGTCT

+2+2+2 -1+2 -2+2+2+2 -2+2+2+2+2+2+2 = 25

Figure 12.4 Comparing two sequences using a dotplot and pairwise alignment. Two sequences may be compared using a dotplot, a two-dimensional graph where the two sequences compared are on different axes and all instances of identity are indicated with a "dot." Any diagonals observed in this plot reflect identity between the two sequences. Three of these are here shown in red. An alignment may intuitively be inferred from a dotplot, but in more general terms, finding the "best" alignment depends on the identification of a best-scoring path through the two-dimensional graph. For the scoring, we apply a scheme with different scores for matches, mismatches, and gaps—in this example +2, −1, and −2, respectively. A mismatch represents a substitution mutation, and a gap reflects a mutational event where one or more nucleotides have been inserted or deleted during evolution. The overall score for the path is obtained by summing the individual scores (bottom). Every path in the graph corresponds to an alignment. There are numerous possible paths, but the alignment with the highest score is considered the optimal alignment. In this case, it is the path as indicated with red and black lines in the graph (three regions of identity are combined). The corresponding alignment and scoring are shown at the bottom.

Figure 12.5 An amino acid substitution matrix is used in the scoring of protein sequence alignments. The matrix BLOSUM62 is depicted. It shows for each pair of amino acids a value reflecting the probability of that particular amino acid substitution. The matrix is exploited in protein alignment algorithms. An example alignment of two sequences with five amino acids (shown below the matrix) shows a scoring based on the BLOSUM62 matrix.

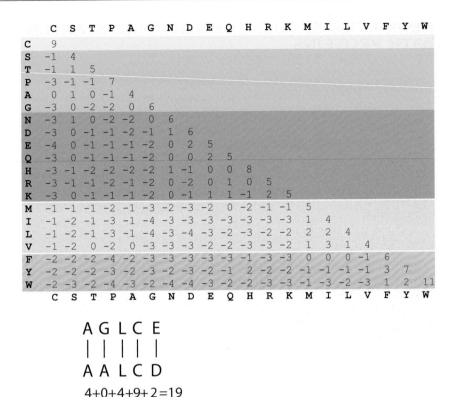

	C	S	T	P	A	G	N	D	E	Q	H	R	K	M	I	L	V	F	Y	W
C	9																			
S	-1	4																		
T	-1	1	5																	
P	-3	-1	-1	7																
A	0	1	0	-1	4															
G	-3	0	-2	-2	0	6														
N	-3	1	0	-2	-2	0	6													
D	-3	0	-1	-1	-2	-1	1	6												
E	-4	0	-1	-1	-1	-2	0	2	5											
Q	-3	0	-1	-1	-1	-2	0	0	2	5										
H	-3	-1	-2	-2	-2	-2	1	-1	0	0	8									
R	-3	-1	-1	-2	-1	-2	0	-2	0	1	0	5								
K	-3	0	-1	-1	-1	-2	0	-1	1	1	-1	2	5							
M	-1	-1	-1	-2	-1	-3	-2	-3	-2	0	-2	-1	-1	5						
I	-1	-2	-1	-3	-1	-4	-3	-3	-3	-3	-3	-3	-3	1	4					
L	-1	-2	-1	-3	-1	-4	-3	-4	-3	-2	-3	-2	-2	2	2	4				
V	-1	-2	0	-2	0	-3	-3	-3	-2	-2	-3	-3	-2	1	3	1	4			
F	-2	-2	-2	-4	-2	-3	-3	-3	-3	-3	-1	-3	-3	0	0	0	-1	6		
Y	-2	-2	-2	-3	-2	-3	-2	-3	-2	-1	2	-2	-1	-1	-1	-1	-3	3	7	
W	-2	-3	-2	-4	-3	-2	-4	-4	-3	-2	-2	-3	-3	-1	-3	-2	-3	1	2	11
	C	S	T	P	A	G	N	D	E	Q	H	R	K	M	I	L	V	F	Y	W

```
A G L C E
| | | | |
A A L C D
4+0+4+9+2=19
```

which is to say that they have a common evolutionary origin. There are two categories of homology. **Orthology** arises as a result of a speciation event. Orthologous proteins carry out the same function in different species. The second category of homology is **paralogy,** which comes about by gene duplication events. Paralogous proteins have similar but not identical functions within an organism. The α- and β-globins (see Figure 12.6) are examples of paralogs—they arose at some point in evolution from an ancestral globin gene by gene duplication and have related but slightly different structures and functions.

As we see in this chapter, alignments are not only useful to identify homologs—there are many other biological applications. Thus, they allow prediction of biological function, the analysis of protein families, comparison of genomes, and investigations of evolution. On the more technical side, alignments are important for genome assembly in shotgun sequencing and in the alignment of DNA sequence reads to a reference genome.

MULTIPLE ALIGNMENTS

In the early days of protein sequence bioinformatics in the 1960s, the phylogenetic analyses required not only proper alignment of two sequences as described above, but alignment of several sequences, a **multiple**

```
α   1 MV-LSPADKTNVKAAWGKVGAHAGEYGAEALERMFLSFPTTKTYFPHF-DLSH-----GSAQVKGHGKKVADALT  68
β   1 MVHLTPEEKSAVTALWGKV--NVDEVGGEALGRLLVVYPWTQRFFESFGDLSTPDAVMGNPKVKAHGKKVLGAFS  73

α  69 NAVAHVDDMPNALSALSDLHAHKLRVDPVNFKLLSHCLLVTLAAHLPAEFTPAVHASLDKFLASVSTVLTSKYR  142
β  74 DGLAHLDNLKGTFATLSELHCDKLHVDPENFRLLGNVLVCVLAHHFGKEFTPPVQAAYQKVVAGVANALAHKYH  147
```

Figure 12.6 Alignment of α- and β-globins. An alignment of human α- and β-globin chains. Coloring of amino acids is based on properties of amino acids and the extent of conservation between the two sequences.

alignment. This kind of alignment was first carried out on a manual basis, but there was an increasing need for computer software for this task. One important computer program was presented in 1988, Clustal. This program and its more recent variants (a current version is named Clustal Omega) have been immensely popular, and up to 2018 were cited about 48,000 times in scientific papers (see Table 12.1). Clustal uses a stepwise protocol, a progressive alignment procedure to achieve the final alignment. Thus, in each step, a single sequence or a subalignment is aligned with a previous sequence or subalignment. Critical for the outcome of this procedure is the order in which the different sequences are aligned. This order is determined by a guide tree computed in a first phase of the Clustal procedure. The guide tree is constructed from a pairwise comparison of all sequences to be aligned.

There are a number of applications to multiple alignments. First, they allow the identification of elements in the sequences that are conserved during evolution. Such elements are often of functional interest. Furthermore, multiple alignments are used in (1) phylogenetic tree construction methods, (2) prediction of protein secondary and three-dimensional structure, and (3) derivation of statistical models of protein families, as further described later. Furthermore, alignments of genomic sequences offer analysis of genome evolution and identification of biologically interesting regions of a genome.

MOLECULAR SEQUENCES ARE COLLECTED IN PUBLIC DATABASES

Whereas protein sequences started to be collected in the 1950s, there were no DNA sequences until the development of gene technology in the late 1970s. In 1977 Frederick Sanger presented his pioneering method of DNA sequencing, and in the 1980s significant amounts of DNA sequence had been collected using this method. Therefore, databases were initiated to collect this information. Thus, GenBank in the United States and the EMBL database in Europe (now European Nucleotide Archive [ENA]) were initiated. These databases are still collecting sequence information and are extensively used by scientists. The amount of data is now quite overwhelming as compared to the early days of sequencing. Thus, the version of Gen-Bank 1982 had 606 sequences containing a total of 680,338 nucleotides, and all of the sequences were printed in a physical book. In comparison, the GenBank version 227 (August 2018) had nearly 1 billion (10^9) sequences and over 3 trillion (3×10^{12}) nucleotides. Printing a book with the current GenBank data is no longer an option.

Protein sequences are also collected in specific databases. One of these is a comprehensive resource named UniProt, the Universal Protein Resource. A subsection of UniProt is Swiss-Prot, a manually annotated (reviewed) protein sequence database with functional information pertaining to each protein. Protein three-dimensional structures as elucidated by x-ray crystallography, NMR, and cryo-electron microscopy are deposited in the RCSB Protein Data Bank.

HUGE COLLECTIONS OF SEQUENCE DATA MAY EFFECTIVELY BE SEARCHED FOR SEQUENCE SIMILARITY

In the 1980s when DNA sequences began to accumulate, there was a need to more effectively search DNA sequence databases for sequence

Alignment between query and database sequence

Figure 12.7 Output of BLAST search. Output from a nucleotide BLAST search where the mRNA of human β-globin (*HBB* gene, blue circle) was searched against the human genome sequence (red circle). Each of the areas in light yellow corresponds to an alignment between the query and a database sequence. Only selected alignments of the output are shown. In each of the alignments, the lines starting "Query" (orange circle) are the input query sequence, and the matching database sequence is in the lines starting with "Sbjct" (green circle). The lines in between indicate identity (|) or nonidentity (space). All of the matching sequences shown here are at chromosome 11 where different genes are located that are related to the *HBB* gene. Thus, not only is the *HBB* gene identified in the search, but also the genes encoding the hemoglobin subunits δ (*HBD*) and ε 1 (*HBE1*).

similarity. The methods developed earlier for pairwise alignments were too slow for this task. The new and faster methods developed were BLAST (Basic Local Alignment Search Tool) and FASTA (acronym based on "fast" and "all," where "all" means both DNA and protein sequences). Both of these methods use a single query sequence (DNA or protein). Alignment between this query sequence and all sequences in the database is attempted, and significant alignments are identified. One reason these methods are faster than the previous pairwise alignment methods is that the first step is an identification of perfect matches of short words between the query and database sequences. These "word hits" correspond to the diagonals of perfect identity in Figure 12.4. From such hits, more extensive alignments are created using the methods for pairwise alignments as previously described. The method with "word hits" is faster, because we are avoiding more extensive exploration of graphs like that in Figure 12.4.

The output from the BLAST and FASTA search is a list of the best matching hits, together with alignments of these hits to the query sequence. BLAST quickly became popular and is also highly ranked in the list of most frequently cited papers (see Table 12.1). An example of output from BLAST is shown in **Figure 12.7**. The results show how genes homologous (paralogous) to *HBB* (*HBD* and *HBE1*, encoding δ- and ε-globin chains, respectively) are identified.

MORE SENSITIVE METHODS TO REVEAL DISTANT EVOLUTIONARY RELATIONSHIPS AMONG PROTEINS

Whereas a method like BLAST has been important in many areas of molecular biology, there was also the discovery in the late 1980s that this type of method is often not sensitive enough to identify the protein relatives with a low degree of sequence similarity. However, there is a solution to this problem. Rather than using a single sequence to query a database, instead a statistical model based on several sequences may be used to achieve better sensitivity. The model is based on an alignment of sequences that we know are evolutionary related based on biochemical work. Such models may be created for DNA as well as protein sequences.

One type of such a model is the position-specific scoring matrix (PSSM). The information and sequence conservation in models like PSSMs may be illustrated with **sequence logos**. The derivation of a sequence logo from a nucleotide sequence multiple alignment is shown in **Figure 12.8**. We used logos in this book to illustrate consensus sequences for transcription factor binding sites (e.g., see Figures 10.2 and 10.9).

The principle of a statistical model based on a multiple alignment is particularly important for proteins. It is implemented in a variety of popular tools—PSI-BLAST, the National Center of Biotechnology Information (NCBI) Conserved Domain Database (CDD), the HMMER software, and Pfam. For instance, PSI-BLAST is a modification of the original BLAST protocol but where a model based on a multiple alignment is used instead of a single sequence to search a database. PSI-BLAST has been extensively used to identify novel members of protein families and to reveal relationships of proteins and protein families that were not previously found.

Figure 12.8 Sequence logos are used to illustrate the extent of sequence conservation in a set of sequences with similar function. The construction of a sequence logo is illustrated with a set of nine nucleotide sequences (A-I) with seven nucleotides in each (1–7). The particular sequences are representatives of the GATA motif recognized by a set of transcription factors (Chapter 10).

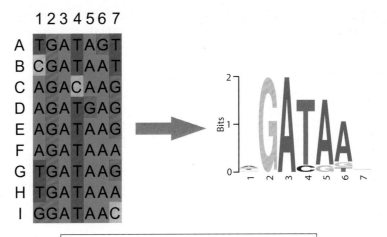

Sequence logo calculations

Starting out from the alignment, the amount of *uncertainty* may be stated as:

$$H_u = -\sum_a f_{u,a} \log_2 f_{u,a}$$

where H_u is the uncertainty at position u and a is one of the four bases in a DNA sequence. $f_{u,a}$ is the frequency of base a in column *u*. *Total information* (total height in the sequence logo) in the position u is represented by the decrease in uncertainty which for DNA is:

$$I_u = \log_2 4 - H_u$$

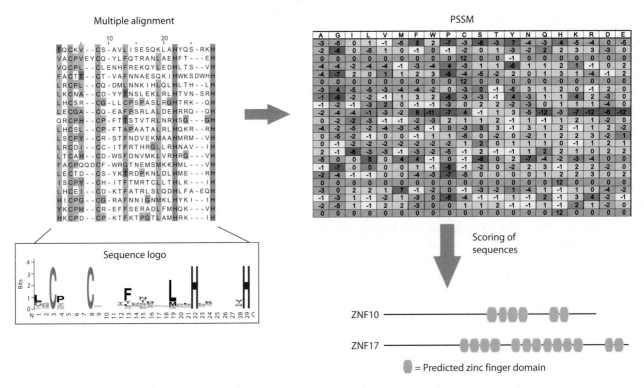

Figure 12.9 A position-specific scoring matrix is used to identify function from amino acid sequence. On top left is shown an alignment of 20 protein sequences known to contain a C2H2 zinc finger motif. Below the multiple alignment is shown the corresponding sequence logo. The multiple alignment is used to create the position-specific scoring matrix (PSSM). The letters on the top are the 20 amino acids, and the rows in the PSSM correspond to the columns (positions) in the multiple alignment. Numbers in the cells reflect the probability of having a certain amino acid in a specific position. Note that the positions with highly conserved residues correspond to large positive numbers. Similar to the scoring with a substitution matrix like BLOSUM62 (Figure 12.5), we may now search any protein sequence for matches to the PSSM and obtain a score. If the score is above a specific threshold value, the matching is considered significant. This is illustrated here with two human zinc finger proteins, ZNF10 and ZNF17, that both are predicted to contain several zinc finger domains.

BIOLOGICAL FUNCTION MAY BE PREDICTED FROM AMINO ACID SEQUENCES

Similar to the NCBI CDD, the Pfam database collects a number of protein families and models of these families. The creation of models (known as profile hidden Markov models or profile HMMs) is based on the software HMMER.

One important application of PSSMs of CDD and the profile HMMs of Pfam is that they may be used to search amino acid sequences for matches. In this way, a previously uncharacterized protein may be examined and tentatively assigned specific structural and functional properties. Important clues as to further experimental studies are thus provided. This is illustrated in **Figure 12.9**, showing how a PSSM may be matched against proteins and predict whether they contain a specific category of **zinc finger** domain (Chapter 10) or not.

Protein domain predictions using CDD and Pfam are often very reliable. Consider, for instance, the structure of the protein c-SRC. The three-dimensional structure has been elucidated experimentally with x-ray crystallography (**Figure 12.10**). It has a kinase domain in addition to SH2 and SH3 domains. All of these domains are accurately predicted by Pfam. This example also illustrates that many proteins are built from more than one domain. These domains are distinct structural units, and they are ordered in a linear manner in the amino acid sequence.

More examples of protein domain predictions are shown in **Figure 12.11**, but now with proteins whose structures and functions are unknown from experimental work. For instance, the human protein encoded by the gene *SBK3* is, like many other human proteins, poorly characterized from a biochemical point of view. However, a prediction using Pfam informs us that the protein has a domain characteristic of protein tyrosine kinases. These enzymes are known to attach a phosphate group to tyrosine of their protein substrates and are typically part of signaling pathways in the cell. From this observation, we may infer that this protein most likely is an important regulatory protein, simply on the basis of its amino acid sequence.

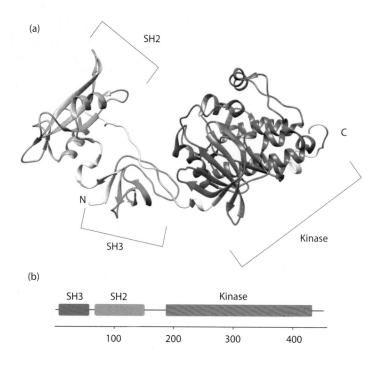

Figure 12.10 Structure of c-SRC protein as compared to computational domain prediction based on amino acid sequence. (a) The three-dimensional structure of the protein c-SRC (also known as proto-oncogene tyrosine-protein kinase Src) is shown. The amino-terminal (N) and carboxylterminal (C) ends of the proteins are indicated. The protein includes an SH2 domain, an SH3 domain, and a tyrosine kinase domain. (b) Colored rectangles represent the domains predicted from the amino acid sequence by Pfam and the HMMER search tool. The three domains of c-SRC are all accurately predicted. (Based on Cowan-Jacob SW et al. 2005. *Structure* 13:861–871. PDB ID: 1Y57.)

Figure 12.11 Protein may be assigned a possible function based on amino acid sequence. Protein domain structures are shown that were predicted using the database Pfam and the HMMER search tool. The four sequences (SBK3, ADCK2, CA109, and ZN814) represent proteins that are poorly character-ized from a biochemical point of view and where domain structure prediction gives clues as to their function.

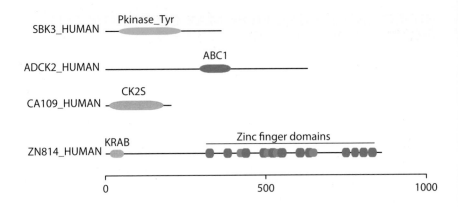

In addition to methods that make use of CDD and Pfam, there are other computational methods that inform us about protein structure and function. Examples are TMHMM and SignalP—methods that are able to accurately predict transmembrane domain helices and N-terminal signal peptides, respectively. Both of these methods allow us to conclude the intracellular location of a protein. For instance, TMHMM predicts whether a protein is located in a membrane or not. About 25% of all human proteins have such a location, and knowing about membrane localization aids in assigning function to a protein.

A NETWORK OF GENE AND PROTEIN FUNCTIONS: GENE ONTOLOGY

Protein sequence databases such as UniProt have a lot of information about proteins, and databases such as Pfam may be used to infer function of proteins. However, for a number of analyses, we need to know more. For instance, we also would like to have information describing how proteins are related in terms of function, such as what proteins are involved in a certain biological process. As an example, consider an experiment to monitor proteins produced in a tumor sample and where these proteins are compared to proteins in a sample of normal healthy cells. In this experiment, we want to identify similarities and differences. Is there, for instance, a set of proteins related to a specific biological process that are more abundant in one category of samples? To answer this question, we need to annotate as many human genes as possible with accurate terms describing biological processes. However, genes and proteins as such are rarely assigned such informative terms. For instance, many proteins have anonymous names, saying nothing about their biological function. To overcome this problem, a project was initiated in 1998 to assign **gene ontology** (**GO**) terms to genes and proteins. The ontology is a unique vocabulary where the terms are arranged in a graph. The root nodes of the graph are the three main GO categories "molecular function," "biological process," and "cellular component." An example of terms parental to the term "protein kinase activity" is shown in **Figure 12.12**. (In this graph the root term "cellular component" is missing, because this term is not linked to "protein kinase activity.")

In order for GO terms to be useful, as many entries as possible in protein and DNA sequence databases need to be assigned one or more GO terms. The sequences of many databases such as Uni-Prot have now been associated with such terms through the work of bioinformaticians.

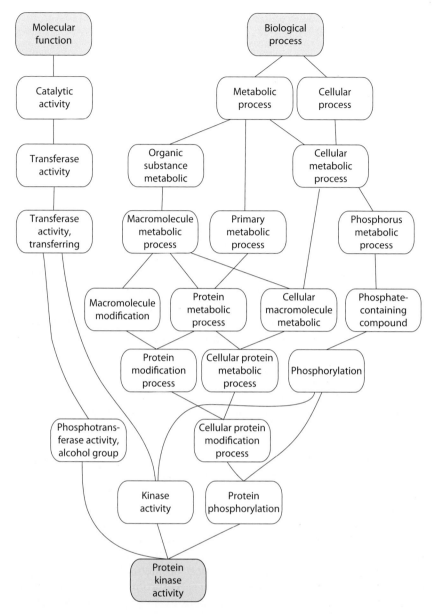

Figure 12.12 Structure and terms in a gene ontology graph. The structure of a gene ontology graph is such that a node may have more than one parent. For instance, the term "protein kinase activity" shown here (light green background at bottom) has three different parents, "phosphotransferase activity, alcohol group," "protein phosphorylation," and "kinase activity." These are under the root terms "molecular function" and "biological process" (shaded gray). The graph is based on a display from the GO browser QuickGO at www.ebi.ac.uk/QuickGO/.

SEQUENCES TO BUILD TREES: PHYLOGENY

We return to questions of molecular evolution. A **phylogenetic tree** is a diagram showing inferred evolutionary relationships between species or between other entities (like in Figure 12.2). **Molecular phylogeny** (or **molecular phylogenetics**) uses DNA, RNA, or protein sequence information to deduce the tree. Remember that already in the 1960s phylogenetic trees were constructed on the basis of amino acid sequences. The methods of phylogeny have since been developed to be more accurate, and there are numerous methods available today. The sequences used for phylogeny

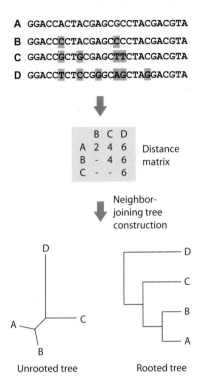

Figure 12.13 From sequences to a phylogenetic tree. An alignment of sequences labeled A–D is shown on top. Sequence differences are highlighted in orange background. By considering every pair of sequences and counting the number of differences, we end up with the distance matrix as shown. The distance matrix may finally be used to construct a tree, for instance using the method of neighbor joining, as shown here. To the left is shown an unrooted tree, which is the actual tree inferred from the neighbor-joining method. To the right is shown a rooted version of the same tree. In this case, the root has been placed between the node D and the closest internal node, a common choice as this is the longest branch.

are either protein or nucleic acid (DNA/RNA) sequences. What type of sequence is used depends on the evolutionary distance being considered. For instance, in the case of human populations, we are considering evolution that has taken place during the last 300,000 years or so. For this situation, we need a frequently mutating sequence, such as **mitochondrial** DNA. A region approximately 1 kb in size in mitochondrial DNA known as the D-loop is particularly variable. Conversely, for the analysis of distant evolutionary relationships, protein sequences or ribosomal RNA sequences are used as they mutate relatively slowly. For instance, to create a tree with species from all domains of life (eukaryotes, bacteria, and archaea), a multiple alignment of protein sequences is often used. For this method to be statistically powerful, each of the sequences used in the alignment has been obtained by merging several proteins that all are universally present in all domains of life.

One simple and popular method of tree construction is **neighbor joining**, presented in the late 1980s. It is an example of a method based on a **distance matrix** (**Figure 12.13**). It has the advantage that it is computationally fast and allows the analysis of thousands of sequences.

Whereas neighbor joining is computationally fast, there are other methods of tree construction that are more accurate. These methods do not rely on a distance matrix but rather analyze information in all columns (positions) of the multiple alignment. They exploit the close relationship between a multiple alignment and evolutionary events (**Figure 12.14**) and use a mathematical model of evolution to take into account parameters such as the relative rates of specific nucleotide or amino acid substitutions.

A wide variety of different biological problems related to evolution may be addressed and visualized by tree construction. A few examples are shown here. First, mitochondrial sequences from different human individuals may be used to examine the relationship between different populations and to infer the history of human migration. A widely accepted model of human evolution is the "Out of Africa" hypothesis, also referred to as "Recent African origin," which states that anatomically modern humans originated in Africa and migrated out of Africa to other continents to replace other human species (**Figure 12.15**). The tree shown in **Figure 12.16** is a neighbor-joining tree based on mitochondrial genome sequences, and it provides support to the idea that the origin of modern humans was in Africa.

The tree in Figure 12.16 shows the chimpanzee as an outgroup. What about the relationship between primates and human? Is the chimpanzee our closest relative? Again, this question may be addressed by molecular

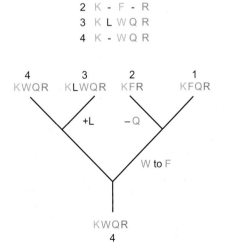

Figure 12.14 Relationship of multiple alignment and phylogenetic tree. On top is shown an alignment of four different protein sequences. In the graph below is shown evolutionary paths consistent with the alignment. The root of the graph is the protein 4. Branches in the graph show evolutionary events, such as the substitution of W with F that leads to protein 1.

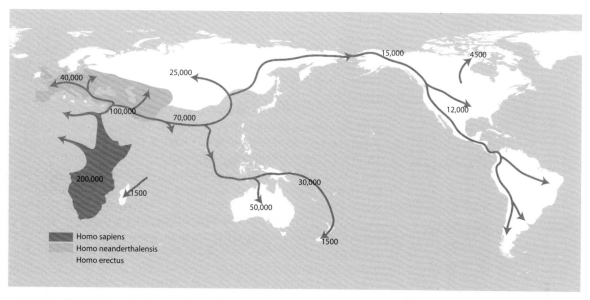

Figure 12.15 Map of early human migrations. Anatomically modern humans migrated out of Africa around 200,000 years ago. They populated different parts of the world and competed out previously existing humans in those parts. Numbers of years ago for different migrations are indicated.

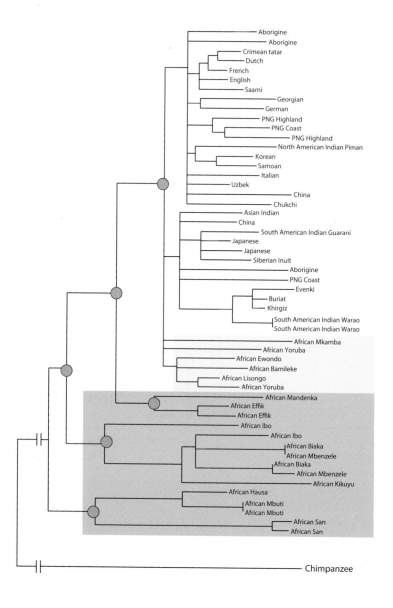

Figure 12.16 The origin of modern humans was in Africa. A neighbor-joining phylogenetic tree based on mitochondrial genome sequences. Nodes with a strong statistical support are shown with orange circles. Sub-Saharan African sequences are shown with a colored background—in one group (green) the most deeply branching sequences are present. The position of the African branches closest to the root of the tree gives strong support for a human origin in Africa. (Based on the results in Ingman M et al. 2000. *Nature* 408(6813):708–713).

phylogeny (**Figure 12.17**). As with our previous example of human individuals, mitochondrial sequences are suitable as a starting point for this analysis. The resulting tree shows that most closely related to humans are chimpanzees and bonobos. These two monkeys are closely related from an evolutionary perspective and diverged from each other around 2 million years ago.

Phylogeny may also be used to examine a particular family of homologous sequences, like proteins. With such an analysis, we may better understand the evolution of a particular kind of gene or protein. The example in **Figure 12.18** is based on an alignment of about 900 animal globin protein sequences, including all human blood cell globins and the related muscle globin myoglobin, but also related sequences from many other vertebrates. It is not a tree to address the question of how biological species are related by evolution. Rather, it is an example of how protein sequences may be classified and how we can trace the origin of specific paralogs. In this case, there are three major groups containing myoglobin (an oxygen-binding protein present in muscle tissue), the β-globin (*HBB* gene) and the α-globin (*HBA* gene), respectively. Within all three groups, all vertebrate species are represented—therefore, the tree shows how these three paralogous proteins diverged early in vertebrate evolution. Paralogs such as the γ-, δ-, and ε-chains of hemoglobin evolved at a later stage during the development of mammals.

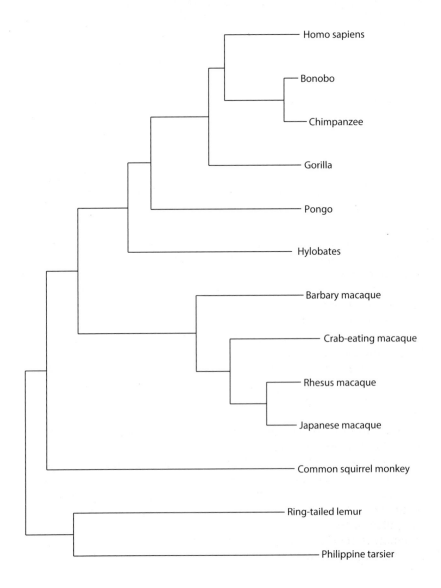

Figure 12.17 Evolutionary relationship of primates. The phylogenetic tree is based on sequences from the D-loop region of the mitochondrial genome. The method of tree construction is neighbor joining. The tree indicates that our closest relatives are chimpanzee and bonobo, followed by gorilla.

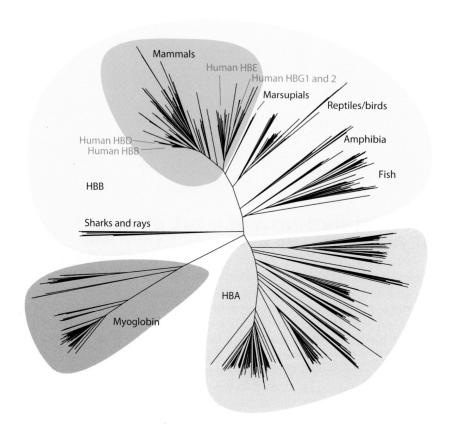

Figure 12.18 Evolutionary relationship of globin proteins. A total of 909 globin protein sequences were used to create a tree with the neighbor-joining method. The proteins included are all vertebrate proteins belonging to the groups of β-globin (*HBB*), α-globin (*HBA*), and myoglobin. The homologs to these three proteins form three distinct groups. Other paralogs, such as the δ- (*HBD*), ε- (*HBE*), and γ-chains (*HBG1* and *HBG2*) of hemoglobin, are found within the group of mammalian *HBB*, showing that these proteins evolved at a later stage during the development of mammals.

Therefore, with this type of analysis, we may conclude on the evolution of a distinct group of proteins, and we can do this only if we already know, by some other method, how the different species are related by evolution.

HUMAN GENOME COMPUTING: PUTTING THE HUMAN GENOME TOGETHER FROM SHORTER SEQUENCES

Reconstruction and analysis of the human genome offer a number of different computational challenges. Most genomes are currently sequenced using a shotgun strategy (Chapter 4). Individual shorter DNA sequence reads are merged into longer sequences using a computational process of **sequence assembly**. In the case of the human genome, more than 100 million sequence reads are required to reconstruct a significant part of the genome. Assembly is seldom straightforward. One significant problem is the presence of repetitive sequences.

The object of the assembly process is to join shorter sequencing reads, thereby creating **contigs**. The joining is based on overlaps between reads, and for this reason one important step of the assembly process is the identification of overlaps. The first assembly programs followed a simple strategy in which the order of merging the reads is based on overlap size. An example is shown in **Figure 12.19**.

One disadvantage of the simple approach in Figure 12.19 is that the assembler can easily be confused by complex repeats, leading to errors in the assembly. In the light of these problems, a better approach is one based on **graphs**. In the example in **Figure 12.20**, a graph (panel [a]) has been computed based on overlaps between the reads. Based on this graph, the assembly is inferred. The graph in panel (b) shows for comparison the correct genome structure.

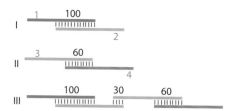

Figure 12.19 Greedy assembly of a genome sequence. The order of joining is based on overlap size. The assembler joins, in order, reads 1 and 2 (overlap = 100 nucleotides), then reads 3 and 4 (overlap = 60 nucleotides), then reads 2 and 3 (overlap = 30 nucleotides), and the final result is a single contig from the four reads provided in the input.

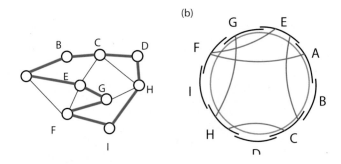

Figure 12.20 Assembly of a genome sequence using a graph. A simple example of a graph method for assembly of a circular genome, for instance a bacterial or a mitochondrial DNA, is depicted. There are nine sequencing reads, labeled A–I. The graph in (a) is computed based on overlaps between the reads. The nodes represent the reads, and a line connects two reads if they overlap. The graph is used to assemble the genome—the green path (in which each node is visited once) through the graph corresponds to the assembly. The correct assembly is shown in (b) with the circular genome (continuous blue line) and sequencing reads A–I (black lines) correctly aligned to the genome. The assembly as inferred from (a) is consistent with the correct assembly in (b). The non-green lines in (a) represent false overlaps resulting from repeats, as illustrated with the red lines in (b).

RECONSTRUCTING A GENOME BY ALIGNMENT OF READS TO THE REFERENCE GENOME

The reference version of the human genome, which is a mosaic from a few individuals, is derived from an assembly process as previously described. But in addition, more than one million human individuals have now been subject to sequencing. The raw data from the DNA sequencing machines are short sequence reads. Using sequence assembly to reconstruct a genome from this data is a demanding process in terms of computational power and complexity. Instead, as a kind of shortcut, the sequence reads may be aligned to the reference genome. An advantage with this method is that the alignment of reads is fast. At the same time, one of the problems with the read alignment procedure is that it mainly identifies single nucleotide variants (SNVs) and short indels, but it is not adequate when it comes to structural variation. Nevertheless, the method has been widely used.

For the task of aligning several million reads to a reference genome, the traditional alignment software (like BLAST, for instance) is inefficient. Instead, specific methods for read alignment were developed. Examples are the aligners named BWA ("Burrows Wheeler Alignment") and Bowtie that were presented in 2009 and have since been used extensively (see Table 12.1). To produce a fast yet accurate alignment, the methods use clever computational tricks, including a data compression algorithm known as the Burrows Wheeler transform.

Once reads have been aligned to the reference genome, the next step is to identify any differences, like SNVs and short insertions or deletions. Typically, a genomic region is covered by numerous sequencing reads. With a tool like the Integrative Genomics Viewer (IGV) developed at the Broad Institute (of MIT and Harvard, Massachusetts), reads aligned to the reference genome may be visualized (**Figure 12.21a**). Any differences as to the reference genome in an individual can be revealed by manual inspection in IGV. An example is shown in **Figure 12.21b**, presenting evidence of a mutation in the FOXP2 gene, associated with a language disorder (Chapter 6).

Manual inspection in a genomic browser is not suitable for a genome-wide identification of variants. Instead, specific computational tools have been developed where the information in the different reads is analyzed to identify any differences with respect to the reference version of the genome. Thereby, the **genotype** of the individual in specific genomic positions is inferred. The procedure of identifying variants is commonly known as **variant calling**.

Not only genomic DNA sequences are mapped to the reference genome. In RNA-Seq experiments, messenger RNAs (mRNAs) are sequenced using a protocol where RNA is reverse transcribed into DNA, followed by

(a)

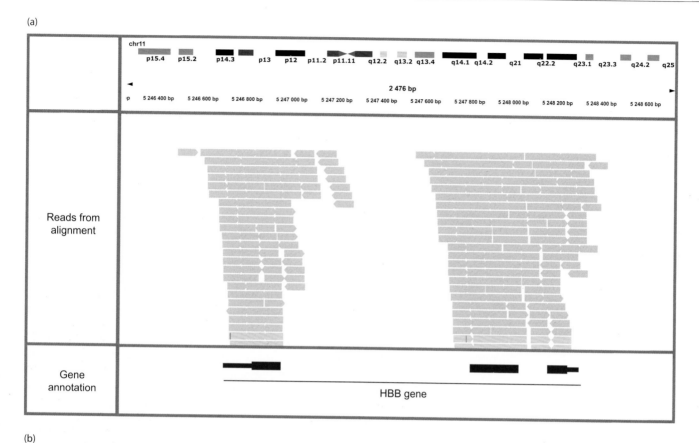

(b)

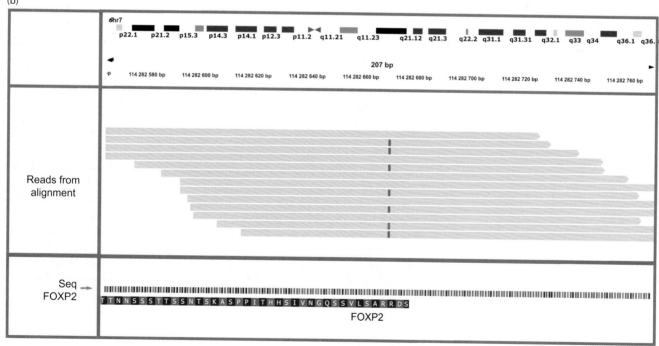

Figure 12.21 Alignment of sequencing reads to the human reference genome. The figure is based on visualization of reads with the Integrative Genomics Viewer (IGV). In both panels the chromosomal localization is shown on top. (a) Reads are shown that are aligned to the β-globin gene *HBB*. Each read is represented by a gray rectangle. Each rectangle is pointed to indicate the strandedness of the read. The exon-intron structure of the *HBB* gene is shown in the lower part under "Gene annotation," where exons are shown with coding sequences (thicker blue rectangles) and UTRs (thinner blue rectangles). Reads were obtained in a project to sequence only the exons, and for this reason the reads primarily match the exonic regions of the genome. (b) A patient with a mutation in the FOXP2 gene was sequenced, and 13 reads have been aligned to an exon of this gene. The mutation is a C > T substitution changing an arginine codon into a stop codon. Among the reads there are seven with a T (red vertical lines), suggesting that the patient is heterozygous for the mutation (genotype is CT). In the lower part is shown the sequence of the reference genome using a color code ("Seq") and the amino acid sequence of the FOXP2 exon.

DNA sequencing. To align such reads that correspond to fully processed mRNA to the human reference genome, we must take into account the exon-intron structure of genes. There are alignment algorithms, such as those of the computer programs TopHat and Hisat, that align reads to the genome while considering splicing.

INFERRING FUNCTIONS OF INDIVIDUAL GENOMIC VARIANTS

The sequence of an individual genome may be inferred from read alignment and variant calling as previously described. Having access to all identified variants, the next step is to attempt to understand their functional consequences. For this interpretation, we need as good **annotation** as possible of the genome. This is to say that we need to know for as many positions as possible in the genome their functional impact. Is the variant in a protein gene? In such a case, is there a predicted amino acid change, and what is the biological effect of that change? Is the variant in a noncoding RNA gene or in an intergenic location? Could it be significant in terms of transcriptional control?

One significant project to assign function to human genes and gene elements is GENCODE. We previously referred to the ENCODE project, which was launched by the National Human Genome Research Institute (NHGRI) in 2003 in an effort to experimentally identify all functional elements in the human genome (Chapter 10). GENCODE was established as part of ENCODE a few years later with the aim to annotate gene features in the human genome.

ANNOVAR is an example of a computational tool that assigns gene localization and, if available, functions to variants. ANNOVAR makes use of numerous database sources, and the annotation of variants may include (1) gene definition systems such as GENCODE, (2) regions conserved between species, (3) transcription factor binding sites, (4) epigenetic data like histone modifications and DNA methylation, (5) dbSNP data, and (6) sites identified by GWASs. At the same time, the human genome is far from perfect in terms of annotation, and the assignment of function to different parts of the genome is still very much a work in progress.

In all cases where novel variants (SNVs) are identified, for instance in the context of inherited disorders and cancer, we want to know the phenotypic outcome of these mutations. Is a certain mutation likely to be associated with a disease, or is it a neutral mutation without any effect? In the instances when a mutation is in a coding sequence, it is of interest to know whether the mutation changes the amino acid sequence or not. In cases in which it does change the amino acid sequence, can the functional consequences be predicted? Methods such as PolyPhen and SIFT predict whether a protein coding sequence nonsynonymous mutation is likely to be benign or deleterious. These methods make use of known structural properties of proteins, annotation from Swiss-Prot, and examine the evolutionary conservation of the particular site in the protein being changed by the mutation.

There are several tools for exploring the annotated human genome. In addition to the UCSC browser that we previously encountered, there is, for instance, the Ensembl browser. Ensembl, a joint project between the European Bioinformatics Institute and the Wellcome Trust Sanger Institute, also aims at the development of a pipeline of computational tools making possible automated annotation of a variety of animal genomes.

COMPARISON TO OTHER SPECIES IMPROVES OUR UNDERSTANDING OF THE HUMAN GENOME

In the field of comparative genomics, genomes from different organisms are being compared. If genomes are to be compared at the level of nucleotide sequences, they need to be sufficiently similar to allow alignment procedures to be meaningful. Methods have been developed that produce extensive multiple alignments. For instance, with such methods, the human genome has been compared to other vertebrate genomes.

A number of questions may be addressed using genome alignments. First, they may be applied in phylogenetic studies. Furthermore, conserved portions may point to regions of the genome that are particularly interesting from a functional point of view. Examples of such regions are exons, noncoding RNAs, and regions that control transcriptional activity (**Figure 12.22**). Yet another application of comparative genomics is in studies of human individual genomic variation. Thus, chimpanzee and other primates' genomes may be used to infer the ancestral variant for a specific human position. Comparative genomics may also be employed to identify human genes that are unusual in terms of recent primate evolution. For instance, *human accelerated regions* (HARs) are conserved throughout vertebrate evolution but are markedly different in humans. Two ncRNAs, HAR1F and HAR1R, expressed in the brain during development, were discovered because they contain such a region characterized by many nucleotide changes in humans as compared to other primates.

We are still far from a complete understanding of the human genome. For instance, we need to identify the complete structures of all genes and

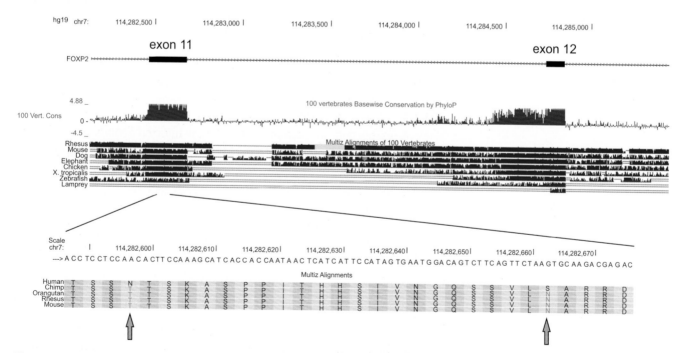

Figure 12.22 Comparing the human genome to those of other animals. A region of the human FOXP2 gene is shown, based on a view with the UCSC browser. A track "100 Vert. Cons" shows the extent of sequence conservation between 100 different vertebrates, and the similarity to a number of individual animals is shown below that track. The extent of conservation overlaps with the exons and with a region upstream of exon 12. At the bottom (a track in the UCSC browser named "Multiz Alignments") is a region of exon 11, showing amino acid sequences in five different animals, including human. The arrows highlight two positions in the FOXP2 protein where the human protein is different from all the other animals.

their transcript variants and identify under what circumstances they are expressed. In addition, as many as possible functional sites in the genome need to be identified, including regulatory signals. The first human genome draft was presented already in the early 2000s, but it will still take a lot of experimental studies and extensive computing to more fully understand the genome.

SUMMARY

- The three major domains of life are bacteria, archaea, and eukaryotes.
- The evolution of species on our planet started more than 4 billion years ago. A tree may be constructed on the basis of molecular sequences to show how all species are related by evolution.
- The sequence similarity of specific proteins that are found in all three domains of life is strong evidence of a common evolutionary origin of all species.
- The content of the human genome may only be understood in the light of evolution.
- The history of sequence bioinformatics started with proteins in the 1960s.
- Sequences are typically compared using a procedure of *alignment*.
- In *multiple alignments*, two or more protein or nucleic acid sequences are aligned.
- GenBank, ENA, and UniProt are three important publically available sequence archives.
- *BLAST* is a widely used computational tool to search sequence databases for sequence similarity.
- A set of evolutionary-related sequences may be used to create a statistical model (a PSSM of profile HMM). This model may be used to find additional members of the sequence family, or it may be used to screen any number of sequences for matches to the model.
- TMHMM and SignalP are methods that predict transmembrane domain helices and N-terminal signal peptides, respectively.
- *Gene ontology* (GO) is a unique vocabulary where terms are arranged in a graph. GO aids in the annotation of the human genome and is useful during analysis of gene expression data.
- A *phylogenetic tree* is a diagram to show inferred evolutionary relationships between species or other objects. Methods of tree construction include *neighbor joining*, which belongs to the category of *distance matrix* methods.
- An important step in the reconstruction of the human genome from individual sequence reads is computational *assembly*. A typical procedure of assembly includes the construction of a graph based on read overlaps.
- Once a reference version of the human genome has been obtained, the sequence of a human individual may be inferred from alignment of reads to the reference genome.
- An essential problem when analyzing the genome of an individual is the *assignment of function to genomic variants*. For this reason, we are in need of as accurate as possible annotation of the human genome at the level of individual nucleotide positions.
- A comparison of the human genome sequence to that of other animals may be used (1) in phylogenetic studies, (2) to identify regions of biological interest, and (3) to identify genes that are unusual in terms of recent primate evolution.

QUESTIONS

1. How can molecular sequences be used to demonstrate the basic principle of evolution that all species are related and have a common origin?
2. Explain the concepts of *homology*, *orthology*, and *paralogy*.
3. What is a *dotplot*?
4. Name some important applications of *multiple sequence alignments*.
5. What is the output of the computer program *BLAST*? For what purposes is the program used?
6. What is the purpose of databases like CDD and Pfam?
7. How many human proteins are located in membranes?
8. Why is it important to assign *gene ontology* terms to genes and proteins?
9. How are multiple alignments used in *distance matrix* methods to create phylogenetic trees?
10. What are the challenges of *sequence assembly*?
11. A human genome sequence may be derived either from (1) sequence assembly or by (2) alignment of reads against the reference genome. Discuss the advantages and disadvantages of these two methods.
12. What is meant by *variant calling*?
13. Think about what we learned so far in this book, and make a list of the various kinds of functional elements we would like to see properly annotated in the human genome.
14. Why is it of interest to compare the human genome to those of other related species?
15. What is a *human accelerated region*?

URLs

NCBI GenBank (www.ncbi.nlm.nih.gov/genbank/)

European Nucleotide Archive (ENA) (www.ebi.ac.uk/ena)

UniProt, the Universal Protein Resource (www.uniprot.org)

RCSB Protein Data Bank (www.rcsb.org/)

HMMER (http://hmmer.org)

Transcription factor database, JASPAR (http://jaspar.genereg.net/)

Pfam database (http:// pfam.xfam.org)

ANNOVAR (http://annovar.openbioinformatics.org)

UCSC browser (http://genome.ucsc.edu/)

QuickGO (www.ebi.ac.uk/QuickGO/)

FURTHER READING

Quote on evolution
Dobzhansky T. 1964. Biology, molecular and organismic. *Am Zool* 4:443–452.

Early days of protein sequences
Dayhoff MO. 1965. *Atlas of protein sequence and structure*. National Biomedical Research Foundation, Silver Spring, MD.
Sanger F, Tuppy H. 1951. The amino-acid sequence in the phenylalanyl chain of insulin. 2. The investigation of peptides from enzymic hydrolysates. *Biochem J* 49(4):481–490.

Sequence alignment algorithms
Henikoff S, Henikoff JG. 1992. Amino acid substitution matrices from protein blocks. *Proc Natl Acad Sci USA* 89(22):10915–10919.

Needleman SB, Wunsch CD. 1970. A general method applicable to the search for similarities in the amino acid sequence of two proteins. *J Mol Biol* 48(3):443–453.
Smith TF, Waterman MS. 1981. Identification of common molecular subsequences. *J Mol Biol* 147(1):195–197.

Multiple alignment software
Sievers F, Wilm A, Dineen D et al. 2011. Fast, scalable generation of high-quality protein multiple sequence alignments using Clustal Omega. *Mol Syst Biol* 7:539.
Thompson JD, Higgins DG, Gibson TJ. 1994. CLUSTAL W: Improving the sensitivity of progressive multiple sequence alignment through sequence weighting, position-specific gap penalties and weight matrix choice. *Nucleic Acids Res* 22(22):4673–4680.

Sequence archives

Benson DA, Cavanaugh M, Clark K et al. 2017. GenBank. *Nucleic Acids Res* 45(1):37–42.

Pundir S, Martin MJ, O'Donovan C. 2017. UniProt protein knowledge-base. *Methods Mol Biol* 1558:41–55.

Toribio AL, Alako B, Amid C et al. 2017. European nucleotide Archive in 2016. *Nucleic Acids Res* 45(D1):D32–D36.

Protein data bank

Rose PW, Prlic A, Altunkaya A et al. 2017. The RCSB protein data bank: Integrative view of protein, gene and 3D structural information. *Nucleic Acids Res* 45(D1):D271–D281.

BLAST

Altschul SF, Gish W, Miller W et al. 1990. Basic local alignment search tool. *J Mol Biol* 215(3):403–410.

Altschul SF, Madden TL, Schaffer AA et al. 1997. Gapped BLAST and PSI-BLAST: A new generation of protein database search programs. *Nucleic Acids Res* 25(17):3389–3402.

PSSMs and profile HMMs

Durbin R, Eddy SR, Krogh A, Mitchison G. 2007. *Biological sequence analysis: Probabilistic models of proteins and nucleic acids.* Cambridge University Press, Cambridge, UK.

Finn RD, Coggill P, Eberhardt RY et al. 2016. The Pfam protein families database: Towards a more sustainable future. *Nucleic Acids Res* 44(1):279–285.

Rose PW, Prlic A, Altunkaya A et al. 2017. The RCSB protein data bank: Integrative view of protein, gene and 3D structural information. *Nucleic Acids Res* 45(D1):D271–D281.

Sequence logos

Crooks GE, Hon G, Chandonia JM, Brenner SE. 2004. WebLogo: A sequence logo generator. *Genome Res* 14(6):1188–1190.

Schneider TD, Stephens RM. 1990. Sequence logos: A new way to display consensus sequences. *Nucleic Acids Res* 18(20):6097–6100.

JASPAR

Mathelier A, Zhao X, Zhang AW et al. 2014. JASPAR 2014: An extensively expanded and updated open-access database of transcription factor binding profiles. *Nucleic Acids Res* 42(Database issue):D142–D147.

InterPro

Finn RD, Attwood TK, Babbitt PC et al. 2017. InterPro in 2017-beyond protein family and domain annotations. *Nucleic Acids Res* 45(D1):D190–D199.

TMHMM

Krogh A, Larsson B, von Heijne G, Sonnhammer EL. 2001. Predicting transmembrane protein topology with a hidden Markov model: Application to complete genomes. *J Mol Biol* 305(3):567–580.

SignalP

Nielsen H, Engelbrecht J, Brunak S, von Heijne G. 1997. Identification of prokaryotic and eukaryotic signal peptides and prediction of their cleavage sites. *Protein Eng* 10(1):1–6.

Gene ontology

Ashburner M, Ball CA, Blake JA et al. 2000. Gene ontology: Tool for the unification of biology. The Gene Ontology Consortium. *Nat Genet* 25(1):25–29.

Gene Ontology Consortium. 2015. Gene Ontology Consortium: Going forward. *Nucleic Acids Res* 43(D1):D1049–D1056.

Neighbor joining

Saitou N, Nei M. 1987. The neighbor-joining method: A new method for reconstructing phylogenetic trees. *Mol Biol Evol* 4(4):406–425.

Phylogeny textbooks

Lemey P, Salemi M, Vandamme A-M. 2014. *The phylogenetic handbook: A practical approach to phylogenetic analysis and hypothesis testing.* Cambridge University Press, Cambridge, UK.

Salemi M, Vandamme A-M. 2007. *The phylogenetic handbook: A practical approach to DNA and protein phylogeny.* Cambridge University Press, Cambridge, UK.

Phylogeny of human populations

Ingman M, Kaessmann H, Paabo S, Gyllensten U. 2000. Mitochondrial genome variation and the origin of modern humans. *Nature* 408(6813):708–713.

Genome assembly review

Simpson JT, Pop M. 2015. The theory and practice of genome sequence assembly. *Annu Rev Genomics Hum Genet* 16:153–172.

Read alignment with the Burrows-Wheeler transform

Langmead B, Trapnell C, Pop M, Salzberg SL. 2009. Ultrafast and memory-efficient alignment of short DNA sequences to the human genome. *Genome Biol* 10(3):R25.

Li H, Durbin R. 2009. Fast and accurate short read alignment with Burrows-Wheeler transform. *Bioinformatics* 25(14):1754–1760.

Alignment of reads corresponding to mRNA to the reference genome

Kim D, Langmead B, Salzberg SL. 2015. HISAT: A fast spliced aligner with low memory requirements. *Nat Methods* 12(4):357–360.

Trapnell C, Pachter L, Salzberg SL. 2009. TopHat: Discovering splice junctions with RNA-Seq. *Bioinformatics* 25(9):1105–1111.

Integrative genomics viewer (IGV)

Robinson JT, Thorvaldsdóttir H, Winckler W et al. 2011. Integrative genomics viewer. *Nat Biotechnol* 29:24.

Thorvaldsdóttir H, Robinson JT, Mesirov JP. 2013. Integrative Genomics Viewer (IGV): High-performance genomics data visualization and exploration. *Brief Bioinform* 14(2):178–192.

GENCODE

Harrow J, Frankish A, Gonzalez JM et al. 2012. GENCODE: The reference human genome annotation for The ENCODE Project. *Genome Res* 22(9):1760–1774.

ANNOVAR

Yang H, Wang K. 2015. Genomic variant annotation and prioritization with ANNOVAR and wANNOVAR. *Nat Protoc* 10(10):1556–1566.

Ensembl

Yates A, Akanni W, Amode MR et al. 2016. Ensembl 2016. *Nucleic Acids Res* 44(D1):D710–D716.

UCSC browser

Kent WJ, Sugnet CW, Furey TS et al. 2002. The human genome browser at UCSC. *Genome Res* 12(6):996–1006.

BLAT, fast alignment of sequences to reference genome

Kent WJ. 2002. BLAT—The BLAST-like alignment tool. *Genome Res* 12(4):656–664.

Human accelerated regions

Pollard KS, Salama SR, King B et al. 2006. Forces shaping the fastest evolving regions in the human genome. *PLOS Genet* 2(10):e168.

Pollard KS, Salama SR, Lambert N et al. 2006. An RNA gene expressed during cortical development evolved rapidly in humans. *Nature* 443(7108):167–172.

Diagnosing the Genome

13

We have seen examples of different functional elements in the human genome, and the importance of these elements is illustrated by mutations that are responsible for monogenic disorders and cancer. Mutations affecting functionally significant regions of the genome may be revealed by DNA sequencing. The dramatic development in this technology has made possible the sequencing of genomic DNA from a large number of healthy individuals, as well as those affected by genetic disease. We now look into more details of how DNA sequencing has been used to identify causative variants in genetic disorders. First, we pay attention to a particular genome, that of a boy named Nicholas Volker.

A DIFFICULT INFLAMMATORY BOWEL DISEASE RESISTS DIAGNOSIS

From the age of 2 years, Nicholas suffered from a mysterious disease. Unusual holes would open in his intestine such that its contents leaked out into a large wound in his abdomen. His disease was similar to Crohn's disease, a type of inflammatory bowel disease, but its standard medications had no effect on Nicholas. Doctors at the Children's Hospital of Wisconsin taking care of Nicholas were indeed puzzled and frustrated—they could not match a known disease to the serious symptoms of Nicholas. The wound in Nicholas's abdomen would not heal, and for this reason he needed frequent surgical care. He underwent about 100 different surgical operations, including one where his colon was removed. By summer 2009 when Nicholas was 4 years old, he had spent more than 300 days in the hospital. He weighed less than 20 pounds, as compared to the 35 pounds expected for a boy of his age. He was greatly suffering from his disease, and there was still no diagnosis.

Alan Mayer, a gastroenterologist at Children's Hospital of Wisconsin and the Medical College of Wisconsin, now proposed that Nicholas' genome was to be sequenced. At this time these institutions had no experience of this type of analysis. As whole genome sequencing would be too costly, it was decided that instead they would go for **whole exome sequencing**—a procedure where only the exons of an individual are analyzed.

THE FAULTY GENE IS IDENTIFIED BY GENOME SEQUENCING

DNA is first isolated from the white blood cells of Nicholas. The DNA is then fragmented and subjected to exome sequencing. The resulting sequences are compared to the human **reference genome**. More than 16,000 differences are recorded. Most of these are either technical artifacts or individual differences not related to disease. The long list of genetic differences needs to be narrowed down.

Only 1,527 of the genetic differences are novel—that is, they have never been observed in an individual before. Of the novel variants, 879 change the amino acid sequence of a protein. Further filtering is based on the idea that the disorder is recessive, and as a consequence, Nicholas is expected to have two copies of the variant—or one copy in case the gene is on the X chromosome. Computational methods are also applied that predict whether a certain **nonsynonymous mutation** is likely to disturb the normal function of the protein. Applying these criteria, only two variants remain. One of them is in the gene *GSTM1*, but this gene seems an unlikely candidate because many individuals in the general population are missing a functional copy of this gene. The filtering process is thus left with one single variant.

The variant is in the gene *XIAP* (X-linked inhibitor of apoptosis), a gene located on the X chromosome. The exact role of the XIAP protein in Nicholas' disease is unclear, but in general this protein has a function to inhibit apoptotic cell death and in addition, it takes part in a number of different signaling pathways. To verify that the mutation is indeed novel, a large number of other individuals are examined with respect to the *XIAP* variant. However, none of them has the variant of Nicholas. A set of laboratory tests confirm that Nicholas' XIAP protein is not functioning as it should. It is concluded that the mutation in the *XIAP* gene is the cause of Nicholas' disease. Where did his mutation come from? It was learned that the disease mutation was carried by Nicholas' mother. The disorder was inherited in an X-linked recessive mode. While Nicholas is lacking a functional *XIAP* gene, his mother is heterozygous for the mutant gene, and for this reason she is healthy.

A CURE FROM TRANSPLANTATION

The gene *XIAP* was previously implicated in a rare inherited disorder named XLP (X-linked lymphoproliferative disease), a disorder that makes the patient unable to fight a common human virus, Epstein-Barr virus (EBV). This disease causes death, typically before the age of 10. The previously known patients had mutations in positions in the gene different from that of Nicholas, but nevertheless it seemed likely that he was at risk of suffering from XLP. Therefore, it was decided that Nicholas should receive a transplant with healthy blood **stem cells** of a donor (for more on stem cells, see Chapter 14.) Such stem cells are in the recipient expected to give rise to immune cells that produce the normal XIAP protein.

In summer 2010 Nicholas receives a transplant of umbilical cord blood from an anonymous donor. A transplantation is a dangerous operation for Nicholas. First, chemotherapy is used to destroy his immune system. Then a new immune system is introduced by transplantation. His body could reject these cells. During the period when his old immune system is eliminated and the new system is building up, he is vulnerable to infections by microorganisms and viruses. The transplantation is not without incident. Nicholas develops encephalitis as a result of virus infection. He also gets a staphylococcal infection. These infections are counteracted with antiviral agents and with antibiotics. Nicholas slowly recovers, and in October he is discharged from the hospital. At this point, the transplantation as such seems to have been successful.

But was Nicholas cured by the transplant? In 2018, 8 years had passed since the transplant. By this time it was clear that Nicholas did not suffer from XLP, and he was also cured from his intestinal disorder.

DIAGNOSIS OF GENETIC ERRORS

Nicholas Volker became the "first child saved by DNA sequencing." In this chapter, we examine more cases where sequencing was critical for correct diagnosis of an inherited disorder. Analysis of DNA is, however, not the only method used to reveal the identity of genetic errors. For instance, many of the currently available screening methods for inherited disorders rely on different biochemical tests that do not involve DNA sequencing. Rather, they monitor the protein product of the gene. One example is sickle cell anemia, where sickle cell hemoglobin may be measured using a number of different separation and quantitation techniques that are based on the unique charge properties of the mutant protein (Chapter 2). In addition, many metabolic disorders may be diagnosed by measurement of enzyme activity. For instance, for the diagnosis of Tay-Sachs disease, the activity of the enzyme hexosaminidase is determined (Chapter 6).

But DNA sequencing is often necessary to obtain a more detailed picture of the underlying genetic defect. For many genetic disorders, clinicians suspect a previously known disorder on the basis of patient symptoms. In such a case, there may be one or a few candidate genes to examine, as well as specific mutations in these genes.

HIGH-THROUGHPUT DNA SEQUENCING METHODS

Whereas genetic testing often is guided by clinical symptoms, there are also cases where a genetic defect is suspected, but the symptoms do not clearly match a previously known disorder. Now and then, clinicians encounter a completely novel genetic disorder. In such cases, we do not know what genes are affected or where mutations are located. One way to address this problem is to sequence the entire genome of the patient.

How is DNA sequenced? Remember that a human genome may be analyzed using a shotgun procedure where fragments of the genome are sequenced individually. The resulting sequence reads are used to reconstruct the whole genome (Chapters 4 and 12). The reads used in the assembly may be obtained with different methods. The first version of the human genome was obtained with Frederick Sanger's protocol for DNA sequencing. But sequencing technology has been developed to a remarkable extent so that it is now much faster and less expensive. Two methods that have been widely used are **reversible terminator sequencing** and **single-molecule real-time (SMRT) sequencing** (**Figures 13.1** and **13.2**), provided by the companies Illumina and PacBio, respectively. Both of the methods make use of DNA replication like in the Sanger method. They monitor incorporation of nucleotides, and the four nucleotides may be distinguished as each is labeled with a fluorophore that generates light with a specific wavelength. A third example of DNA sequencing technology is **nanopore sequencing** (Oxford Nanopore Technologies). This method does not involve DNA replication but relies on the measurement of changes

Figure 13.1 Reversible terminator sequencing. (a) In the Illumina reversible terminator sequencing protocol, the DNA to be sequenced is first amplified many times to make possible monitoring of nucleotide incorporation. As a result, the flow cell where the sequencing reaction is taking place contains millions of clusters of DNA, where in each cluster the DNA sequences are identical. All DNA molecules are attached to the surface of the flow cell, and their 3′ ends are available for polynucleotide chain elongation. (b) A DNA molecule in one of the clusters in the flow cell is shown. Required for replication are four dNTPs, DNA polymerase, and, for the first sequencing cycle, a primer. The detailed structure of a newly incorporated nucleotide is shown to the right. Like the other dNTPs that are used during the sequencing, it is modified at its 3′ end (red bar) to prevent further elongation of the polynucleotide chain. Furthermore, it contains a fluorophore that is specific to the nucleotide. Once the nucleotide is incorporated, the fluorophore emits a light signal that is recorded. The base is identified based on the color characteristic of the fluorophore. (c) The incorporated nucleotide is then deblocked at its 3′ OH, and the fluorophore is removed. DNA is now ready for incorporation of a new nucleotide. A new cycle is next initiated by introducing a new set of labeled dNTPs. (d) The final output of the sequencing procedure is a set of recorded images for the flow cell. Here are shown four different images, each representing a sequencing cycle. Each color in an image is representing a specific nucleotide. For each cluster, the sequence may be read from the colors. Reversible terminator sequencing is used by sequencing machines marketed by Illumina, Inc.

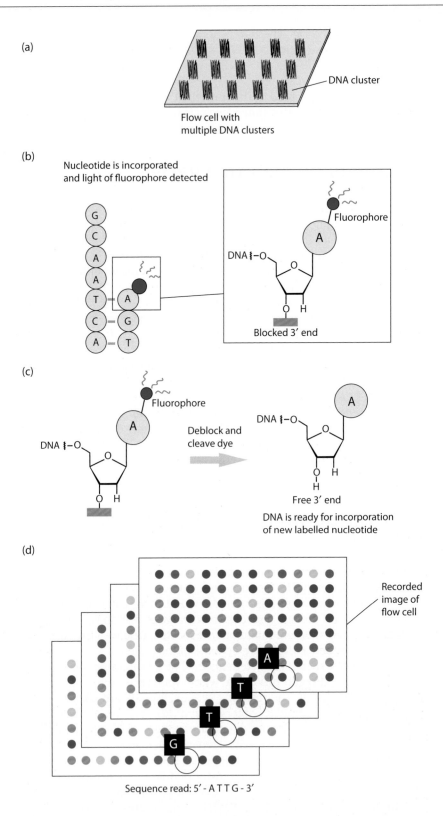

in electrical current density when a DNA molecule is transported across a small orifice, a nanopore. Oxford Nanopore Technologies markets the MinION, a highly portable sequencing device (Figure 5.10).

None of the DNA sequencing methods generate a completely accurate base sequence. The error rate in Illumina sequencing is about 10^{-3}, where the most common error is a **substitution**. The SMRT and nanopore

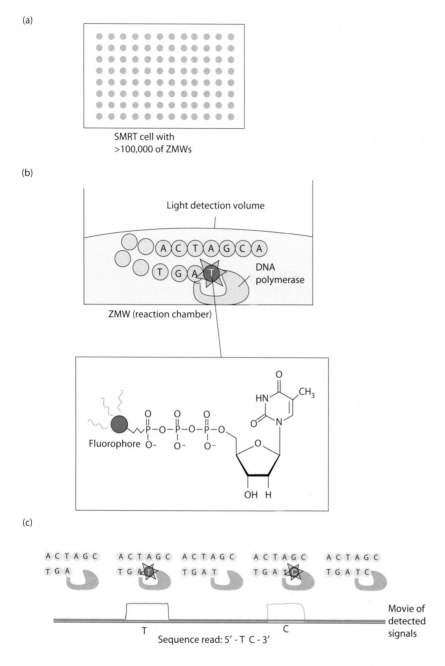

(a)

SMRT cell with
>100,000 of ZMWs

(b)

Light detection volume

DNA
polymerase

ZMW (reaction chamber)

Fluorophore

(c)

Sequence read: 5' - T C - 3'

Movie of
detected
signals

Figure 13.2 Single-molecule real-time (SMRT) sequencing. (a) A SMRT cell has more than 100,000 reaction chambers, "zero-mode waveguides" (ZMWs). (b) In each ZMW is monitored the replication of a single molecule. A single DNA polymerase is fixed on the bottom of the ZMW. The reaction chamber is constructed such that it allows measurement of emitted light restricted to a minute volume (10^{-21} liters, light yellow area) at the very bottom of the chamber where the polymerase reaction is taking place. Thereby, only nucleotides interacting with the DNA polymerase will be detected. As in Illumina sequencing, dNTPs are used with a fluorophore, but the fluorophore is chemically linked to the terminal phosphate of the nucleotide instead of the base (nucleotide structure is shown within rectangle). Each of the nucleotides is labeled with a different colored fluorophore. (c) The replication of DNA in SMRT technology is shown. When a dNTP enters the catalytic site of the polymerase, it stays there for a short while and a light pulse is produced. When the nucleotide is incorporated into DNA, the terminal phosphates and the fluorophore are cleaved off, and a new dNTP may enter into the polymerase active site. During the whole process of DNA synthesis, all light pulses are recorded to form a "movie," from which the nucleotide sequence may be inferred. SMRT is a product of Pacific Biosciences of California, Inc.

methods are associated with even higher error rates. Because of errors during DNA sequencing, there is a need for sequencing every region of DNA multiple times. The sequencing **coverage** is the number of reads that include a given nucleotide (for multiple reads mapping to a genome region, see also Figure 12.21).

Whereas Illumina sequencing has been used extensively, one significant advantage of the SMRT and nanopore technologies is that they produce long sequences, several thousands of nucleotides long,[1] as compared to reversible terminator Illumina sequencing that typically generates read lengths up to about 200 nucleotides. Obtaining long sequences is particularly helpful in the sequencing of repetitive regions, where too short reads make the assembly process difficult (Chapter 12).

---

[1]In 2017 was reported the first read longer than one million bases, a read from chromosome 19 obtained by nanopore sequencing.

One significant improvement as compared to the early days of Sanger sequencing is that we now have the ability to sequence a large number of DNA fragments in parallel. Thus, the sequencing procedures are carried out with millions of different DNA fragments at the same time, so in one single experiment a vast amount of sequence is generated. The output from the Illumina and PacBio sequencing machines is truly remarkable—in the order of 200 and 20 gigabases/day, respectively. This means that raw sequence data for an equivalent of several human genomes may be generated in one day only. At the same time, we will most likely in the future see further development in the area of DNA sequencing to make it even more powerful.

WHOLE GENOME AND EXOME SEQUENCING

Remember that the human genome is some three billion letters. Sequencing the entire genome—**whole genome sequencing (WGS)**—of a patient with a suspected genetic disease is certainly possible but not always necessary. Instead, a common procedure is to sequence only the regions of the genome that are the exons. This technique is known as **whole exome sequencing (WES)**.

The motivation for exome sequencing is that a majority of monogenic disorders are estimated to be the result of exonic mutations. In exome sequencing, sequences immediately flanking the exons are also included. In this way, the most important **splicing** mutations will be covered. Furthermore, sequencing the exons is less expensive and time-consuming than the whole genome. This is because the exons of protein genes correspond to only about 3% of the whole genome (Chapter 4). The data obtained from WES are also easier to analyze using computational tools. The disadvantage with exome sequencing is obviously that we will be missing the majority of the noncoding parts of the genome.

The basic procedure of WGS was outlined in Chapters 4 and 12, in which first small pieces of genomic DNA are sequenced and then the short sequence reads are used to reconstruct the whole genome using a computational procedure. An overview of WES is shown in (**Figure 13.3**). The genomic DNA to be analyzed is first sheared into numerous smaller pieces, similar to what happens in WGS. The next step is an enrichment for the exonic regions. Thus, the genomic DNA fragments are allowed to hybridize (form duplex structures by base-pairing) to short DNA or RNA sequences that all are from the exonic regions of the human genome. These short sequences have been synthesized based on known human exon sequences. The genomic DNA fragments of the sample that bind to the exonic bait sequences are then isolated, amplified with polymerase chain reaction, and subjected to DNA sequencing. Finally, the large collection of sequences obtained is aligned to the reference genome, and all differences are identified (see also Chapter 12).

A large number of individuals have been subject to exome sequencing, and much of this information is available to researchers. For instance, the Genome Aggregation Database (gnomAD) collects exomes from a wide variety of sequencing projects and is in 2018 accounting for about 123,000 different exomes.[2] These exomes constitute a rich source of information for studies of human genetic information in protein coding regions. For instance, variants that are likely to destroy the function

---

[2]The number of exomes is guaranteed to grow in the near future. For instance, the UK Biobank has set a goal to sequence 500,000 exomes by 2020.

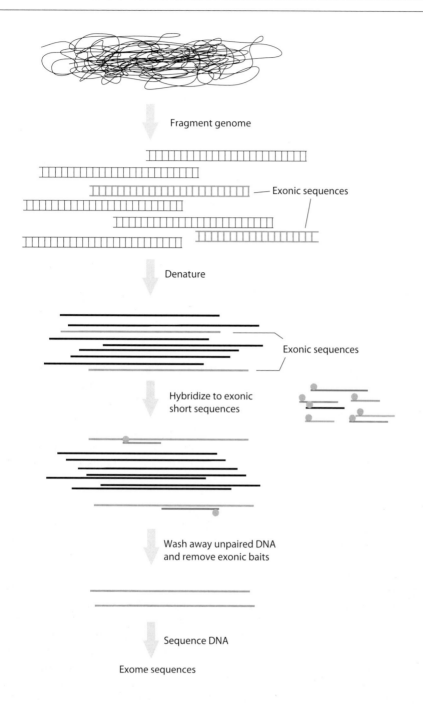

Fragment genome

Exonic sequences

Denature

Exonic sequences

Hybridize to exonic
short sequences

Wash away unpaired DNA
and remove exonic baits

Sequence DNA

Exome sequences

Figure 13.3 Whole exome sequencing. Genomic DNA to be analyzed is sheared into smaller pieces. The DNA fragments are then allowed to hybridize to short sequences that are representing the exonic regions of the human genome. These short sequences have been synthesized based on known human exon sequences and are labeled with biotin (gray circles). Biotin has a strong affinity for the protein streptavidin. A common purification procedure for biotin-labeled molecules involves streptavidin-coated magnetic beads. Thus, the pieces of genomic DNA binding to the biotin-labeled sequences are attracted to a magnet, whereas the unpaired DNA fragments can be washed away. The paired DNA fragments may be finally isolated by raising the temperature, so that they are discharged from the biotin-labeled molecules.

of the protein gene, such as nonsense mutations, may be identified. By comparing the expected and observed numbers of variants that cause the production of truncated proteins, it has been possible to identify genes that are sensitive to such variants. These genes are likely to have critical functions.

DIAGNOSIS OF A NEUROMUSCULAR DISORDER: ANOTHER SUCCESS STORY OF GENOME SEQUENCING

The case of Nicholas Volker has been followed by many other instances where genome sequencing has been used to diagnose a rare disorder. As one example, sequencing helped to clarify a rare neuromuscular disorder.

Figure 13.4 Biosynthesis of dopamine and serotonin. Important genes involved in dopa-responsive dystonia are those that encode the enzymes GTP cyclohydrolase (GCH1), tyrosine hydroxylase (TH), and sepiapterin reductase (SPR).

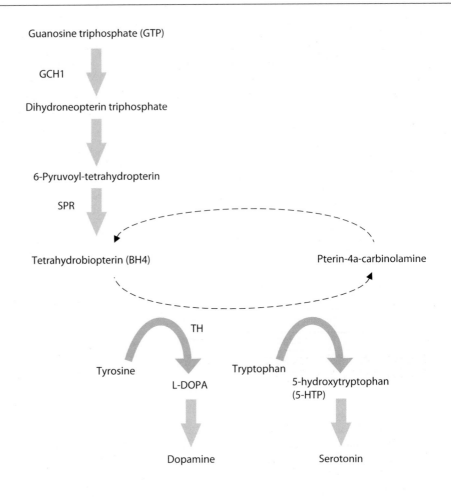

The twins Noah and Alexis Beery were born in 1996. At the age of 5 years, they were diagnosed with a genetic disorder named dopa-responsive dystonia (DRD). A characteristic symptom is abnormal movements. The standard therapy for DRD is to supply L-DOPA, the same drug that is used to treat another movement disorder, Parkinson's disease. Many of the symptoms of Noah and Alexis were indeed alleviated by L-DOPA. However, at the age of 13, Alexis developed a serious cough and breathing problem. Doctors did not have a clue as to what was causing her condition.

The major genes involved in DRD are those that encode GTP cyclohydrolase (*GCH1*), tyrosine hydroxylase (*TH*), and sepiapterin reductase (*SPR*). The enzyme products of *GCH1* and *SPR* are both involved in a pathway for synthesis of the neurotransmitters dopamine, noradrenaline, and serotonin. Tyrosine hydroxylase is specifically involved in the pathway leading to dopamine (**Figure 13.4**).

To identify the responsible gene, it was decided that the genomes of Alexis and Noah were to be analyzed with WGS. The father of the twins, Joe Beery, was a chief information officer at Life Technologies, a biotechnology company. The company helped fund a study where the genomes were analyzed using Life's sequencing technology.

A research team at Baylor College of Medicine, Texas, examined the whole genomes of the twins to identify causative genetic variants. The disorder was suspected to be recessive, as there were no signs of the disease in the immediate family history. From all the variants, the scientists therefore focused on previously unknown **missense** variants in coding regions that were consistent with a recessive mode of inheritance. This means that the parents should have genotype Aa and the twins aa, where "A" is the healthy and "a" the disease allele. Among the final candidates, only three of the identified genes had such mutations present in both twins. One of the genes was *SPR*, a gene previously associated with DRD. It turned out that this was the gene pertinent to the disorder. The twins were found to have not one but two different mutations in this gene. They inherited a missense variant from their father and a **nonsense** variant from their mother (**Figures 13.5** and **13.6**). In the language of genetics, the twins are **compound heterozygotes**—that is, for the disease to manifest, there are two different mutant alleles in the same gene.

The sepiapterin reductase encoded by the *SPR* gene is required for formation of both dopamine and serotonin (see Figure 13.4). The twins already received L-DOPA therapy for dopamine production. Could also the production of serotonin be improved with a drug? A therapy with 5-HTP (5-hydroxytryptophan), a substrate in the formation of serotonin, was initiated in both twins. Their symptoms were dramatically reduced as a result of the combination therapy with L-DOPA and 5-HTP. Therefore, this story is another example when genome sequencing was able to improve the lives of affected individuals. However, it must be emphasized at this stage that there are also many cases of rare disorders where diagnosis has not led to a successful therapy.

GENOME SEQUENCING TO IDENTIFY GENETIC ERRORS

How efficient are WGS and WES sequencing in identifying the cause of a monogenic disorder? As we have seen, a large number of differences are identified when comparing a patient genome to the reference version of the human genome. The large number is because of individual variation, and because errors during DNA sequencing generate false-positive differences. The long list of differences needs to be filtered using different criteria. For instance, we may focus on a set of specific genes suspected to be associated with the symptoms of the patient. Sequencing of parents and siblings to the patient is often helpful. It should also be noted that once a single likely causative mutation has been found, it must be verified using

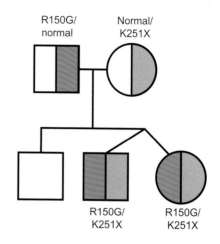

Figure 13.5 Pedigree showing inheritance of dopa-responsive dystonia. The parents are both heterozygous for a recessive allele (red/orange), and the twins inherited one mutation in the *SPR* gene from their mother and one from their father and are affected by the disorder. The pedigree also shows an unaffected sibling of the twins.

Figure 13.6 Twins Noah and Alexis Beery, with their parents. (Courtesy of Retta Beery.)

other methods. At any rate, genome sequencing will not always identify the causative mutation. There are now several studies in which a larger number of people with a suspected genetic disorder have been analyzed with genome sequencing. The reported percentage of patients that are successfully diagnosed is in the range between 25% and 55%.

A less expensive alternative to WES and WGS is to sequence a large panel of genes that are known to be disease associated. One such commercially available panel is composed of exonic regions from around 4,800 different genes, much less sequence data as compared to the whole exome. In one specific study using such a panel, 20 newborns and children admitted to the Children's Hospital of Eastern Ontario, Canada, suspected to have a genetic disorder were screened. For eight of these, sequencing of the 4,800 genes provided a diagnosis.

MANY PEOPLE ARE AFFECTED BY A RARE DISORDER

The diseases of Nicholas Volker and the Beery twins are examples of rare disorders with a genetic background. The definition of a "rare" disease is slightly different in different parts of the world. In Europe, a disease is rare when it affects fewer than 1 in 2,000. In the United States, a disease is considered rare when it affects fewer than 200,000 US citizens at any given time. There are 30 million US citizens, 30 million EU citizens, and more than 300 million people worldwide who are estimated to have a rare disease. About 6,000–7,000 different such diseases are known today. It is thought that 80% of them have a genetic background, and for this reason, DNA sequencing is indeed an important method for discovering their molecular background.

RARE GENETIC DEFECTS ARE BEING ANALYZED AT A LARGE SCALE

Initial attempts to identify the cause of rare inherited disorders first examined individual cases, like those that we encountered above. Additional examples are listed in **Table 13.1**. However, as WGS and WES have become routine procedures, the analysis of rare disorders has now entered a phase where a large number of genomes may be analyzed by a single laboratory.

In one study, 722 individuals with inherited retinal disease (IRD) were examined with genome sequencing. IRD describes a group of conditions that is highly heterogeneous with respect to genotype and phenotype. More than 250 genes have been implicated in IRD. The majority of the 722 individuals had no prior genetic diagnosis explaining the retinal disease. With the help of WGS, the likely cause of the disease was identified in 56% of the individuals. Most of the pathogenic mutations were substitutions and indels, but in 30 individuals the causative mutation was a structural variant (typically a deletion), not identifiable with WES.

In the event a newborn is suspected to carry a genetic disorder and where the standard diagnostics fail, it is now possible to carry out genome sequencing. In Children's Mercy Hospital in Kansas City, Kansas, a team set up a protocol for rapid WGS. Thus, the whole procedure of DNA sequencing, alignment of reads to the reference genome, and variant calling is carried out within 26 hours or even less to allow as early diagnosis as possible for newborns. A large number of acutely ill children have now been diagnosed using this protocol. Examples of successful diagnosis are shown in

Table 13.1 Disorders diagnosed with whole genome or exome sequencing

Individual	Gene	Chromosome	Mutation	Category of mutation	Inheritance	WES/WGS	Name of disorder
James Lupski[a,e]	SH3TC2	chr5	A > G, G > A	Missense, nonsense	One variant from father, one from mother	WGS	Charcot-Marie-Tooth Disease
Nicholas Volker[f]	XIAP	chrX	G > A	Missense	Inherited from mother	WES	Inflammatory bowel disease
Noah and Alexis Beery[g]	SPR	chr2	A > T, A > G	Nonsense, missense	One variant from father, one from mother	WGS	Dopa-responsive dystonia (DRD)
Bea Rienhoff[b,h]	TGFB3	chr14	G > A	Missense	*De novo* mutation, not present in parents	WES	Loeys-Dietz syndrome-5
Adam Foye[c,i]	TTN	chr2	G > C, G > A	Splicing, splicing	One variant from father, one from mother	WGS	Centronuclear myopathy
Lily Grossman[d,j]	ADCY5	chr3	C > T	Missense	*De novo* mutation, not present in parents	WGS	Dyskinesia

Note: For the disorders of Nicholas Volker and the Beery twins, the reader is referred to the text.

a James Lupski is a scientist at the Baylor College of Medicine, Texas. He is affected by the genetic disease Charcot-Marie-Tooth (CMT), and in his research he has studied this disease as well as several others.

b It was realized soon after her birth that Bea Rienhoff had a disorder with a genetic background. Her symptoms included contracted fingers, widely spaced eyes, and a failure to gain weight. Doctors were unable to diagnose her condition. Bea is a daughter of Hugh Rienhoff, a biotech entrepreneur with a background in clinical genetics. Although Bea was otherwise healthy, her father was anxious that she would develop heart problems later in life, and wanted to find a relevant diagnosis. He bought laboratory equipment in order to study his daughter's DNA himself—an example of "do-it-yourself biology." Exome sequencing finally found the causative mutation.

c At the age of 13 months, Adam Foye received the diagnosis of centronuclear myopathy (CNM). However, his condition was not based on changes in any of the 13 genes previously associated with CNM. About 10 years later, he was one of three children whose genomes were sequenced by 23 research teams in the CLARITY Challenge, an international competition designed by Boston Children's Hospital. The goal of the competition was to "identify best methods and practices for the analysis, interpretation and reporting of individuals' DNA sequence data, to provide the most meaningful results to clinicians, patients and families." The winning team identified the genomic changes behind Adam's disease, two mutations in the *TTN* gene, a gene encoding the huge muscle protein titin.

d Lilly Grossman suffers from a genetic disorder that causes involuntary movements and weakness. During her childhood, the nights were particularly difficult for her as dramatic attacks of tremor would cause her to wake up constantly. Her parents were frustrated that a diagnosis for the disease was lacking. They had heard about the genome sequence of the Beery twins, and when Lilly was 15 years old, the family was happy to enter into a project at the Scripps Institute called IDIOM (Idiopathic Diseases of Man). This is a clinical project designed to use WGS to help determine the genetic cause of idiopathic diseases—rare conditions that defy a diagnosis or are unresponsive to standard treatments. Sequencing of Lilly's genome revealed a mutation in the *ADCY5* gene that scientists believe is causing her tremors. A drug called Diamox (acetazolamide) was known to help another family with mutation in the same gene, and Lilly's symptoms were somewhat reduced when she was administered this drug. In this context, it is also interesting to note that the identification of Lilly's mutation led to the discovery of additional patients with mutations in the same gene (Chen, D.H. et al., Neurology, 2015. 85(23): p. 2026–35).

e Lupski JR, Reid JG, Gonzaga-Jauregui C et al. 2010. Whole-genome sequencing in a patient with Charcot-Marie-Tooth neuropathy. *N Engl J Med* 362(13):1181–1191.

f Worthey EA, Mayer AN, Syverson GD et al. 2011. Making a definitive diagnosis: Successful clinical application of whole exome sequencing in a child with intractable inflammatory bowel disease. *Genet Med* 13(3):255–262.

g Bainbridge MN, Wiszniewski W, Murdock DR et al. 2011. Wholegenome sequencing for optimized patient management. *Sci Transl Med* 3(87):87re83.

h Rienhoff HYJr, Yeo CY, Morissette R et al. 2013. A mutation in *TGFB3* associated with a syndrome of low muscle mass, growth retardation, distal arthrogryposis and clinical features overlapping with Marfan and Loeys-Dietz syndrome. *Am J Med Genet A* 161A(8):2040–2046.

i Brownstein CA, Beggs AH, Homer N et al. 2014. An international effort towards developing standards for best practices in analysis, interpretation and reporting of clinical genome sequencing results in the CLARITY Challenge. *Genome Biol* 15(3):R53.

j Chen YZ, Friedman JR, Chen DH et al. 2014. Gain-of-function ADCY5 mutations in familial dyskinesia with facial myokymia. *Ann Neurol* 75(4):542–549.

Table 13.2. A US program called Newborn Sequencing in Genomic Medicine and Public Health (NSIGHT) is aimed at exploring the possibilities of genome-scale sequencing in newborns. One interesting observation from recent genome sequencing of newborns with a suspected genetic defect is that in more than half of the children with a diagnosis, the disease is caused by a ***de novo* mutation**, meaning that it came about during the production of a parental sperm or egg cell, or early in the fertilized egg. Such a mutation is not identified when testing the DNA of the parents.

An ambitious genome sequencing project was initiated in 2012 in the United Kingdom, the 100,000 Genomes Project. Included in the project are about 25,000 cancer patients (both cancer and normal tissue from each patient, in total, 50,000 genomes) and about 17,000 patients with rare disease (patient genome and genomes of two relatives, in all 50,000 genomes). The project is run by the company Genomics England, which is set up and owned by the UK Department of Health. In November 2018 an impressive 92,000 genomes had been sequenced, and results are being returned to

Table 13.2 Whole genome sequencing of children in neonatal and pediatric intensive care units

Causal gene	Disease	Inheritance pattern	Chromosome coordinates of causal variants (hg19)	Causal variants
GJB2	Keratitis-ichthyosis-deafness syndrome	Autosomal dominant, de novo	13:20763634–20763636	In-frame deletion
BRAT1	Lethal neonatal rigidity and multifocal seizure syndrome	Autosomal recessive, homozygous	7:2583573–2583574	Frameshift insertion
MMP21	Heterotaxy	Autosomal recessive, compound heterozygous	10:127462732–127462732; 10:127460915–127466819	Frameshift deletion; Exon-13 deletion
PRF1	Familial hemophagocytic lymphohistiocytosis type 2	Autosomal recessive, compound heterozygous	10:72358167–72358167; 10:72360387–72360387	Missense; missense
ABCC8	Familial hyperinsulinism type 1	Autosomal dominant	11:17424218–17424218	Missense
PTPN11	Noonan syndrome	Autosomal dominant, de novo	12:112915523–112915523	Missense
PTPN11	Noonan syndrome	Autosomal dominant, de novo	12:112926258–112926258	Missense
SCN2A	Epileptic encephalopathy	Autosomal dominant, de novo	2:166245193–166245193	Missense

Source: Based on information in Willig LK et al. 2015. *Lancet. Resp Med* 3(5):377–387.

Note: At Children's Mercy Hospital in Kansas City, Missouri, 35 acutely ill children with suspected genetic disorder were studied. Of these, 20 were successfully diagnosed using rapid whole genome sequencing, out of which 8 are shown here.

participating individuals and their clinical teams. Furthermore, the National Health Service (NHS) in England has announced that the 100,000 Genomes Project is to be expanded to incorporate the whole genome sequencing of 1 million patients.

It should be noted that in order to identify the causes of rare genetic disorders, we benefit not only from sequencing the affected individuals and their parents. Also the sequencing of healthy individuals, such as represented in the gnomAD dataset referred to previously, is helpful to clarify the role of disease mutations.

We see from the cases mentioned that the causative genetic change is not identified in all patients affected by a rare genetic disorder. It should also be noted that for most of the cases where the defect is known, this does not lead to effective treatment. Nevertheless, there is no doubt that there are significant advantages of genome sequencing as a diagnostic tool. In Chapter 14, we examine more closely how to make use of a genetic diagnosis as we look into the prospects for successful therapy.

SUMMARY

- In human *WGS*, the whole genome of an individual is subject to DNA sequencing. In *WES*, only the exome is sequenced.
- The methods used in finding the genetic cause of an inherited disorder depend on whether the patient is suspected to have a previously known disorder or not. For well-known disorders, there are biochemical tests or methods based on DNA sequencing of a gene or a limited number of genes. In case there is no clue as to the genetic background based on clinical observations and available tests for individual genes, WES or WGS are possible alternatives.
- High-throughput DNA sequencing methods include reversible terminator sequencing and SMRT sequencing. Both of these methods make use of DNA replication *in vitro* and provide generation of sequences from a large number of DNA templates in parallel.
- WGS or WES is able to identify the genetic cause of a suspected monogenic disorder in only 25%–55% of cases. The low success rate is partly because it is difficult to identify the disease-causing variant against a background of a large number of genetic variants that are unique to an individual.

- At least 4% of all people are estimated to have a "rare" disorder—that is, a disorder unique to an individual or to a group of close relatives. About 80% of these rare disorders are suspected to have a genetic cause.
- Rare genetic defects are being analyzed at a large scale, for instance, in the UK 100,000 Genomes Project. In addition, independently from rare diseases, millions of human genomes are in the next few years being sequenced worldwide, projects in which genetic variants are being inferred and compared to phenotypic data.

QUESTIONS

1. If the genome of an individual with a suspected genetic disease is sequenced, what are major technical challenges when it comes to identifying the causative mutation?
2. When it comes to diagnosis of genetic disorders, what are the alternatives to genome sequencing? Under what circumstances is genome sequencing the preferred alternative?
3. What are the similarities and differences between *reversible terminator sequencing* and *SMRT sequencing*? How does *nanopore sequencing* compare to the other two methods?
4. What is the best possible error rate in DNA sequencing methods?
5. What are the advantages and disadvantages of *WGS* and *WES*?
6. What are the experimental steps involved in *WES*?
7. What is meant by a *rare disorder*?
8. What is a *compound heterozygote*?
9. What is a de *novo mutation*?
10. What is the objective of the UK *100,000 Genomes Project*?

URLs

gnomAD (http://gnomad.broadinstitute.org)

Genomics England (www.genomicsengland.co.uk)

On WGS and WES to analyze Mendelian disorders (http://massgenomics org/2017/01/exome-whole-genome-disorders.html)

Rare disorders

National Human Genome Research Institute (www.genome.gov/)

Global Genes, RARE List (https://globalgenes.org/rarelist/)

National Organization for Rare Disorders (NORD) (https://rarediseases.org/)

Rare Disease Day (www.rarediseaseday.org/article/what-is-a-rare-disease)

FURTHER READING

Case of Nicholas Volker

Johnson M, Gallagher K. 2017. *One in a billion: The story of Nic Volker and the dawn of genomic medicine.* Simon & Schuster Paperbacks, New York.

Worthey EA, Mayer AN, Syverson GD et al. 2011. Making a definitive diagnosis: Successful clinical application of whole exome sequencing in a child with intractable inflammatory bowel disease. *Genet Med* 13(3):255–262.

Sequencing of individual genomes

Bainbridge MN, Wiszniewski W, Murdock DR et al. 2011. Whole-genome sequencing for optimized patient management. *Sci Transl Med* 3(87):87re83.

Brownstein CA, Beggs AH, Homer N et al. 2014. An international effort towards developing standards for best practices in analysis, interpretation and reporting of clinical genome sequencing results in the CLARITY Challenge. *Genome Biol* 15(3):R53.

Chen YZ, Friedman JR, Chen DH et al. 2014. Gain-of-function ADCY5 mutations in familial dyskinesia with facial myokymia. *Ann Neurol* 75(4):542–549.

Lupski JR, Reid JG, Gonzaga-Jauregui C et al. 2010. Whole-genome sequencing in a patient with Charcot-Marie-Tooth neuropathy. *N Engl J Med* 362(13):1181–1191.

Rienhoff HY Jr, Yeo CY, Morissette R et al. 2013. A mutation in *TGFB3* associated with a syndrome of low muscle mass, growth retardation, distal arthrogryposis and clinical features overlapping with Marfan and Loeys-Dietz syndrome. *Am J Med Genet A* 161A(8):2040–2046.

Whole exome sequencing

Ng SB, Turner EH, Robertson PD et al. 2009. Targeted capture and massively parallel sequencing of 12 human exomes. *Nature* 461(7261):272–276.

Ng SB, Buckingham KJ, Lee C et al. 2010. Exome sequencing identifies the cause of a Mendelian disorder. *Nat Genet* 42(1):30–35.

DNA sequencing review

Shendure J, Balasubramanian S, Church GM et al. 2017. DNA sequencing at 40: Past, present and future. *Nature* 550:345.

SMRT sequencing of human genome

Chaisson MJ, Huddleston J, Dennis MY et al. 2015. Resolving the complexity of the human genome using single-molecule sequencing. *Nature* 517(7536):608–611.

Mendelian disorders and genome sequencing

Gilissen C, Hoischen A, Brunner HG, Veltman JA. 2011. Unlocking Mendelian disease using exome sequencing. *Genome Biol* 12(9):228.

Yang Y, Muzny DM, Reid JG et al. 2013. Clinical whole-exome sequencing for the diagnosis of Mendelian disorders. *N Engl J Med* 369(16):1502–1511.

Genetic variation from genome sequencing and gnomAD database

Lek M, Karczewski KJ, Minikel EV et al. 2016. Analysis of protein-coding genetic variation in 60,706 humans. *Nature* 536(7616):285–291.

MacArthur DG, Balasubramanian S, Frankish A et al. 2012. A systematic survey of loss-of-function variants in human protein-coding genes. *Science* 335(6070):823–828.

Newborn genome sequencing

Berg JS, Agrawal PB, Bailey DB Jr. et al. 2017. Newborn sequencing in genomic medicine and public health. *Pediatrics* 139(2).

Daoud H, Luco SM, Li R et al. 2016. Next-generation sequencing for diagnosis of rare diseases in the neonatal intensive care unit. *CMAJ* 188(11):E254–E260.

Miller NA, Farrow EG, Gibson M et al. 2015. A 26-hour system of highly sensitive whole genome sequencing for emergency management of genetic diseases. *Genome Med* 7(1):100.

Petrikin JE, Cakici JA, Clark MM et al. 2018. The NSIGHT1-randomized controlled trial: Rapid whole-genome sequencing for accelerated etiologic diagnosis in critically ill infants. *NPJ Genomic Medicine* 3:6.

Saunders CJ, Miller NA, Soden SE et al. 2012. Rapid whole-genome sequencing for genetic disease diagnosis in neonatal intensive care units. *Sci Transl Med* 4(154):154ra135.

Wright CF, FitzPatrick DR, Firth HV. 2018. Paediatric genomics: Diagnosing rare disease in children. *Nat Rev Genet* 19:253.

Willig LK, Petrikin JE, Smith LD et al. 2015. Whole-genome sequencing for identification of Mendelian disorders in critically ill infants: A retrospective analysis of diagnostic and clinical findings. *Lancet Respir Med* 3(5):377–387.

WGS/WES and inherited retinal disease

Carss KJ, Arno G, Erwood M et al. 2017. Comprehensive rare variant analysis via whole-genome sequencing to determine the molecular pathology of inherited retinal disease. *Am J Hum Genet* 100(1):75–90.

Correcting Genome Errors   14

G iven that many inherited diseases as well as cancer are based on
fatal changes in the genomic DNA sequence, what are our possibilities of "correcting" these errors? This question brings us to the
subject of **gene therapy**. Gene therapy refers to procedures where DNA or
RNA is delivered to patients to treat a disease. For instance, if a patient has
a mutated and defective gene, a possible treatment is the introduction of a
normal and healthy copy of the gene. But we also discuss other methods
that may compensate for the genetic disorder. For instance, DNA and RNA
may be administered to patients to modify the expression of a gene.

In principle, we are today able to correct errors and to introduce
healthy genes into sick individuals, but there are a number of difficult technical problems. For this reason, gene therapy has so far been used at a
limited scale. But as technology is now advancing quickly, it seems likely
that it will soon be in much wider use. For instance, we discuss powerful methods that were recently developed for the purpose of changing the
genomic sequence *in vivo*.

GERMLINE AND SOMATIC CELL THERAPY

We first turn to procedures that change the genome and later in this chapter discuss methodology related to changing the expression of a gene.
With regard to genomic changes, we need to distinguish between two different categories—**germline** and **somatic** therapy. In germline therapy,
a genetic change is introduced into a sperm or egg cell, into a fertilized
egg (zygote), or at an early stage in embryonic development. This type
of technology was applied to an animal already in 1983 when mice were
genetically modified to produce a human version of a growth hormone. A
consequence of germline therapy is that the genetic change will be transmitted to every cell of the mature body. Also, all descendents of the modified individual will carry the genetic change. Although germline therapy is
technically feasible, there are a number of safety issues. We want to make
sure that there is a high degree of specificity to the genetic alteration. For
instance, we do not want changes to occur in positions in the genome that

are not desired. In addition, we need to be sure that the genetic modification that we have accomplished does not have undesired long-term consequences. For these reasons, scientists have typically been against human germline therapy.

In somatic cell therapy, only a subset of the cells in the body is genetically changed. An important advantage in terms of safety is that the DNA changes are not transferred to the offspring. Therefore, this form of cell therapy is less controversial from an ethical point of view. At the same time, there are technical problems. For instance, the gene that we want to introduce or change should reach a specific type of target cell or tissue. It is sometimes difficult to achieve this specificity, and there are cells that may not even be accessible to gene transfer.

ADMINISTERING A FUNCTIONAL GENE TO A PATIENT BY TRANSPLANTATION OF HEMATOPOIETIC STEM CELLS

Early attempts to cure a genetic disease included the use of **stem cells** from a healthy donor. For instance, in the 1950s, experiments started to cure patients with cancer. Stem cells are cells that are able to differentiate into different kinds of cells. There are different types of stem cells in the human body. Hematopoietic stem cells have been widely used to restore a normal gene into a patient with a genetic disorder. The hematopoietic stem cells of bone marrow develop into all kinds of blood cells, such as red blood cells and different types of immune cells. Bone marrow has a remarkable capacity to produce blood cells—it can generate hundreds of billions of new blood cells every day. Another useful source of hematopoietic stem cells is blood of the umbilical cord.

A large number of disorders may be treated by administering to the patient hematopoietic stem cells originating from an healthy individual. Examples are leukemia, immune disorders, different blood disorders, and a range of metabolic disorders. The stem cells are introduced into the patient by intravenous injection, and these cells will then populate the bone marrow and produce new cells. A basic requirement for this therapy to work is that the gene that is deficient in the patient is expressed by any of the cells derived from the hematopoietic stem cells. It should be noted that this type of stem cell therapy is not strictly "gene therapy," because that particular term refers to procedures relating to specific DNA or RNA sequences, procedures we see several examples of later in this chapter.

A number of patients with a genetic disorder have been successfully cured with bone marrow or cord blood transplantation. However, a problem with transplants is that foreign cells are typically recognized by the immune system, and this results in rejection of the foreign cells. To avoid this rejection, stem cells should originate from a donor who is compatible with the patient in terms of HLA type. HLA is short for "human leukocyte antigen," molecules that are part of the immune system. HLAs are on the surface of cells, and the immune system uses these to distinguish between self and foreign invaders. A donor should ideally have the same HLA antigens on the surface of the cells as those of the recipient. However, the chances of finding a matched donor are small because of the diversity of HLA antigens. If an individual has a certain HLA type, there are only 42 out of 1 billion human individuals with exactly that HLA type. A child is unlikely to have the same HLA set as one of its parents—therefore, a parent can rarely be a donor of bone marrow. An alternative option to cure a child with an inherited disorder, although unusual, is to give birth to a "savior sibling" as discussed in the next section.

CORRECTING A GENETIC DISORDER WITH THE HELP OF A "SAVIOR SIBLING"

In October 1995, Laurie and Allen Strongin gave birth to a boy named Henry. Immediately after birth, it was discovered he had one extra thumb and four different heart problems. After a few weeks, he was diagnosed as having Fanconi's anemia, a rare genetic disorder.

Fanconi's anemia (FA) was first described by Guido Fanconi, a Swiss pediatrician, in 1927. It is a rare disease—about 1,000 individuals are affected by the disease worldwide, and it occurs in about 1 per 130,000 births. It has a somewhat higher frequency among Ashkenazi Jews. Both of Henry's parents belong to this group.

The disease may be caused by mutations in many different genes. There are 17 different FA genes, all involved in DNA repair. When DNA is damaged, a complex is normally assembled in the nucleus, which is composed of eight proteins, FANCA, -B, -C, -E, -F, -G, -L, and -M. When one of these is defective, DNA repair is much less efficient. Fanconi patients also have a risk of developing cancers, for instance, acute myeloid leukemia (AML).

Henry's particular form of the disease is severe and normally leads to death at a young age. The mutation is in an intron of the gene *FANCC* and has the effect that one of the exons of the gene is skipped during splicing, resulting in a change in reading frame during translation.

FA is a recessive disorder—that is, you are healthy as long as you have at least one copy of the healthy gene. The Strongin's are both carriers of the mutated *FANCC* gene. For the couple, there was a 25% chance they get an affected child.

Could the life of Henry be saved? In principle, the normal FANCC gene could be introduced by supplying Henry with cells producing the normal protein. A bone marrow transplantation would be one way to accomplish this. For this procedure to work, the donor and Henry must have the same HLA type. The chances of finding a matched donor are small, as explained. In May 1996, Henry's parents were contacted by doctors suggesting a novel approach using a combination of *in vitro* fertilization (IVF) and preimplantation genetic diagnosis (PGD). The idea was to give birth to a child who could save the life of Henry. This was the first time IVF and PGD were used in this manner.

The attempts to produce a child to save the life of Henry were initiated in December 1997. The procedure started with subjecting Laurie Strongin to a hormone treatment to stimulate the production of eggs. Eggs were then collected and fertilized with sperm in a laboratory. The fertilized eggs were analyzed with respect to the disease gene as well as to the HLA type. Selected eggs were finally introduced into Laurie's uterus.

A number of criteria need to be met for a successful pregnancy giving birth to a healthy child. First, the fertilized egg needs to be the right HLA type. The probability of a sibling having the same HLA type is about 25%. Second, the disease gene must be completely absent so that the egg is homozygous for the normal gene—this happens with a probability of 25% (both parents are heterozygous, see method of Figure 2.4). And finally, the implanted egg must not be aborted but give rise to a successful pregnancy. For Laura Strongin, things did not work out well with the attempts to create a healthy "savior sibling" to Henry. During 4 years, there were nine attempts—in all Laura received 353 injections and produced 198 eggs. The cost of these experiments was about $135,000. In the cases where a healthy fertilized egg was produced, the pregnancy failed, or there was a miscarriage.

Eventually there was a pressure with time, as Henry needed a bone marrow transplantation before he got too sick. It was decided that Henry would receive a transplant from an anonymous donor. This attempt to save Henry was initiated in July 2000 when he was 4.5 years old. Unfortunately, Henry got worse; he needed to go to the hospital several times, needed blood transfusions, and he rapidly lost weight. Henry died in a hospital in December 2002. The cause of death was a lung infection resulting from the fungus Aspergillus—this infection is not treatable and is a common cause of death in Fanconi patients.

Whereas the heroic attempts to save Henry ended tragically, there were other cases where the combination of PDG and IFV turned out more successful. Molly Nash was another child suffering from Fanconi anemia. Her parents, Lisa and Jack Nash, considered PGD. The first attempt resulted in miscarriage. In the second attempt, pregnancy failed. In the third attempt, a larger number of eggs were retrieved; only one was a match, but this was successful, and Lisa Nash became pregnant. In August 2000, Adam was born. Using stem cells derived from Adam's umbilical cord, Molly's disease could be cured. The transplant did not cure her Fanconi anemia, but it did prevent leukemia from being developed. Adam has been called "the world's first savior sibling" and "the first designer baby."

Another child rescued by a "savior sibling" is the UK resident Charlie Whitaker. He suffered from a rare blood disorder named Diamond-Black-fan anemia. This disorder prevented him from producing red blood cells, and he needed regular blood transfusions. Treatment was not allowed in the United Kingdom, so the parents looked for help in the United States. Charlie's younger brother Jamie was born as a result of IVF-PGD, and stem cells from Jamie could be used to cure Charlie.

VIRAL VECTORS FOR GENE THERAPY

As it is difficult to find a perfect donor for stem cell transplantation, other methods have been considered for delivery of a healthy gene. Viruses have an effective machinery to introduce their DNA into host cells, and for this reason, they have become popular tools in gene therapy.

Different categories of viruses have been used. **Retroviruses** have their genomes in the form of an RNA molecule, unlike most other living systems where the genome is made up of DNA. Retroviruses have a mode of propagation that includes the reverse transcription of their RNA genome into a DNA copy, a copy that then may be inserted into the host genome (**Figure 14.1**). The advantage with retroviruses is that they will stably integrate their genetic information into the host cell. This means that all descendents of the genetically changed host cell will contain the viral sequences. At the same time, retroviral vectors integrate into several sites in the genome in an uncontrolled manner, as we see later in this chapter.

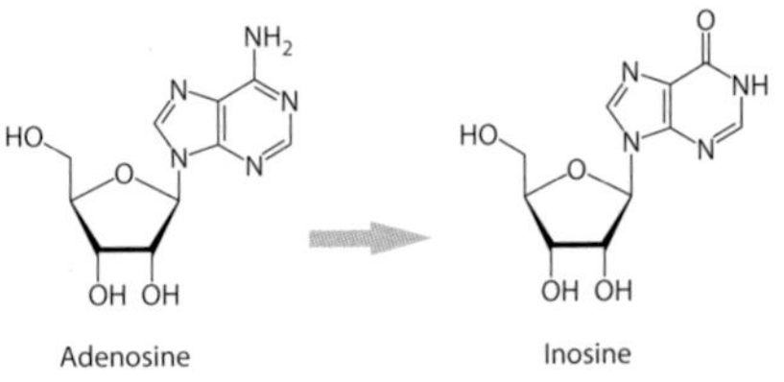

FIRST CASE OF GENE THERAPY: CURING ADENOSINE DEAMINASE DEFICIENCY

The history of gene therapy begins with the successful treatment of a girl, Ashanti DeSilva. She suffers from the inherited disorder known as adenosine deaminase severe combined immunodeficiency (ADA-SCID). The disease is caused by a defect in the gene encoding the enzyme adenosine deaminase (ADA), an enzyme responsible for a reaction where adenosine is metabolized into inosine (**Figure 14.2**).

If the ADA enzyme is deficient, the nucleotide dATP will accumulate. In turn, dATP will inhibit ribonucleotide reductase, an enzyme-producing deoxyribonucleotides (dNTPs), the building blocks of DNA. When dNTPs are

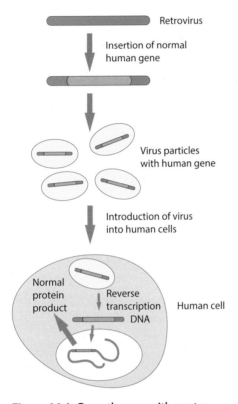

Figure 14.1 Gene therapy with a retrovirus as vector. A normal copy of a human gene under its own promoter is inserted into a modified retrovirus genome. The genome is packaged into virus particles, and these are allowed to transduce patient cells, such as bone marrow cells. DNA is integrated into the genome of patient cells and produces functional protein.

Figure 14.2 Reaction catalyzed by the enzyme adenosine deaminase. Adenosine deaminase (ADA) deaminates adenosine to form inosine. This enzyme is deficient in the disorder adenosine deaminase severe combined immunodeficiency (ADA-SCID).

limiting, cell division is impaired. A high expression of ADA is required by the lymphocytes and other immune cells. Therefore, the immune system is compromised when ADA is deficient.

In 1990 Ashanti DeSilva received gene therapy. The healthy ADA gene was introduced with the help of a retrovirus into her own T cells (cells important in cell-mediated immunity). First, T cells were withdrawn from her immune system and grown in culture in the laboratory after having been mixed with a retrovirus carrying the ADA gene. The resulting T cells, now carrying the healthy ADA gene, were then reintroduced into Ashanti by injection. Normal ADA protein was now produced by her. She received additional injections over a period of 2 years. As a precaution, to complement this treatment, she also received therapy in the form of enzyme protein. In 2003 she was examined again. About 20% of her lymphocytes still produced the healthy ADA protein, 10 years after treatment. As Ashanti also received ADA in the form of protein, one cannot make definite conclusions on how effective the gene therapy was. But it was nevertheless encouraging that ADA was produced as a result of the gene therapy.

Whereas Ashanti DeSilva is reported as a success story, there were also significant backlashes during the history of gene therapy. For instance, there was an experiment with patients with ornithine transcarbamylase deficiency (OTCD). Clinical trials were initiated in 1997. One of the patients was 18-year-old Jesse Gelsinger. He was treated with an adenovirus vector carrying the gene encoding ornithine transcarbamylase. Most unfortunately, his immune system reacted strongly against the viral vector used to carry the gene into his cells. This led to multiple organ failure and brain death. Gelsinger was the first patient to die as a result of a gene therapy trial.

WITH LENTIVIRAL VECTORS LEUKEMIA IS AVOIDED

In the case of Ashanti DeSilva's therapy, the type of retrovirus used was derived from a murine leukemia virus, a type of **gamma retrovirus**. Since then, gamma retroviruses have been successfully used several times for gene therapy. However, for some diseases (SCID-X1, chronic granulomatosis, and Wiskott-Aldrich syndrome), patients frequently developed leukemia as a result of the gene therapy. This is because the virus has a tendency to insert nonrandomly in the genome as it prefers promoters and enhancers of highly expressed genes. Furthermore, long terminal repeat (LTR) regions present in the vector contain promoter or enhancer elements that tend to drive the expression of a host gene located next to the LTR region. As a result, **oncogenes** in the host genome may easily be activated.

To avoid this problem, **lentiviruses** have instead been exploited more recently. The human immunodeficiency virus, HIV, is one well-known example of such a virus. In the lentiviral DNA constructs used in gene therapy, the gene of interest has its own promoter, but the promoter sequences of the virus LTR have been removed. Thereby, the problem of activating other genes is avoided. Another advantage of lentiviruses is that they are able to also target nondividing cells. Lentiviral vectors were, for instance, used in experiments to cure patients with the Wiskott-Aldrich syndrome (WAS). In this disease blood coagulation and the immune system are defective. Responsible for the disease are mutations in WAS, a gene encoding WASP, a protein regulating the cytoskeleton. The protein is expressed in blood cells, including immune cells. For gene therapy, the wild-type WAS gene was inserted into a lentiviral vector where the gene was under the control of its normal promoter. Bone marrow cells of the patients were collected, and these cells were then subjected to the lentiviral vector construct. The resulting cells were reinfused intravenously back into the patients. The results showed that integration of the viral construct was different as compared to the previously used gamma-retroviral vectors—integration into the genome was more random, and there was no activation of oncogenes.

GENE THERAPY WITH ADENO-ASSOCIATED VIRUS

As an alternative to retroviruses, adeno-associated viruses (AAVs) have been used in gene therapy. Whereas many retroviruses infect mainly dividing cells, AAVs are able to also infect nondividing cells. Furthermore, they do not integrate into the host DNA. AAVs have been shown to be effective in different forms of somatic cell therapy. For instance, promising results were reported in 2017 in a clinical trial where an AAV vector containing a factor IX gene was infused in patients with hemophilia B (Chapter 9). The same year an AAV-based therapy for inherited retinal dystrophy was approved in by the US Food and Drug Administration. The purpose of the therapy is to introduce a healthy copy of the gene *RPE65* into the eyes of patients with mutations in this gene. This was one of the first gene therapies in the United States approved for a genetic disease—incidentally a costly therapy, estimated at US$850,000. A 13-year-old boy in the United States was the first patient to receive this therapy in 2018. A third example of an AAV-based therapy is directed against **spinal muscular atrophy** 1 (SMA1). The disorder leads to loss of motor neurons and muscle wasting. It is a fatal disorder, being the most common genetic cause of infant death. It most often leads to death before age 2. The cause of SMA1 is a defective gene so that sufficient amounts of protein SMN cannot be produced. In one clinical study, an AAV vector with the *SMN* gene was intravenously introduced into 15 affected children patients. The outcome was dramatic—after 20 months all children were alive and had much improved motor functions as compared to untreated SMA1 patients. For more on SMA gene therapy, see also section "An oligonucleotide therapy for spinal muscular atrophy" in this chapter.

CANCER GENE THERAPY WITH p53

Gene therapy may also be used for the treatment of cancer. For instance, the gene most commonly mutated in cancers is the *TP53* gene. The gene encodes p53, a protein that plays an important role in blocking the formation of tumors. Important mechanisms of p53 include activation of DNA repair proteins when DNA is damaged, arrest during cell division to allow DNA repair, and initiation of **apoptosis** when DNA cannot be adequately repaired. When the *TP53* gene is mutated, it leads to a high frequency of mutations in the genome, thereby facilitating the development of cancer.

Could normal p53 protein be introduced to cancer cells to overcome the harmful effect of *TP53* mutations? A group of Chinese scientists developed a method for the treatment of cancer that was the first gene therapy product to be approved for clinical use in humans. The drug, Gendicine, is an adenovirus that is engineered to express the normal p53 protein. It was approved in 2003 by the Chinese State Food and Drug Administration to treat head and neck squamous cell carcinoma, one of the cancers where p53 is known to be mutated.

CANCER GENE THERAPY BY BOOSTING IMMUNE RESPONSE: CAR-T CELL THERAPY

A recent and promising gene therapy for cancer is one where immune cells are engineered to kill cancer cells. The therapy is referred to as **CAR-T cell therapy**.

As a first step in this method, blood is withdrawn from the patient. A procedure is applied for enrichment of the immune cells known as T cells. The cells are then genetically engineered with the help of a retrovirus to express a chimeric antigen receptor (CAR) on their surface. Finally, the modified T cells are administered to the patient. The novel T-cell receptors will recognize a specific antigen present on the surface of cancer cells, and upon this recognition the T cell will be activated to destroy the cancer cell.

One application of this technology is a product named axicabtagene ciloleucel (produced by the company Kite Pharma), designed to treat lymphomas. In this therapy, the CAR targets the antigen CD19, a protein expressed on the cell surface of B-cell lymphomas and leukemias. In 2017 promising results were reported in a study of patients with aggressive non-Hodgkin's lymphoma. More than a third of the 101 patients on the trial were still in complete remission at 6 months.

GENOME EDITING: CRISPR/Cas9

The gene therapy experiments with viral vectors illustrate how a healthy copy of the gene is supplied to compensate for a deficient gene. Thus, in these methods there are no changes made to the sequence of the erroneous gene. However, there are also methods available that can accomplish these kinds of alterations. Methods to change a DNA sequence (site-directed mutagenesis) were developed already in the 1980s. Mutations were created in a piece of DNA in the test tube, and then the modified DNA was reintroduced into an organism. The aim of such studies was to examine phenotypic effects of mutations. But more recently, much more powerful methods have been developed to edit a genome sequence, methods that allow editing directly *in vivo*. Most notably, there is a technology known as CRISPR/Cas9, discovered and developed by Emmanuelle Charpentier and Jennifer Doudna. It started to be used around 2013 and in the last years we have seen a dramatic increase in the number of applications of this technology.

The CRISPR/Cas9 methodology is based on a bacterial defense system, a system protecting bacteria from foreign DNA, such as plasmids and phages. Bacteria are slightly different when it comes to the molecular components of this defense system. For editing of just any genome, including the human genome, researchers have exploited a simple system that is present in the bacterium *Streptococcus pyogenes* (**Figure 14.3**). In the bacterium, the first step of the defense mechanism is recognition of the invading DNA. How does the bacterial machinery distinguish between its own DNA and foreign DNA? Part of the answer is that the host DNA contains more frequently short sequences known as Chi sites, and these sequences are recognized as "self." After a foreign DNA has been recognized as "non-self," pieces of it are cut out and inserted into a locus in the bacterial genome referred to as "clustered regularly interspaced short palindromic repeats" (CRISPR). This locus is then transcribed and processed into smaller crRNA molecules. An additional RNA, tracrRNA, base pairs to a region in crRNA. The resulting RNA structure binds to the Cas9 protein, and this complex, in turn, is able to recognize foreign DNA with a sequence complementary to the crRNA. The Cas9 protein, a nuclease enzyme, will cleave both

Figure 14.3 Bacterial immunity based on CRISPR/Cas9 in *Streptococcus pyogenes*. The first step in *S. pyogenes* CRISPR/Cas9 immunity is the acquisition of a fragment (cyan) of an invasive DNA into the CRISPR locus. This fragment is adjacent to a PAM sequence ("protospacer adjacent motif," yellow). It is cut out by Cas proteins (Cas1, Cas2, and Csn2) and introduced into the CRISPR array as a new spacer. DNA elements previously identified as foreign DNA and that are within the array are represented by the red and green rectangles. The CRISPR array is part of a bacterial locus also containing the gene for a tracrRNA (trans-activating crRNA) and four different Cas proteins. In a second step, the CRISPR array is transcribed and processed to form crRNA molecules, each of them carrying a piece of foreign DNA. The tracrRNA gene from the CRISPR locus is also transcribed, and the RNA (orange) forms complexes with the crRNAs. In the third step, an invading DNA is recognized and destroyed. The crRNA/tracrRNA structure will recognize the DNA based on sequence complementarity to the crRNA component (base-pairing as shown between two cyan rectangles), and the recruited Cas9 protein (a nuclease) will cleave the DNA next to a PAM motif (position of scissors).

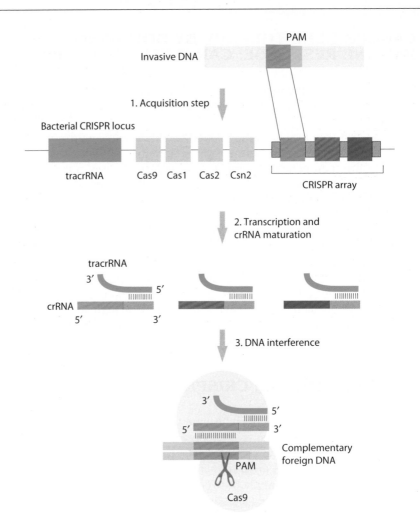

strands of DNA, leading to degradation of the foreign DNA. The recognition and cleavage of invading DNA by the crRNA/tracrRNA/Cas9 complex are referred to as the DNA interference step of the CRISPR defense system.

For genome engineering experiments, the DNA interference component of the CRISPR defense system is exploited. The crRNA and tracrRNA have been combined into a single transcript known as a sgRNA (single guide RNA). This RNA in combination with Cas9 can be used to target any site in a genome and there generate a double-strand break. The break is then repaired, either by error-prone mechanisms like nonhomologous end joining (NHEJ, Figure 5.9) or by creating a specific sequence at this site using a homologous repair template.

As the CRISPR/Cas9 system has only two components, the Cas9 enzyme and the sgRNA, it offers a convenient and effective system to edit a genome sequence. The genes or RNAs encoding the Cas9 and sgRNA components of the CRISPR system may be delivered to an animal in different ways—with the help of a virus, in the form of a lipid nanoparticle, or even by direct injection of DNA or RNA.

CRISPR/Cas9 AND SICKLE CELL ANEMIA

We now examine a specific example of how CRISPR/Cas9 is used to correct a genomic error. We have discussed disorders related to erroneous β-globin, such as sickle cell anemia and β-thalassemia. Could these disorders be treated using a gene editing methodology like CRISPR? As a step to develop therapies, a research team has demonstrated that the CRISPR/Cas9 system may be used to correct mutations in human hematopoietic stem cells, mutations such as the Glu6Val mutation of sickle cell anemia (**Figure 14.4**). The CRISPR/Cas9 system was delivered with an adeno-associated virus vector. This system included a guide RNA directed against the site of the Glu6Val mutation in the β-globin gene. In addition, a DNA molecule was supplied with the nucleotide sequence of the normal β-globin gene, allowing homology directed repair of the double-stranded break introduced by Cas9. In this manner, the Glu6Val mutation could be corrected. The principle of homology directed repair is outlined in **Figure 14.5**.

There are additional attempts to develop an efficient gene therapy for sickle cell anemia. In Chapter 10 we touched upon the fact that some sickle cell disease patients have milder symptoms of the disease as they have higher levels of fetal hemoglobin (HbF). The higher expression of HbF is caused by genetic variants that reduce the expression of *BCL11A*, a regulator of HbF. This observation is now exploited as a gene therapy for both sickle cell anemia and β-thalassemia. From 2018 β-thalassemia patients in Europe are part of a trial in which CRISPR is used to reduce the expression of *BCL11A* by genetically modifying an enhancer of that gene. The result

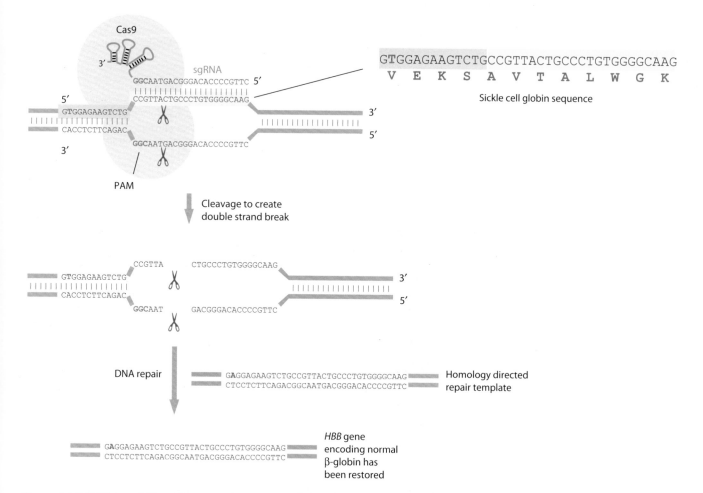

Figure 14.4 Editing a sickle cell genome sequence with CRISPR/Cas9. CRISPR/Cas9 was used to correct the Glu6Val mutation responsible for sickle cell anemia. A sgRNA was designed to be complementary to a portion of the *HBB* gene. The complementary region is highlighted with yellow background, and an upstream portion of the sickle cell gene containing the Glu6Val mutation is with a green background. The mutant nucleotide T and resulting valine (V) are shown in red. Cas9 will create a double-strand break (DSB) close to a PAM motif (red GGC sequence). The DSB may be repaired using different mechanisms, but here is shown homology directed repair. It uses a template as shown where the nucleotide A of the normal *HBB* gene is highlighted in green. The repair template sequence will replace the original patient sequence, and the normal *HBB* gene will be restored.

Figure 14.5 Homology directed repair of a double-strand break in DNA. A double-strand break in DNA may be repaired using homology directed repair. On top is shown a DNA which has been damaged by a double-strand break together with the corresponding homologous undamaged DNA. The first step in the repair pathway is the trimming of ends to generate 3′ tails. In the next step, a free 3′ end invades the homologous DNA on the basis of strand complementarity. A branch point is thus formed, and this branch point can move along the DNA. The 3′ end of the invading strand is extended as guided by the new template, while the branch point is able to migrate. The newly synthesized DNA then pairs with the top strand. The top strand is finally synthesized using the repair template DNA sequence. The green portions show regions of DNA synthesized by the repair pathway and that have the sequence of the repair template DNA.

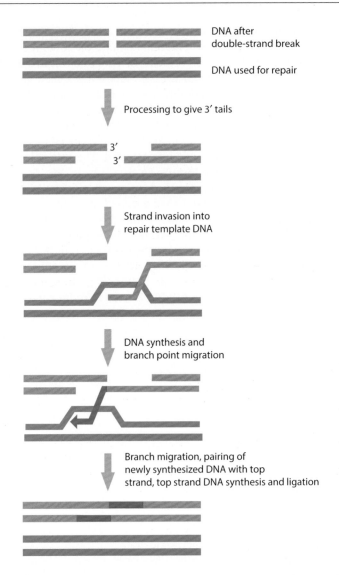

DNA after double-strand break

DNA used for repair

Processing to give 3′ tails

Strand invasion into repair template DNA

DNA synthesis and branch point migration

Branch migration, pairing of newly synthesized DNA with top strand, top strand DNA synthesis and ligation

is efficient expression of HbF, compensating for the lack of normal hemoglobin. In a procedure developed by the company CRISPR Therapeutics, hematopoietic stem cells are withdrawn from the patients, modified using the CRISPR protocol, and finally reintroduced into the patient. The same strategy will be attempted for sickle cell anemia.

THE NUMBER OF CRISPR APPLICATIONS IS GROWING RAPIDLY

The CRISPR/Cas9 technology has the potential to cure inherited disorders and cancer as it allows the correction of DNA sequence errors. There are variants of the technology that allow a more precise editing as compared to the method outlined in Figure 14.4. Thus, editing of genomic sequences is possible without cleavage of double-stranded DNA, thereby eliminating the requirement for double-stranded DNA break repair. A method where an A•T base pair is converted to a G•C pair is shown in **Figure 14.6**.

In addition, there are many other applications of CRISPR/Cas9 in basic biomedical research. For instance, it allows studies of gene function by disrupting genes. In this type of approach, multiple sites in a genome may be examined in a single experiment by using libraries of sgRNAs. Furthermore, CRISPR components allow targeting of double-stranded DNA,

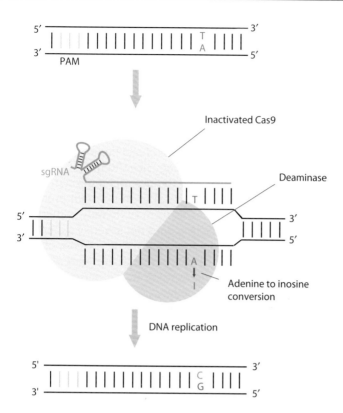

Inactivated Cas9

sgRNA

Deaminase

Adenine to inosine
conversion

DNA replication

Figure 14.6 Base editing without DNA cleavage. An A•T base pair is converted to a G•C pair with a system derived from the components of CRISPR/Cas9. The first step is the binding of Cas9 and a sgRNA such that a bubble of single-stranded DNA is formed. Cas9 is catalytically inactive, preventing cleavage of DNA. It is fused to a deaminase enzyme, which is able to deaminate adenine to form inosine. During replication of DNA, the inosine is read as guanine, eventually leading to replacement of A•T with G•C.

and by modification of the system, CRISPR can be used to cut RNA as well as single-stranded DNA at specific sites. Moreover, CRISPR is developed to identify viral and other pathogenic sequences, showing its potential as a simple and efficient *diagnostic* tool.

MUTATIONS IN DUCHENNE MUSCULAR DYSTROPHY AND THE EFFECT OF EXON ELIMINATION

To evaluate the use of CRISPR for human disease, a number of studies have been carried out in model organisms such as the mouse. As a first example, we consider a mouse model of Duchenne muscle dystrophy (DMD), a human disease characterized by progressive muscle weakness and wasting. DMD is a relatively common inherited disorder. Approximately 1 in 3,300 boys are affected. Responsible for the disease are mutations in a gene present on the X chromosome. The gene, also named *DMD*, is huge with about 2 million base pairs and 81 exons. The gene encodes dystrophin, a protein whose biological function is to act as a link between cytoskeleton actin filaments and proteins located on the inside of the muscle fiber's cellular membrane.

There are many different mutations that cause DMD. About 60% of cases of the disease are due to deletions of one or more exons. The typical outcome of such a deletion is that the normal reading frame of the messenger RNA (mRNA) product is disrupted, leading to premature stop codons. Remember from Chapter 4, Figure 4.19b, that the ends of exons need to be compatible with respect to codons to preserve the translation reading frame. The exon ends need to be of the same phases (0, 1, or 2) as further illustrated in **Figure 14.7**.

As an example of the principles in Figure 14.7, we consider a specific common disease variant of DMD where a genomic region with exons 49–50 is deleted. In this case, exons 48 and 51 are not compatible with respect to

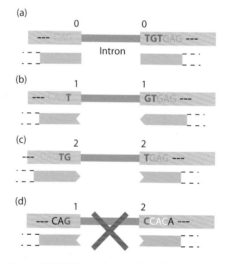

Figure 14.7 The relationship of translation reading frame to splicing. Four examples of the joining of two exons are shown, each with a different configuration as to the translation reading frame. (a) The joining of two exons where both exons are of phase 0, meaning that one codon in one exon is directly combined with another codon in the other exon. To graphically illustrate phase 0, the green rectangles representing the exons end with straight lines. (b) Both exons are of phase 1, meaning that the first nucleotide in a codon is provided by one exon and the next two nucleotides of the codon are in the second exon. Phase 1 is also shown with green rectangles with broken lines that point to the left. (c) As in (a) and (b), but exons are of phase 2, that is the first two nucleotides in a codon are provided by one exon, and the third nucleotide of the codon is in the second exon. Phase 2 is shown with green rectangles with broken lines that point to the right. (d) Whenever the phases of two exons do not match, in this case the phases are 1 and 2, the result is a shift in translation reading frame.

the reading frame. Therefore, the deletion of exons 49 and 50 gives rise to a shift in reading frame (**Figure 14.8a** and **b** where we use the same graphical representation of exon end phases as in Figure 14.7).

From Figure 14.8, it is also evident that the reading frame may be restored by removing additional exons. For instance, if we were to avoid also the exon 51, the translation reading frame will be restored as the 3′ end of exon 48 is compatible with the 5′ end of the exon 52 (**Figure 14.8c**). This means that in the case DMD is caused by a deletion of exons 49–50, a possible therapy is provided by removal of additional exons. Such deletions may be accomplished with methodology like CRISPR.

It should be kept in mind that the dystrophin protein expressed as a result of such a therapy is not identical to the normal protein, as exonic material is missing. Required in this form of therapy is therefore that the protein resulting from exon deletion does not severely suffer from the lack of sequence specified by the eliminated exons. In the case of DMD, it fortunately turns out that some parts of dystrophin are more dispensable than others. In fact, it has been demonstrated that a person can live a rather normal life with as much as 46% of the *DMD* coding sequence deleted.

In addition to human DMD, there is a mouse model of this disease, isolated and characterized in 1989. The actual mutation responsible for the phenotype is a point mutation converting a glutamine codon to a stop

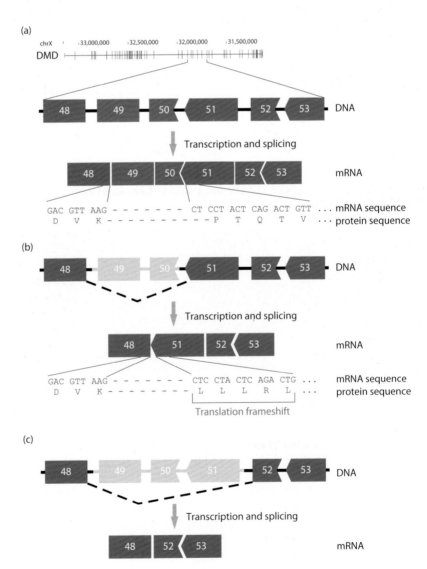

Figure 14.8 DMD exon structure and relationship to translation reading frame. (a) On top is shown the *DMD* gene structure with exons in green. Below is shown details of exons 48–53 at the level of DNA, RNA, and protein. Exons are shown as rectangles where straight or pointed lines at the ends of the exons reflect the intron phases, as in Figure 14.7. At bottom is shown the mRNA and protein sequences of the 3′ end of exon 48 and the 5′ end of exon 51. (b) The two exons 49 and 50 are deleted at the genomic level. Splicing of exons 48 and 51 is shown with a dashed line. The mRNA resulting from splicing is now read by the translation machinery so that a shift of reading frame occurs at the 5′ end of exon 51. (c) The correct reading frame may be restored by deletion of exon 51. The protein product is now lacking the sequences encoded by the exons 49–51 but is otherwise normal. Chromosomal positions refer to human genome version GRCh37/hg19.

codon. The mutation is in exon 23 of the mouse *DMD* gene. There is no equivalent human mutation; the mouse mutation just serves as a model for low expression of dystrophin and a phenotype similar to the human DMD disease. To cure mice carrying this mutant gene, an exon deletion method as previously outlined has been attempted as described in the next section.

RESTORING DYSTROPHIN EXPRESSION IN DMD BY ELIMINATION OF EXONS

Using the mouse model of DMD as described, different research groups showed in 2015 that muscle function was significantly improved in these animals when deletion of exon 23 was accomplished with the help of CRISPR. Exon 23 may be removed without altering the translation reading frame of *DMD*. In the absence of this exon, the amino acid sequence of the protein product is identical to the normal protein except that the amino sequence of exon 23 (a total of 72 amino acids) is lacking.

To create the deletion, two different CRISPR guide RNAs were introduced into the mutant mice with the aid of an AAV. The guide RNAs were constructed to specify cleavage sites flanking exon 23 (**Figure 14.9**). A fragment was excised as a result of the two cuts and ends joined with NHEJ. (Remember that NHEJ will generate a mixture of repair products, but in this case, the repair sites are in intronic regions that are not critical for splicing or any other important *DMD* gene function).

Many DMD patients have mutations that cause a shift of reading frame during translation. In one experiment, DMD patient myoblasts (stem cells that are muscle cell progenitors) grown in culture were genetically altered with CRISPR/Cas9 to introduce mutations. By creating either small deletions or by deletion of one or more exons, the correct reading frame could be restored. As a result, dystrophin expression was observed in the myoblasts. There was also significant expression of human dystrophin in mice that had been transplanted with the cured human cells.

CRISPR/Cas9 APPLIED TO PATIENTS

So far, we discussed the use of CRISPR in model organisms, or with cultured human cells. But in 2016 the first patient was treated with the help of CRISPR technology. In China, a patient with aggressive lung cancer received cells modified with CRISPR. Immune cells were removed from the patient's blood, and a gene named *PDCD1* that encodes "programmed cell death protein 1" (also called PD-1) was inactivated. The patient cells lacking active PD-1 were cultured to increase the number of cells and then reintroduced to the patient. The normal functions of PD-1 are to downregulate the immune response and promote self-tolerance. The absence of PD-1 activates the immune system to attack cancer cells.

There are additional trials to use CRISPR/Cas9 on patients. In 2017 an attempt was initiated in China (First Affiliated Hospital of Sun Yat-Sen)

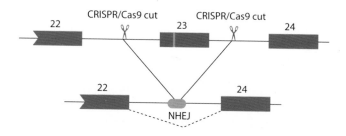

Figure 14.9 CRISPR deletion of mouse DMD exon. Exons 23–25 (green) of a mouse mutant *DMD* gene are shown. Exon 24 has a nonsense (stop-gain) mutation (orange bar). To eliminate this exon, two CRISPR/Cas9 guide RNAs are directed against sites flanking the exon 24. A region containing exon 23 will be excised, and the distal ends are repaired using nonhomologous end joining (NHEJ). In the DNA construct after repair, the exons 22 and 24 are joined through splicing (dashed line).

to destroy human papillomavirus (HPV), a virus that causes cancer. A gel containing the components of the CRISPR/Cas9 machinery is applied to the cervix. The idea is that in HPV-infected cells, the virus will be selectively targeted and destroyed, while normal cells will not be affected by the therapy. This was in fact the first attempt to edit cells while they are still inside the body.

Experiments have also been carried out with editing of human preimplantation embryos with CRISPR/Cas9. In these cases, embryos are used that are not intended for establishing a pregnancy. For instance, US scientists have reported on the successful correction of a mutation in the gene *MYBPC3*, a mutation responsible for hypertrophic cardiomyopathy, a disease where the heart muscle thickens. Similarly, a Chinese research group has reported correction of a β-thalassemia mutant (A > G 28 nucleotides upstream of the transcription start site) in human embryos. In this case, a base editor was used as outlined in Figure 14.6.

GENE THERAPY TO CHANGE GENE EXPRESSION WITH OLIGONUCLEOTIDES

The previous examples of gene therapy with CRISPR/Cas9 are cases where the genome is edited. We now turn to a different kind of strategy for gene therapy where the disease gene is left unchanged and instead expression of a disease-related gene is altered with short synthetic DNA or RNA molecules, **oligonucleotides**. The examples to follow demonstrate various uses:

1. Allele-specific downregulation of a mutant gene (Huntington's disease).
2. Masking of an erroneous splice site (Usher syndrome).
3. The enhancement of a specific splice product by reducing the effect of a splicing inhibitory element (spinal muscular atrophy).
4. Elimination of phenotype by exon skipping (DMD).

The exogenous nucleic acids may act using different mechanisms. In the next section, we consider **RNA interference** (Chapter 8).

ARTIFICIAL MICRORNAs EXPLOITED AS A THERAPY FOR HUNTINGTON'S DISEASE

Huntington's disease (HD) (Chapter 7) is caused by an expansion of CAG repeats in the gene *HTT*. In a dominant inherited disorder like HD, where the mutated gene copy is toxic, a possible strategy for therapy is to reduce the level of the RNA expressed from the mutated gene. This could, for instance, be achieved using RNA interference.

In one approach for HD gene therapy, reduced expression of the mutant *HTT* gene has been attempted with artificial **miRNAs** that are complementary to the mutant gene but not to the normal *HTT* gene. The major difference between normal and mutant *HTT* is the expansion of repeats, but this difference is difficult to exploit if we desire specific downregulation of mutant *HTT*. Another possibility is to make use of the fact that HD patients are heterozygous in specific positions of the *HTT* gene. For instance, consider the single nucleotide polymorphism (SNP) rs362331 in exon 50 of the *HTT* gene. About 40% of all HD patients are heterozygous with respect to this SNP, one copy of the gene having C and one copy T in this position. For instance, one patient may have C in this position in the mutant version of *HTT*. In such a case, an artificial miRNA may be constructed as shown in

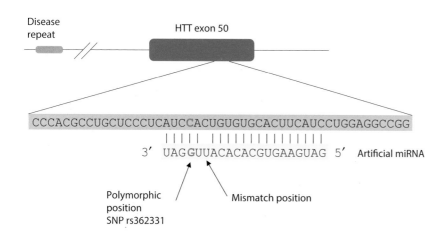

Figure 14.10 Artificial miRNA as a tool to downregulate mutant huntingtin. The gene *HTT* encodes the huntingtin protein. The nucleotide sequence of a portion of *HTT* exon 50 is shown. This exon is part of the disease variant of *HTT*. A miRNA is designed to be complementary to the exon 50 region. The position of SNP rs362331 is highlighted in red, and the position of a mismatch in the miRNA is indicated.

Figure 14.10. In this miRNA, an additional mutation has been introduced. This is to improve discrimination between normal and mutant *HTT* transcripts by decreasing the stability of the RNA-RNA hybrid formed between the miRNA and the normal *HTT* transcript. The method with an artificial miRNA has so far been tested only on cultured cells but might be applicable to patients.

OLIGONUCLEOTIDES TO MASK AN ERRONEOUS SPLICE SITE: USHER SYNDROME

Not only may oligonucleotides be used to initiate cleavage at specific locations in mRNAs, as described for HD and artificial miRNAs. They may also be exploited as agents to block a specific site in an mRNA. In this case, DNA oligonucleotide sequences are used instead of RNA. Examples are found among diseases related to splicing. In Chapter 9 we saw examples of mechanisms regulating splicing, and that many different regions in a precursor mRNA may be involved (Figure 9.10). Hence, **exon skipping** may be achieved through masking of a 5′ splice site or by interfering with an exonic splicing enhancer (ESE). As another alternative, exon inclusion may be favored by masking an intronic splicing silencer (ISS).

If we consider an inherited disease where a mutation introduces an erroneous splice site, one idea for therapy is to mask this site with a DNA oligonucleotide. An example of such as disease is Usher syndrome. We previously encountered this inherited cause of deafness in the context of splicing defects in Chapter 9. All cases of Usher type IC disease among Acadians are caused by a mutation G > A in exon 3 of the *USH1C* gene (Chapter 9, Figure 9.13). The mutation is in a coding sequence region but introduces a novel 5′ splice site.

The effect of oligonucleotide therapy has been examined in a mouse model of Usher syndrome. The mouse equivalent of the human *USH1C* gene was engineered to contain a portion of the human gene, including the G > A mutation. The mice examined were homozygous for the mutation. These mice are deaf and have severe problems with balance and spatial orientation. This is demonstrated by the fact that they do not react to high-amplitude sound and that they show a characteristic head-tossing and circling behavior.

Researchers wanted to examine if oligonucleotides could mask the erroneous splice site. Therefore, oligonucleotides were constructed like one named ASO-29 as shown in **Figure 14.11**. They pair with a region of the mRNA precursor that contains the mutated site.

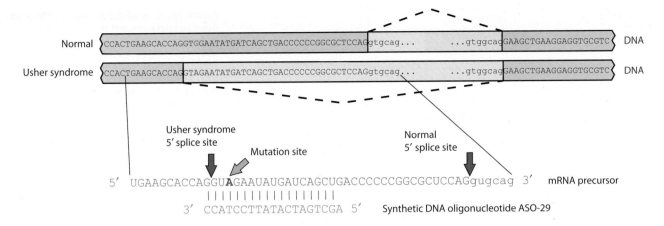

Figure 14.11 **A DNA oligonucleotide masks an erroneous splice site in Usher syndrome.** On top is a portion of the human gene *USH1C*, with exons 3 and 4 (green background) and the intervening intron 3 (yellow background). The normal sequence is shown as well as that representing a form of Usher syndrome with the substitution mutation G > A in exon 3 (G in cyan and A in red). The mutation introduces a novel 5′ splice site. The splice events in normal individuals and in Usher syndrome are shown as dashed lines. Below is shown more details of the disease splice site and its relationship to the normal splice site at the level of the RNA transcript. In a mouse model of the disorder, the mouse gene homologous to *USH1C* was engineered to contain a piece of human *USH1C*, including the portion shown here with exons 3 and 4. The oligonucleotide ASO-29 (sequence in red) is pairing to the disease 5′ splice site and is thereby able to mask this site for the splicing machinery.

Such oligonucleotides were injected into the body cavity (peritoneum) of newborn mice. As controls, mice were used that were untreated or treated with an oligonucleotide with a slightly modified sequence that did not allow a perfect match to the splice site. When the mice had reached adult stage, they were examined. Animals treated with ASO-29 were apparent healthy, they moved around in a normal manner and had improved hearing as compared to the untreated mice. The studies of mouse movement with and without therapy with ASO-29 are shown in **Figure 14.12**. This figure also demonstrates that movement was normal in mice heterozygous for the mutation.

This study illustrates that oligonucleotides may be used to counteract a genetic defect *in vivo*, at least in mice. At the same time, there are some significant problems applying this technology for Usher syndrome therapy in humans. For instance, whereas hearing in the mouse matures after birth, it develops in humans in the uterus, and for this reason any intervention has to be at the prenatal stage.

SPLICING PLAYS A CRITICAL ROLE IN SPINAL MUSCULAR ATROPHY

We previously saw that children affected by spinal muscular atrophy 1 (SMA1) could be cured by gene therapy, where a gene encoding the SMN protein is supplied. We now turn to a different gene therapy for SMA disorders.

Figure 14.12 **Normal movement behavior in mice with *USH1C* mutation is restored by an antisense oligonucleotide.** Three graphs show how mice move in a cage. To the left is shown the abnormal circling behavior of a mouse homozygous for a mutation G > A in exon 3 of the *USH1C* gene (216AA). The middle graph shows the same 216AA mouse as to the left but treated with an oligonucleotide and showing normal movement. The third graph represents a heterozygous mouse (216GA) with normal movement behavior. (From Lentz JJ et al. 2013. *Nat Med* 19:345–350. With permission from Springer Nature.)

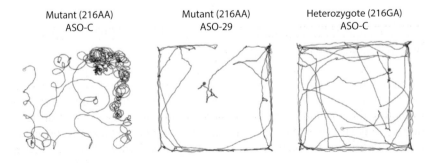

Mutant (216AA)
ASO-C

Mutant (216AA)
ASO-29

Heterozygote (216GA)
ASO-C

(a)

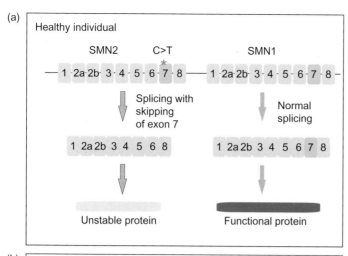

(b)

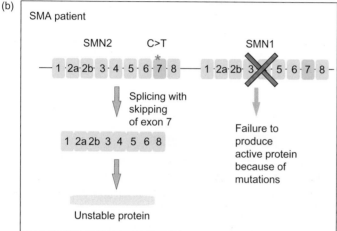

Figure 14.13 The production of SMN protein in health and disease. (a) In a healthy individual, SMN protein is produced from the *SMN1* gene, while the *SMN2* gene has a mutation (C > T, red star) that leads to skipping of exon 7 for a majority of transcripts. Therefore, only a small amount of normal SMN protein is produced from the *SMN2* gene. (b) In a patient with spinal muscular atrophy (SMA), the expression of protein from *SMN1* is defective and at the same time the amount of protein produced from gene *SMN2* is not sufficient.

Two genes, *SMN1* and *SMN2*, encode the protein SMN ("survival of motor neurons") and both genes specify the same amino acid sequence. The two genes are close to each other on chromosome 5 and are related to the disorder SMA1. A difference between the genes is that in *SMN2* there is a nucleotide substitution C > T in the sixth position of exon 7. This variant is within an ESE and has the effect that only 10%–20% of all transcripts are spliced normally to give rise to the SMN protein. The rest (80%–90%) is alternatively spliced so that exon 7 is skipped. This splice product gives rise to a truncated protein (SMNΔ7), a variant of the protein that is unstable and only partially functional (**Figure 14.13a**).

The molecular background to SMA is that the *SMN1* gene is mutated such that a functional SMN protein is not produced from this gene (**Figure 14.13b**). SMA is a recessive disorder and is manifested when a functional *SMN1* gene is completely absent. In such a case, only the gene *SMN2* is able to produce functional SMN protein, but the level of expression is not sufficient.

AN OLIGONUCLEOTIDE THERAPY FOR SPINAL MUSCULAR ATROPHY

An oligonucleotide therapy has been developed to compensate for the deficiency in the *SMN1* gene of SMA patients. The idea behind this therapy is to modify the splicing of the *SMN2* gene so that a larger amount of functional protein is produced from this gene. But rather than targeting the actual C > T substitution characteristic of this gene, the therapy exploits another

Figure 14.14 Enhancing the expression of normal SMN protein in spinal muscular atrophy. The *SMN2* intron 7 contains an intronic splicing silencer (ISS, yellow background), giving rise to skipping of exon 7. Nusinersen is a drug that is an oligonucleotide complementary to the ISS element. It interferes with the ISS and promotes exon 7 inclusion.

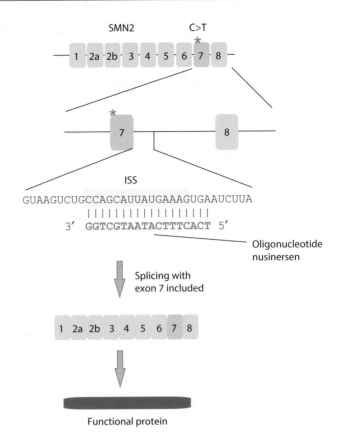

property of the *SMN2* gene. Careful studies of *SMN2* have revealed that in addition to the C > T substitution in exon 7, there is an additional element involved in the skipping of exon 7—an ISS located toward the 5′ end of the intron that is between exons 7 and 8. A drug named nusinersen has been developed, which is an oligonucleotide complementary to this ISS element (**Figure 14.14**). The oligonucleotide is chemically modified (a phosphorothioate derivative) as compared to DNA (**Figure 14.15**). These modifications are to render it more resistant to nucleases that degrade exogenous DNA.

The effect of nusinersen is to reduce exon skipping and thereby produce an mRNA encoding normal SMN protein. It is believed that there are at least two different mechanisms behind the action of nusinersen. First, it blocks the binding of a ribonucleoprotein named hnRNP A1, which

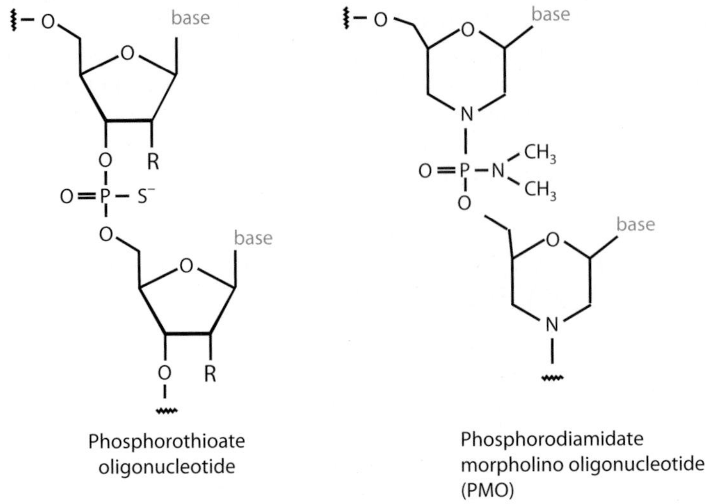

Figure 14.15 Derivatives of natural nucleic acids employed as therapeutic oligonucleotides. In the phosphorothioate derivatives, 2′-hydroxy groups of the sugar rings are replaced with 2′-O-2-methoxyethyl groups (R), and the phosphate linkages are replaced with phosphorothioate linkages. In the phosphorodiamidate morpholino oligomer (PMO), the sugar residues are replaced with methylenemorpholine rings, and the phosphate linkages are replaced with phosphorodiamidate linkages.

is responsible for the splicing silencing. Second, it causes changes in RNA secondary structure and prevents an inhibitory interaction with distant downstream sequences within intron 7.

Nusinersen has been shown to be efficient and safe in Phase 3 trials. In 2016 it was approved by the US Food and Drug Administration (FDA) as a treatment for SMA, and it is also used elsewhere in the world. The oligonucleotide made history as the first drug for SMA. It has been reported that nusinersen will be among the most expensive drugs. Thus, the first year of treatment was estimated to cost US$700,000.

OLIGONUCLEOTIDES TO PRODUCE EXON SKIPPING IN DUCHENNE MUSCLE DYSTROPHY

Remember that many cases of DMD are caused by exon deletions or other mutations that alter the reading frame. We again consider the specific variant of DMD where exons 49–50 are deleted. The deletion of exons gives rise to a shift in reading frame, and we discussed how CRISPR may be used to eliminate additional exons to restore the reading frame (see Figure 14.8). But the elimination of exons may also be accomplished at the level of splicing with an oligonucleotide therapy.

To induce skipping of exon 51, the drug eteplirsen (Exondys 51) has been developed. Its likely mechanism is to block the binding of an ESE protein (**Figure 14.16**). It is a phosphorodiamidate morpholino 30-nucleotide oligonucleotide (PMO) (see Figure 14.15). PMOs are, like the phosphorothioate oligonucleotides, more resistant to nucleases than normal DNA. In addition, the absence of charge makes them less likely to interact nonspecifically with proteins. Eteplirsen was found to be efficient in clinical trials and was approved as a drug to treat DMD by the FDA in 2016. As with nusinersen, the treatment is costly—about US$300,000 for 1 year of treatment.

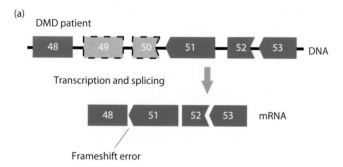

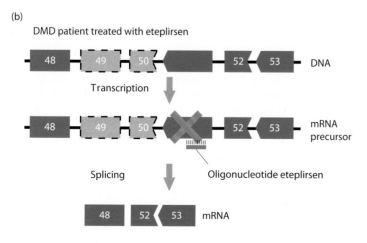

Figure 14.16 Curing DMD with exon skipping. (a) A DMD patient with a genomic deletion that eliminates the exons 49 and 50. The combination of exons 48 and 51 gives rise to a shift in reading frame. (b) The same patient as in (a) but treated with eteplirsen, an oligonucleotide pairing with a region in exon 51 (blue rectangle). This drug has the effect that exon 51 is skipped. As a result, exons 48 and 52 are joined, two exons that are compatible in terms of reading frame. A protein is produced from the spliced RNA that is lacking the amino acid sequence as specified by exons 49–51, but is otherwise intact and sufficiently functional. Eteplirsen has the nucleotide sequence 5'-CTCCAACATCAAGGAAGATGGCATTTCT-3.

The technology of exon skipping described for DMD assumes that the protein of interest is reasonably functional without one or more exons. It should be kept in mind that for many proteins, these criteria are not met.

IMPROVING THE FUNCTION OF A SICK PROTEIN

In addition to gene therapy, there are other methods for treatment of genetic disorders. As one example, we consider here a therapy that makes no attempt to alter the genome or to modify gene expression. Rather, the focus is on the erroneous protein product and modifying its structure such that its function is improved. This strategy has been attempted in the case of cystic fibrosis. The protein that is deficient in cystic fibrosis is CFTR, the cystic fibrosis transmembrane conductance regulator that we discussed in Chapter 6. This protein forms a channel in the membrane of cells and functions as a transporter of chloride ions. It is important for generation of a functional mucus layer that is necessary for effective clearance of bacteria and debris at surfaces such as those of the airway system. When the protein is defective, cells that line the passageways of the lungs and other organs produce mucus that is thicker and stickier than normal, giving rise to inflammation and destruction of the lungs.

The most common mutation in the *CFTR* gene is a deletion of three bases, resulting in a deletion of F508 in the protein (Figure 6.17). This mutation, F508del, is present in 90% of all cystic fibrosis patients. It results in aberrant folding and degradation of CFTR, and the small fraction of the mutant protein that reaches the cell membrane has reduced activity. Another mutation is G551D, the third most common mutation in cystic fibrosis. This mutation has the effect that the chloride channel is less likely to be open. Could substances be found to improve the function of these mutant CFTR proteins? The company Vertex Pharmaceuticals screened about 230,000 compounds to find one that improved the behavior of the G551D mutant CFTR. One effective compound, ivacaftor, was identified, which increases CFTR channel activity. In patients who receive this drug, the decline in lung function slows. The price for the compound is high though, about $294,000 a year. Another drug, lumacaftor, is able to improve folding of the mutant CFTR protein. Since 2015 it is used in combination with ivacaftor to treat patients who are homozygous for the F508del mutation.

After 2015, additional compounds began being evaluated by multiple companies to further improve the folding of the F508del CFTR. A combination of three compounds, one channel opener and two that aid in folding, is in late clinical trials in 2018. This combination is expected to provide efficient therapy, also for patients with one F508del allele.

The type of cystic fibrosis therapy that exploits the effects of compounds on the structure and function of mutant CFTR is of interest, because there are difficulties with gene therapy methods designed to modify the *CFTR* gene or its expression. The cells responsible for production of the CFTR protein are difficult to reach with viral vectors or nucleic acid oligonucleotides, as they are protected by a thick mucus layer. In this situation, the use of small molecules to improve the function of the sick protein seems promising for a majority of these patients.

SUMMARY

- In *gene therapy*, an individual is delivered a nucleic acid to treat a disease. There are two major categories of gene therapy, one in which the genome is modified and one in which gene expression is altered.

- In *germline* therapy, a genetic change is introduced into a sperm or egg cell, or at an early stage in embryonic development. This type of change is transmitted to all descendents of the modified individual. In *somatic* therapy, a genetic change is introduced into a subset of somatic cells, a modification not transmitted to future generations.
- One of the first methods to correct genetic disorders was through transplantation of hematopoietic stem cells derived from a healthy donor. A problem with transplantation of hematopoietic stem cells is that it is difficult to find a donor of the right HLA type. In rare cases, this problem has been addressed using a "savior sibling."
- For gene therapy, vector constructs based on retroviruses, lentiviruses and AAVs are commonly used to deliver a healthy gene.
- In CAR-T cell therapy, immune cells are engineered to kill cancer cells.
- CRISPR/Cas9 is a system for *in vivo* genome editing, which is based on a bacterial defense system. In essence, the Cas9 nuclease generates a double-strand break at the genomic site as specified by a guide RNA molecule, and the double-strand break is then repaired to generate any desired sequence.
- Cases where gene therapy is used to alter gene expression include (1) allele-specific downregulation of a mutant gene, (2) masking of an erroneous splice site, (3) enhancement of a specific splice product by reducing the effect of a splicing inhibitory element, and (4) elimination of phenotype by exon skipping. All of these methods make use of synthetic oligonucleotides to target a specific DNA or RNA region.
- Some cells are difficult to target with somatic gene therapy. In such cases, an alternative therapy is to target the mutant protein responsible for the disease phenotype. This type of therapy has been exploited in cystic fibrosis.

QUESTIONS

1. What is meant by *gene therapy*?
2. What types of disorders may be treated with transplantation of hematopoietic stem cells?
3. What are advantages and disadvantages of *gamma retroviruses, lentiviruses*, and *AAVs* in the context of gene therapy?
4. What examples are there of *cancer gene therapy*?
5. What is the purpose of the CRISPR/Cas9 system in the bacterium *S. pyogenes*? What is the function of the Cas9 protein, the tracrRNA, and the crRNA, respectively?
6. What changes have been made to the bacterial CRISPR/Cas9 system for use in gene editing technology?
7. Using the CRISPR/Cas9 method, a cut is accomplished in the DNA to be edited. What short sequence close to the site of cleavage is required in the original CRISPR/Cas9 method?
8. What properties of the *DMD* gene and its product allow the use of *exon skipping* as a means of therapy?
9. Consider the exons 48–53 of the *DMD* gene as depicted in Figure 14.8. Which of the following exon combinations give rise to a shift in reading frame?
   a. 48-49-51-52
   b. 48-49-50-53
   c. 48-49-52-53
10. Discuss the different mechanisms of action for oligonucleotides used in gene therapy.

11. Explain how a SNP may be exploited in gene therapy for Huntington's disease.
12. What is the target for the oligonucleotide nusinersen used in spinal muscular atrophy therapy?
13. Why are *synthetic analogs* of oligonucleotides used for gene therapy?
14. In cases when it is difficult to deliver a gene or a therapeutic oligonucleotide to a specific tissue, what other therapy methods are available?

FURTHER READING

Early germline modification in mice

Palmiter RD, Norstedt G, Gelinas RE et al. 1983. Metallothionein-human GH fusion genes stimulate growth of mice. *Science* 222(4625):809–814.

Fanconi's anemia

Verlinsky Y, Rechitsky S, Schoolcraft W et al. 2001. Preimplantation diagnosis for Fanconi anemia combined with HLA matching. *JAMA* 285(24):3130–3133.
Whitney MA, Saito H, Jakobs PM et al. 1993. A common mutation in the *FACC* gene causes Fanconi anaemia in Ashkenazi Jews. *Nat Genet* 4(2):202–205.

Savior siblings

Steinbock B. 2003. Using preimplantation genetic diagnosis to save a sibling: The story of Molly and Adam Nash. *Ethical issues in modern medicine*. eds Steinbock BA, Arras JD, London AJ (McGraw-Hill, New York), 6th ed, pp. 544–545.
Strongin L. 2010. *Saving Henry, a mother's journey*. Hyperion, New York, 1st ed, pp. xii, 271 p.

Gene therapy, overviews

Dunbar CE, High KA, Joung JK et al. 2018. Gene therapy comes of age. *Science* 359(6372).
Kaiser J. 2011. Clinical research. Gene therapists celebrate a decade of progress. *Science* 334(6052):29–30.
Naldini L. 2015. Gene therapy returns to centre stage. *Nature* 526(7573):351–360.

ADA gene therapy

Hoggatt J. 2016. Gene therapy for "Bubble Boy" disease. *Cell* 166(2):263.
Muul LM, Tuschong LM, Soenen SL et al. 2003. Persistence and expression of the adenosine deaminase gene for 12 years and immune reaction to gene transfer components: Long-term results of the first clinical gene therapy trial. *Blood* 101(7):2563–2569.

Gene therapy with lentiviral vectors

Aiuti A, Biasco L, Scaramuzza S et al. 2013. Lentiviral hematopoietic stem cell gene therapy in patients with Wiskott-Aldrich syndrome. *Science* 341(6148):1233151.
Biffi A, Montini E, Lorioli L et al. 2013. Lentiviral hematopoietic stem cell gene therapy benefits metachromatic leukodystrophy. *Science* 341(6148):1233158.
King A. 2016. Gene therapy: A new chapter. *Nature* 537:S158.
Naldini L, Trono D, Verma IM. 2016. Lentiviral vectors, two decades later. *Science* 353(6304):1101–1102.
Persons DA. 2010. Lentiviral vector gene therapy: Effective and safe? *Mol Ther* 8(5):861–862.

Gene therapy with adeno-associated viruses

George LA, Sullivan SK, Giermasz A et al. 2017. Hemophilia B gene therapy with a high-specific-activity factor IX variant. *N Engl J Med* 377(23):2215–2227.
Lillicrap D. 2017. FIX It in one go: Enhanced factor IX gene therapy for hemophilia B. *Cell* 171(7):1478–1480.
Mendell JR, Al-Zaidy S, Shell R et al. 2017. Single-dose gene-replacement therapy for spinal muscular atrophy. *N Engl J Med* 377(18):1713–1722.

Cancer gene therapy

Amer MH. 2014. Gene therapy for cancer: Present status and future perspective. *Mol Cell Ther* 2:27.
Fesnak AD, June CH, Levine BL. 2016. Engineered T cells: The promise and challenges of cancer immunotherapy. *Nat Rev Cancer* 16(9):566–581.
Pan JJ, Zhang SW, Chen CB et al. 2009. Effect of recombinant adenovirus-p53 combined with radiotherapy on long-term prognosis of advanced nasopharyngeal carcinoma. *Am J Clin Oncol* 27(5):799–804.
Sadelain M. 2017. CD19 CAR T cells. *Cell* 171(7):1471.

Hemophilia B gene therapy

Barzel A, Paulk NK, Shi Y et al. 2015a. Promoterless gene targeting without nucleases ameliorates haemophilia B in mice. *Nature* 517(7534):360–364.
Porteus M. 2017. Closing in on treatment for hemophilia B. *N Engl J Med* 377(23):2274–2275.
van den Berg HM. 2017. A cure for hemophilia within reach. *N Engl J Med* 377(26):2592–2593.

Early mutagenesis

Kunkel TA. 1985. Rapid and efficient site-specific mutagenesis without phenotypic selection. *Proc Natl Acad Sci USA* 82(2):488–492.

Therapeutic genome editing

Porteus MH. 2015. Towards a new era in medicine: Therapeutic genome editing. *Genome Biol* 16(1):286.
Reardon S. 2015. Leukaemia success heralds wave of gene-editing therapies. *Nature* 527(7577):146–147.

A mouse model of sickle cell anemia and gene therapy

Hardison RC, Blobel GA. 2013. Genetics. GWAS to therapy by genome edits? *Science* 342(6155):206–207.
Xu J, Peng C, Sankaran VG et al. 2011. Correction of sickle cell disease in adult mice by interference with fetal hemoglobin silencing. *Science* 334(6058):993–996.

CRISPR/Cas9 and applications

Bourzac K. 2017. Gene therapy: Erasing sickle-cell disease. *Nature* 549(7673):S28–S30.

Chen JS, Ma E, Harrington LB et al. 2017. CRISPR-Cas12a target binding unleashes single-stranded DNase activity. *bioRxiv*.

Chertow DS. 2018. Next-generation diagnostics with CRISPR. *Science* 360(6387):381–382.

Cyranoski D. 2016. CRISPR gene-editing tested in a person for the first time. *Nature* 539(7630):479.

Dever DP, Bak RO, Reinisch A et al. 2016. CRISPR/Cas9 β-globin gene targeting in human haematopoietic stem cells. *Nature* 539(7629):384–389.

Doudna J. 2015. Genome-editing revolution: My whirlwind year with CRISPR. *Nature* 528(7583):469–471.

Doudna JA, Gersbach CA. 2015. Genome editing: The end of the beginning. *Genome Biol* 16:292.

Gaudelli NM, Komor AC, Rees HA et al. 2017. Programmable base editing of A·T to G·C in genomic DNA without DNA cleavage. *Nature* 551:464.

Gootenberg JS, Abudayyeh OO, Kellner MJ et al. 2018. Multiplexed and portable nucleic acid detection platform with Cas13, Cas12a, and Csm6. *Science* 360(6387):439–444.

Jiang F, Doudna JA. 2017. CRISPR-Cas9 structures and mechanisms. *Annu Rev Biophys* 46:505–529.

Komor AC, Badran AH, Liu DR. 2017. CRISPR-based technologies for the manipulation of eukaryotic genomes. *Cell* 168(1–2):20–36.

Konermann S, Lotfy P, Brideau NJ et al. 2018. Transcriptome engineering with RNA-targeting type VI-D CRISPR effectors. *Cell* 173(3):665–676.

Ledford H. 2015. Mini enzyme moves gene editing closer to the clinic. *Nature* 520(7545):18.

Ledford H. 2016. CRISPR: Gene editing is just the beginning. *Nature* 531(7593):156–159.

Levy A, Goren MG, Yosef I et al. 2015. CRISPR adaptation biases explain preference for acquisition of foreign DNA. *Nature* 520(7548):505–510.

Ma H, Marti-Gutierrez N, Park S-W et al. 2017. Correction of a pathogenic gene mutation in human embryos. *Nature* 548:413.

Niu Y, Shen B, Cui Y et al. 2014. Generation of gene-modified cynomolgus monkey via Cas9/RNA-mediated gene targeting in one-cell embryos. *Cell* 156(4):836–843.

Travis J. 2015. Making the cut. *Science* 350(6267):1456–1457.

Wright Addison V, Nuñez James K, Doudna Jennifer A. 2015. Biology and applications of CRISPR systems: Harnessing nature's toolbox for genome engineering. *Cell* 164(1):29–44.

Wu Y, Liang D, Wang Y et al. 2013. Correction of a genetic disease in mouse via use of CRISPR-Cas9. *Cell Stem Cell* 13(6):659–662.

Xiao-Jie L, Hui-Ying X, Zun-Ping K et al. 2015. CRISPR-Cas9: A new and promising player in gene therapy. *J Med Genet* 52(5):289–296.

Xie F, Ye L, Chang JC et al. 2014. Seamless gene correction of beta-thalassemia mutations in patient-specific iPSCs using CRISPR/Cas9 and piggyBac. *Genome Res* 24(9):1526–1533.

Yin H, Xue W, Chen S et al. 2014. Genome editing with Cas9 in adult mice corrects a disease mutation and phenotype. *Nat Biotechnol* 32(6):551–553.

CRISPR/Cas9 and Duchenne muscular dystrophy

Long C, Amoasii L, Mireault AA et al. 2016. Postnatal genome editing partially restores dystrophin expression in a mouse model of muscular dystrophy. *Science* 351(6271):400–403.

Nelson CE, Hakim CH, Ousterout DG et al. 2016. In vivo genome editing improves muscle function in a mouse model of Duchenne muscular dystrophy. *Science* 351(6271):403–407.

Ousterout DG, Kabadi AM, Thakore PI et al. 2015. Multiplex CRISPR/Cas9-based genome editing for correction of dystrophin mutations that cause Duchenne muscular dystrophy. *Nat Commun* 6:6244.

Tabebordbar M, Zhu K, Cheng JK et al. 2015. In vivo gene editing in dystrophic mouse muscle and muscle stem cells. *Science*.

Hemophilia B and gene therapy in mouse models

Barzel A, Paulk NK, Shi Y et al. 2015b. Promoterless gene targeting without nucleases ameliorates haemophilia B in mice. *Nature* 517(7534):360–364.

Li H, Haurigot V, Doyon Y et al. 2011. In vivo genome editing restores haemostasis in a mouse model of haemophilia. *Nature* 475(7355):217–221.

PolyQ disorders and oligonucleotide therapy

Fiszer A, Krzyzosiak WJ. 2014. Oligonucleotide-based strategies to combat polyglutamine diseases. *Nucleic Acids Res* 42(11):6787–6810.

Huntington's disease gene therapy

Glorioso JC, Cohen JB, Carlisle DL et al. 2015. Moving toward a gene therapy for Huntington's disease. *Gene Ther* 22(12):931–933.

Miniarikova J, Zanella I, Huseinovic A et al. 2016. Design, characterization, and lead selection of therapeutic miRNAs targeting huntingtin for development of gene therapy for Huntington's disease. *Mol Ther Nucleic Acids* 5:e297.

van Bilsen PH, Jaspers L, Lombardi MS et al. 2008. Identification and allele-specific silencing of the mutant huntingtin allele in Huntington's disease patient-derived fibroblasts. *Hum Gene Ther* 19(7):710–719.

Zhang Y, Engelman J, Friedlander RM. 2009. Allele-specific silencing of mutant Huntington's disease gene. *J Neurochem* 108(1):82–90.

Therapy based on modification of splicing

Arechavala-Gomeza V, Khoo B, Aartsma-Rus A. 2014. Splicing modulation therapy in the treatment of genetic diseases. *Clin Genet* 7:245–252.

Spitali P, Aartsma-Rus A. 2012. Splice modulating therapies for human disease. *Cell* 148(6):1085–1088.

Usher type 1C disease and oligonucleotide therapy

Ebermann I, Lopez I, Bitner-Glindzicz M et al. 2007. Deafblindness in French Canadians from Quebec: A predominant founder mutation in the USH1C gene provides the first genetic link with the Acadian population. *Genome Biol* 8(4):R47.

Lentz JJ, Jodelka FM, Hinrich AJ et al. 2013. Rescue of hearing and vestibular function by antisense oligonucleotides in a mouse model of human deafness. *Nat Med* 19(3):345–350.

Ouyang XM, Hejtmancik JF, Jacobson SG et al. 2003. USH1C: A rare cause of USH1 in a non-Acadian population and a founder effect of the Acadian allele. *Clin Genet* 63(2):150–153.

Ouyang XM, Yan D, Du LL et al. 2005. Characterization of Usher syndrome type I gene mutations in an Usher syndrome patient population. *Hum Genet* 116(4):292–299.

Spinal muscular atrophy and oligonucleotide therapy

Cartegni L, Krainer AR. 2002. Disruption of an SF2/ASF-dependent exonic splicing enhancer in SMN2 causes spinal muscular atrophy in the absence of SMN1. *Nat Genet* 30(4):377–384.

Chiriboga CA, Swoboda KJ, Darras BT et al. 2016. Results from a phase 1 study of nusinersen (ISIS-SMN[Rx]) in children with spinal muscular atrophy. *Neurology* 86(10):890–897.

Corey DR. 2017. Nusinersen, an antisense oligonucleotide drug for spinal muscular atrophy. *Nat Neurosci* 20(4):497–499.

Finkel RS, Chiriboga CA, Vajsar J et al. 2016. Treatment of infantile-onset spinal muscular atrophy with nusinersen: A phase 2, open-label, dose-escalation study. *Lancet* 388(10063):3017–3026.

Naryshkin NA, Weetall M, Dakka A et al. 2014. Motor neuron disease. SMN2 splicing modifiers improve motor function and longevity in mice with spinal muscular atrophy. *Science* 345(6197):688–693.

Ottesen EW. 2017. ISS-N1 makes the first FDA-approved drug for spinal muscular atrophy. *Translational Neurosci* 8:1–6.

DMD and oligonucleotide therapy

Goyenvalle A, Griffith G, Babbs A et al. 2015. Functional correction in mouse models of muscular dystrophy using exon-skipping tricyclo-DNA oligomers. *Nat Med* 21(3):270–275.

Lim KR, Maruyama R, Yokota T. 2017. Eteplirsen in the treatment of Duchenne muscular dystrophy. *Drug Des Dev Ther* 11:533–545.

Young CS, Pyle AD. 2016. Exon skipping therapy. *Cell* 167(5):1144.

Cystic fibrosis therapy

Brodsky JL, Frizzell RA. 2015. A combination therapy for cystic fibrosis. *Cell* 163(1):17.

Deeks ED. 2016. Lumacaftor/ivacaftor: A review in cystic fibrosis. *Drugs* 76(12):1191–1201.

Epilogue

We have examined the information content of the human genome with an emphasis on its inherent logic based on the letters A, T, C, and G. These letters and combinations of them have different functional meanings, depending on the context in which they are located in the genome. Our look into the operation of the different DNA sequence motifs has highlighted important biochemical functions of the human genome, including the complex flow of genetic information from DNA to protein. We have also seen how changes in the genome sequence may have phenotypic consequences—in some cases, they result in serious disease. In this final chapter, some of the findings discussed in this book are summarized. Individual genetic information, the analysis of genetic disorders, as well as gene therapy were examined in Chapters 5, 13, and 14. Here we briefly discuss some accompanying ethical issues.

DNA SEQUENCES HAVE A VARIETY OF FUNCTIONAL ROLES

Throughout this book, we discussed different DNA sequence elements of the human genome. These are typically short motifs, with sizes ranging from two (like a CpG site), three (such as a codon), up to about 10–20 nucleotides. Motifs like these that are part of a protein coding gene are summarized in **Figure 15.1**. Some of the elements, like transcription factor binding sites, are operational at the level of DNA, whereas others such as splicing signals operate at the RNA level. Figure 15.1 shows the nature of a protein coding gene, but many of the elements of such a gene are also found in the genes of noncoding RNAs.

It is important to realize that all DNA and RNA sequence elements of functional significance have one or the other binding partner. These are either nucleic acids, DNA or RNA, or they are proteins. One example of nucleic acid interactions is when mRNA codons are being read by tRNA anticodons during translation (**Figure 15.2a**). In such instances, a substitution mutation can change a codon such that a different amino acid is incorporated into protein. Other examples where a sequence encoded in

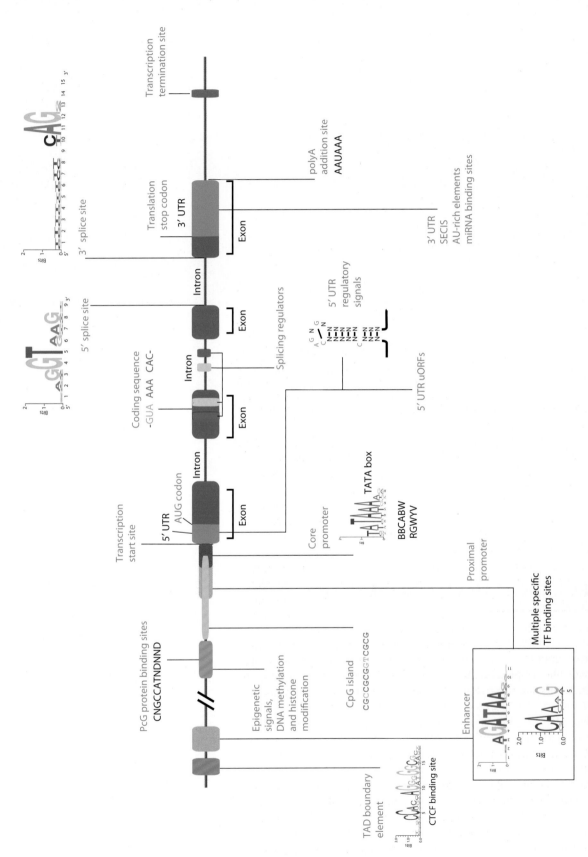

Figure 15.1 A gene contains multiple functional DNA sequence elements. The organization of a human protein coding gene is shown with important elements shown as colored rectangles and with the names of the elements in red. The figure summarizes key elements discussed throughout the book. Conserved sequences are typically shown as sequence logos.

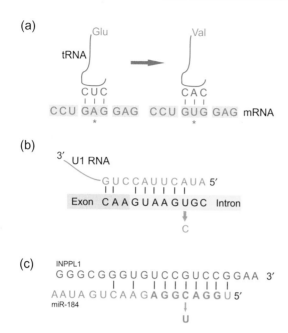

(a)

Glu → Val

tRNA

CUC → CAC

CCU GAG GAG   CCU GUG GAG   mRNA

(b)

3′ U1 RNA

GUCCAUUCAUA 5′

Exon CAAGUAAGUGC Intron

↓
C

(c)

INPPL1
GGGCGGGUGUCCGUCCGGAA 3′

AAUAGUCAAGAGGCAGGU 5′
miR-184

↓
U

Figure 15.2 RNA interactions based on base-pairing. (a) To the left is shown a glutamate tRNA (green) with anticodon CUC reading a glutamate GAG mRNA codon. The mRNA sequence is in red. The star indicates a position in mRNA mutated in sickle cell anemia. To the right is shown how the normal codon GAG has been altered to GUG in sickle cell anemia and is now read by a valine tRNA with anticodon CAC. (b) The 5′ splice site is recognized by the U1 RNA (red) through base-pairing. Here is shown the recognition of one of the 5′ splice sites of the gene *IKBKAP* with the exonic part in light green and the intronic region with a light yellow background. A mutation T > C (U > C at the RNA level, arrow) at the sixth position of the 5′ end of the intron characteristic of familial dysautonomia disrupts the pairing between U1 RNA and the mRNA precursor. (This is the same mutation as shown in Figure 9.14a.) (c) The miR-184 miRNA targets the INPPL1 mRNA for degradation in a mechanism based on base-pairing. A mutation C > T (C > U at the RNA level, arrow) disrupts the interaction of the miRNA seed region (bold letters) with the mRNA target sequence (same mutation as in Figure 8.20a).

the genome is recognized by a nucleic acid based on sequence complementarity is when a spliceosomal RNA interacts with an mRNA precursor or when a microRNA (miRNA) targets an mRNA (**Figure 15.2b,c**). All of the interactions in Figure 15.2 are examples of RNA-RNA interactions, but we also encounter situations of interactions between DNA and RNA. One example is the DNA-RNA hybrid formed in the expression of the *FMR1* gene as discussed in Chapter 10 (Figure 10.29).

However, proteins are also highly significant binding partners to DNA or RNA molecules. A specific protein-DNA interaction is illustrated in **Figure 15.3**, showing the reading of DNA bases through an interaction of specific amino acids in the protein GLI1, a zinc finger transcription factor. We have seen many other examples of DNA and transcription factor interactions such as in Figures 10.3 and 10.5. This type of specific binding is quantitatively significant, as there are some 1,600 different transcription factors encoded by the human genome. In addition, there are proteins other than transcription factors that recognize specific sequences and structural features of DNA. For instance, a range of DNA repair proteins interact with DNA and are able to recognize specific DNA damage. The proteins known as TRF1 and TRF2 interact with telomeres of chromosomes and regulate their length. Furthermore, a number of proteins interact with DNA in the context of epigenetic regulation.

There are also many proteins that bind to specific RNAs. For instance, there are iron regulatory proteins that interact with the iron response element of specific mRNAs related to iron metabolism (Figure 8.4a). In addition, numerous proteins bind to RNA in the context of splicing and the regulation of mRNA translation (Figures 8.11 and 8.15).

Whereas there are simple rules for the pairing of one nucleic acid to another, the recognition of a nucleic acid by a protein is a more complex biochemical story. Billions of years of evolution have created a myriad of ways in which a protein may specifically bind to a nucleic acid. Individual recognition may typically be understood only if we have access to the detailed three-dimensional structure of the protein-nucleic acid complex. As discussed in Chapter 10, elements in DNA that are important for protein recognition are the actual base sequence as well as DNA shape.

When we are to predict the outcome of a genomic change like a nucleotide substitution, the accuracy of that prediction is dependent on the nature of the region being mutated. First, consider a substitution mutation affecting an

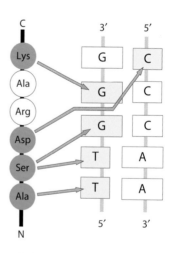

C

3′   5′

Lys   G   C

Ala   G   C

Arg   G   C

Asp   T   A

Ser   T   A

Ala

5′   3′

N

Figure 15.3 A DNA-protein interaction. The transcription factor GLI1 is a zinc finger protein that binds in a sequence-specific manner to DNA. The amino acid sequence of a portion of one of the zinc fingers of the protein is shown to the left, and a region of DNA specifically interacting with the protein is to the right. The amino acids in orange interact with specific bases in the double-stranded DNA as indicated by the arrows.

RNA-RNA interaction. If it affects a protein coding region we can reliably predict the consequences of such a mutation using the genetic code (see Figure 15.2a). In the case of splicing and miRNA action, the rules are less precise because the complementarity of the two RNA strands may not be perfect, and there might be proteins that influence the efficiency of pairing. Finally, consider a DNA or RNA sequence whose only interaction partner is a protein. If this nucleic acid sequence is affected by a substitution mutation, the outcome with respect to protein binding is often problematic to predict. For instance, many mutations in enhancer or promoter regions are hard to evaluate in terms of protein binding efficiency.

We have also seen examples of disorders where the gene elements as shown in Figure 15.1 are mutated. These disorders are summarized in **Table 15.1**. (For additional information on disorders, traits, and mutations, the reader is reminded of Tables 5.1, 5.2, 13.1, and 13.2.) Genetic disorders illustrate the biological significance of specific DNA sequences and highlight molecular mechanisms related to the genome and the flow of genetic information in the cell. At the same time, the molecular mechanisms responsible for the phenotype of many disorders remain poorly understood.

We have seen that studies of genetic disorders effectively highlight the role of the four letters of our genome for the human phenotype. However, it is important to note that a phenotype is also formed by other factors, as discussed in the next section.

YOUR GENOME IS NOT YOUR FATE

The US science fiction movie *Gattaca* (yes, the name is based on the four nucleotides of DNA), launched in 1997, is about a world in a "not too distant future" where genetic modification is being used to create "optimal" individuals. The main character of the movie was born without using this technology, and for this reason he is part of a group whose members are "inferior" or "invalids." He is discriminated against as compared to another group of people being genetically "valid." The movie brings to mind ethical topics related to human genetics. For instance, how deterministic should you regard the genome—or put another way, how much of your destiny is in your genes?

In specific cases, there is indeed a strong genetic contribution to the life of an individual. Consider for instance a serious disorder like Huntington's disease. If an individual has enough copies of the CAG repeat of the *HTT* gene, the disease will certainly manifest itself. Similarly, an individual will suffer from cystic fibrosis if homozygous for the F508del mutation. However, there are many common disorders with a genetic component, but where the genetic contribution is not as strong as in the monogenic disorders. Rather, these disorders come about by an interplay between multiple genetic variants as well as nongenetic factors—mainly environmental factors and lifestyle. Examples of common multifactorial disorders are cancer, Alzheimer's disease, diabetes, and heart disease/stroke. All of these have a prevalence of 1/10 or more and are therefore much more common than any of the inherited monogenic disorders. As with multifactorial disorders, many nondisease traits, such as height, physical strength, and intelligence, are the result of interplay of genetic variants and nongenetic factors. The role of factors not directly associated with the actual nucleotide sequence of the genome is also illustrated by examples of monozygotic ("identical") twins that are not phenotypically identical in spite of their identical genomes.

We reach the conclusion that for many disorders and nondisease traits, your genome is certainly not your fate. Rather, lifestyle and environmental factors are of critical importance.

Table 15.1 Inherited disorders and genomic DNA elements

Disease/trait	Gene	Genomic element mutated	Details of mutation	dbSNP identifier
Sickle cell anemia	HBB	Coding sequence	A > T = > Missense	rs334
Tay-Sachs	HEXA	Coding sequence	G > A = > Missense	rs28941770
Hereditary hemochromatosis	HFE	Coding sequence	G > A = > Missense	rs1800562
Brunner syndrome	MAOA	Coding sequence	CAG = > TAG Stopgain	rs72554632
Speech-language disorder 1, SPCH1	FOXP2	Coding sequence	G > A = > Missense	rs121908377
Speech-language disorder 1, SPCH1	FOXP2	Coding sequence	CGA = > TGA Stop-gain	rs121908378
Hurler-Scheie syndrome	IDUA	Coding sequence	TGA Stop = > GGA Gly Stop-loss	rs387906504
Tay-Sachs	HEXA	Coding sequence	Insertion of TATC, Frameshift	rs387906309
Resistance to HIV infection	CCR5	Coding sequence	32 nucleotide deletion, Frameshift	rs333
Cystic fibrosis	CFTR	Coding sequence	Inframe deletion	rs121908745
Huntington's disease	HTT	Coding sequence	CAG repeat expansion	
Spinocerebellar ataxia 1	ATXN1	Coding sequence	CAG repeat expansion	
Spinocerebellar ataxia 2	ATXN2	Coding sequence	CAG repeat expansion	
Spinocerebellar ataxia 3	ATXN3	Coding sequence	CAG repeat expansion	
Spinocerebellar ataxia 7	ATXN7	Coding sequence	CAG repeat expansion	
Hereditary hyperferritinemia cataract syndrome	FTL	5′ UTR	Substitutions and indel mutations	
Marie Unna hereditary hypotrichosis	HR	5′ UTR	T > C = > uORF ATG codon mutated	rs387906382
Fragile X	FMR1	3′ UTR	T > C = > affecting AU-rich element	
Myotonic dystrophy 1	DMPK	3′ UTR	CTG repeat expansion	
Keratoconus	miR-184	miRNA seed region	C > T = > affecting binding to mRNA 3′ UTR?	
Hemophilia B originating from Queen Victoria of the United Kingdom	F9	Intron	A > G, affecting splicing	rs398122990
Usher syndrome 1C	USH1C	Exon	G > A, affecting splicing	rs151045328
Familial dysautonomia (FD)	IKBKAP	Intron	T > C, affecting splicing	rs111033171
Hutchinson-Gilford progeria syndrome (HGPS)	LMNA	Exon	C > T, affecting splicing	rs58596362
Spinal muscular atrophy (SMA)	SMN2	Exon	C > T, affecting splicing	rs4916
β-Thalassemia	HBB	Core promoter	Substitution mutations affecting TATA box or DPE	
δ-Thalassemia	HBD	Proximal promoter	T > C, disrupts GATA-1 binding site	
Many forms of cancer	TERT	Proximal promoter	Two C > T mutations, disrupt GABP binding site	
X-linked sideroblastic anemia (XLSA)	ALAS2	Enhancer	Substitution mutations affecting binding of GATA-1	
Enhanced expression of HbF in sickle cell disorder	BCL11A	Enhancer	G > T, affects binding of TAL1 and GATA-1	rs1427407
Elevated risk for Parkinson's disease	SNCA	Enhancer	A > G, affects binding of EMX2 and NKX6-1	rs356168
Various developmental disorders	EPHA4 and others	Enhancer	Genomic rearrangements, changes context of enhancer	
Fragile X	FMR1	Promoter	CGG repeat expansion in promoter region	
β-Plus thalassemia	HBB	Polyadenylation signal	Substitution mutations within the motif AAUAAA	

(Continued)

Table 15.1 (*Continued*) **Inherited disorders and genomic DNA elements**

Disease/trait	Gene	Genomic element mutated	Details of mutation	dbSNP identifier
MELAS	*MT-TL1*	Mitochondrial ncRNA gene	A > G, A3243G, multiple effects on tRNA structure and function	rs199474657
Cartilage-hair hypoplasia	*RMRP*	Nuclear ncRNA gene	Various substitution and indel mutations	
Prader–Willi syndrome	Gene cluster including *SNORD115* and *SNORD116*	SnoRNA genes	Large deletion	
Spinocerebellar ataxia 8	LncRNA *ATXN8OS*	LncRNA	CTG repeat expansion	
Exfoliation syndrome (XFS)	LncRNA *LOXL1-AS1*	LncRNA	Various mutations	
Susceptibility for celiac disease	LncRNA lnc13	LncRNA	C > T, affecting RNA-protein interactions	rs917997
Facioscapulohumeral muscular dystropy (FSHD)	*DUX4/DBET*	Region including *DUX4/DBET*	Reduced number of D4Z4 repeats	
Angelman syndrome	*UBE3A*	Region including *UBE3A*	Large deletion	

Note: Summary of diseases and traits discussed in this book.

The phenotypic consequences of single nucleotide polymorphism variants, discussed in Chapter 5, are also related to the discussion on genetic fate. Thus, it must be borne in mind that under certain circumstances a variant may be beneficial, but in other situations, the same variant is neutral or bad. An example that we discussed early in the book was the genetic variant conferring the sickle cell trait. It is responsible for poor oxygen transport, but it may also offer a selective advantage in areas affected by malaria. In addition, there is the significant principle that there is a complex interplay between different variants within an individual genome. Thus, the effect of a certain genomic variant is dependent on the presence of other variants, so whether it is good, bad, or neutral depends on its genetic background.

WHO HAS ACCESS TO A PERSON'S GENOME?

We turn to ethical issues that appear when individuals are subject to limited or extensive genetic analysis. One of them is about privacy. In case your genome is sequenced for a clinical or private purpose, who should have access to the information? Some argue that genome sequence information should be kept private to avoid misuse. An opposing standpoint is that genomes are to be shared. For instance, sharing the information with researchers has the advantage that it may be exploited for scientific purposes to make important conclusions as to human genetic variation and its medical significance. An example where this principle has been applied is the UK Biobank, where genetic data (single nucleotide polymorphisms [SNPs]) as well as health data of no less than 500,000 individuals have been made available to researchers.

When sharing genome information, there are cases where people are willing to expose their identity. There are, for instance, a number of scientists who have made their genomes publically available. But a more common situation is that people whose genomes are sequenced remain anonymous. This is, for instance, the case for the UK Biobank data where the contributors have been de-identified. But in this type of situation, can we preserve anonymity? Unfortunately, there are examples where the identities of individuals in genome sequencing projects have been revealed. For instance, it was shown in 2013 that publically available genomic sequences of anonymous people could be used to search genetic information of public genealogy databases and in this way surnames were identified. In combination with other public resources, the identities of a large number of individuals could be elucidated. From this study it is clear that it may be difficult to hide the identity of a person, given that we have access to the full genome sequence or extensive genomic variant data.

In 2018 the California police used a novel method in forensics. They wanted to find the identity of a serial killer responsible for more than 50 rapes and 12 murders carried out in the 1970s and 1980s. Based on DNA from crime scenes, they uploaded a DNA profile (SNP data) to a site named GEDmatch.com. This is a site where people with a genealogy interest may upload their genome data and in this way identify novel family relationships. However, the police used this freely available website to identify relatives to the serial killer and constructed a family tree based on this information. By filtering on the basis of parameters such as sex and place of residence, a single suspect could be identified. Finally, DNA from the suspect was retrieved and was found to perfectly match the DNA of the crime scenes. Although successful from a police point of view, this story raises the question of how much control people have over information they have given to public databases. For instance, knowing about how the police now have made use of the data, do all users of GEDmatch accept that the information they uploaded could potentially be used to identify criminal relatives?

Not only is genome analysis carried out in the clinic, we have also seen that there are private companies that offer such services ("direct-to-consumer" genetic testing), for instance, screening for SNPs (Chapter 5). Can we be sure that the genomic information does not fall into the wrong hands and is being misused? Could consumers make inappropriate or dangerous decisions based on their DNA test results? There is a risk that companies want to make money from an over-interpretation of genetic variant information and that thereby misleading information is presented to the client. How are we protected from such misinformation? Should the activities of commercial genetic companies be regulated? In the United States, the company 23andMe has been under control of the US Food and Drug Administration (FDA). In 2017 the FDA allowed the company to market tests that assess genetic risks for 10 disorders, including Parkinson's and late-onset Alzheimer's diseases. And in 2018 the FDA approved three *BRCA1/BRCA2* mutations, the variants 185delAG and 5382insC in the *BRCA1* gene and the 6174delT variant in the *BRCA2* gene.

The "invalids" of the *Gattaca* movie were subject to discrimination. Could this happen in real life? Some people fear that genetic discrimination could be employed by insurance companies. What decisions should insurance companies make on the basis of genetic information? Can a certain genetic makeup leave you uninsured? Furthermore, could an employer discriminate against you based on your genome sequence? In the United States, the Genetic Information Nondiscrimination Act (GINA) of 2008 is designed to protect people from genetic discrimination. For instance, it prevents employers from using genetic information when making an employment decision. However, many jurisdictions already permit insurance companies to access certain genetic information in specific circumstances. The extent to which this is ethically defensible is still hotly contested.

HOW MUCH DO YOU WANT TO KNOW ABOUT YOUR OWN GENOME?

Testing for inherited disorders started already some 50 years ago when newborns were screened for PKU (phenylketonuria), a disease where the amino acid phenylalanine is accumulated in the blood. The motivation for the screening is that if untreated, PKU may result in mental retardation. In the era of genomics we now see a lot more possibilities in terms of analysis of genetic disorders. As a consequence, there are a number of ethical issues that arise.

For instance, what newborns should be tested genetically or have their genome sequences determined? It is now theoretically possible to analyze the full genome of every newborn. But is it meaningful? Would it help to identify genetic disease? Two research projects argue against newborn screening in this way. One study demonstrated that for some 20% of the newborns, exome sequencing missed a disorder identifiable with biochemical tests. Exome sequencing, not to mention whole genome sequencing, is also costly. In another study, many parents were uninterested in having their newborns sequenced. They were concerned with privacy, with receiving alarming results, and with insurance discrimination issues. A conclusion reached in 2016 by the American Society of Human Genetics (ASGH) is that only babies with an undiagnosed illness should be subject to genome sequencing.

When a newborn is subject to screening, a sample of blood is withdrawn for analysis. This brings us to another ethical topic, that of consent. For instance, if parents have agreed to genome analysis of their newborn, it will be important that the blood sample is not being used for other purposes that are beyond the analysis specified in the agreement. The issue of consent and possible misuse of samples apply of course not only to newborns but to any case where a biological sample is collected for research or clinical purpose.

Also in the context of adult genetic screening we are facing ethical problems. Some of the dilemmas arose already at a time before whole genome analysis, when genetic tests for individual disorders became available. In terms of testing for individual diseases, consider for instance a young person with relatives that are affected by Huntington's disease (HD), a disease for which there is right now no effective therapy. If that person has the disease allele, he or she will develop the disease later in life. Does such an individual want to know about his or her genetic predisposition? There is no simple answer. Studies have shown that there are individuals at risk for HD that choose to undergo testing, whereas others do not.

Decisions about genetic screening are different for diseases where there is therapy available. As one example, consider screening for mutations in the *BRCA1* and *BRCA2* genes. Individuals with certain mutations in these genes have a 45%–90% probability of developing breast cancer. There are possible advantages of knowing about these mutations. Unlike the case with HD, there is preventive therapy for breast cancer available, and a more frequent surveillance can reduce the disease risk. The actress Angelina Jolie received information about mutations in her *BRCA1* gene. She made headlines when she had her breasts and ovaries removed on the basis of this information. The situation with heritable factors increasing the risk of breast cancer still brings up the question—does an individual want to know about any of these mutations? It is a nontrivial question, and it is up to the patient to make an informed decision.

Genetic analysis including genome sequencing sometimes involves less specific phenotypic implications of mutations. For this reason, it is not always clear what kind of genomic information should be shared with patients. For instance, if the probability of developing a specific disease is 10% above average under a certain condition, do patients understand this information and are they helped by it?

There is another concern in the context of genome sequencing. Consider an individual with a suspected inherited disorder whose genome is being investigated to find the causative mutation. What if in the analysis of the genome a mutation that is responsible for another serious disease is identified? Should in such a case the patient be informed about this secondary finding? A similar question appears in projects where a large number of genomes are being sequenced. In Iceland about one-third of the population has had their genomes analyzed by the private company deCODE. The genetic makeup of many other Icelanders may be inferred from this data because of the extensive genealogical information that has been collected. The most common inherited risk factor for breast cancer in Iceland is a mutation named 999del5 in the *BRCA2* gene. If the information is available to the company, should carriers of this allele be told about the mutation? Kari Stefansson, the founder of deCODE, argues that they should, but many others think not.

Furthermore, if an individual gets to know information crucial to a disease, should all close relatives be informed and analyzed? Should members of a family share their genome information among themselves? As a special issue not related to disease, when genomes of a whole family are investigated, do the family members want to know about unexpected paternity?

SELECTING AGAINST DISEASE MUTATIONS

So far we have discussed ethical issues that are concerned with genome information. We now turn to issues related to the application of genetic information. For instance, in the event a couple wants to give birth to a child, what are the options to avoid a mutation causing a serious disease? First, prenatal diagnosis may be a cause for abortion. In many countries, such a diagnosis may be carried out in the clinic for a limited number of disorders, and on the basis of the results, the parents can decide if pregnancy should be terminated. This is controversial though. In case you allow the termination of pregnancy, what are the diseases that you think are adequate for such an operation? Furthermore, who decides which traits are normal and which constitute a disorder? Even more controversial are abortions that are based on prenatal diagnosis for conditions other than disease. One example is the selection of sex. For instance, in China selective abortion has contributed to a strong gender bias in the population.

Another option to avoid a disease mutation is the procedure with *in vitro* fertilization (IVF) and preimplantation genetic diagnosis (PGD) as described in Chapter 14. With this technology, a number of eggs fertilized in the laboratory are genetically analyzed and appropriate eggs with a specific genotype are then reintroduced into the mother's uterus. The motivation for PGD may be that one or both of the parents suspect they are carriers of a specific disease mutation, and they want to avoid the mutation being transmitted to their children. We also saw in Chapter 14 that PGD in rare cases has been used to achieve a "savior sibling."

In the United Kingdom, PGD is available for nearly 400 disorders, but the regulation of the technology is different in different countries. The application of PGD is growing most rapidly in China—there are estimates that China's use of PGD outpaces that in the United States (2017). One clinic in China recorded 41,000 IVF procedures in 2016. A factor contributing to this growth is that in the population there is very little religious and ethical concern about PGD. Furthermore, the recent legislation in China that allows mothers to have two children has led many older women to seek fertility treatment.

The Chinese clinics licensed to do PGD are allowed to use it only to avoid serious disease or to assist infertility treatment. Sex selection with PGD is forbidden. Also, "enhancement" applications are not allowed. For instance, many Asians are not able to effectively process alcohol because they carry a specific genomic variant. It is reported that some Chinese parents ask to have a baby without this variant—they want their sons to take part in business lunches where alcohol is served. However, PGD clinics say no to this type of request not related to disease.

In addition to abortion and IVF/PGD, a third alternative to eliminate a serious disease mutation is to apply a novel genome editing procedure such as CRISPR/Cas9 (Chapter 14). In principle, such technology could be used to directly eliminate a disease mutation at the germline level and thereby cure an inherited disorder. CRISPR/Cas9 is not yet being used for heritable genome editing because there are major safety concerns. However, they may be contemplated in the near future, and it is therefore of importance to consider what kind of experiments or therapy should be allowed. This is discussed in some more detail in the next section.

ETHICS OF GERMLINE GENE THERAPY

We can distinguish between two categories of germline genetic modification. On the one hand, there are experiments carried out on germline cells or fertilized eggs *in vitro*. These experiments are used for the purpose of research and do not give rise to a human being. An example of such a study is from 2017 where the role of the transcription factor OCT4 in early development was studied by introducing mutations in this gene with the help of CRISPR/Cas9. The embryos used in this study were donated by couples who had undergone IVF treatment. The embryos were allowed to develop in the laboratory for only a few days. This type of study is mainly to address fundamental scientific questions related to early development, but experiments of this nature may also aid to improve culture conditions in future IVF treatments. In addition, they may help to study technical details of the gene-editing procedure. For instance, off-target effects may be evaluated. Nevertheless, editing human embryos *in vitro* raises ethical and regulatory issues.

On the other hand, a procedure to introduce germline genetic changes may be intended for use in establishing a pregnancy, in which case a new human being is born who is able to transmit the genetic changes to future generations. This later therapy may be referred to as "heritable germline editing." It has not yet been applied to humans, but as it soon may become technically possible and safe, it is important to think about the ethical issues connected to it. Under what circumstances should it be implemented, and what are the benefits and risks?

As to the purpose of germline editing, it could first be used to prevent or treat disease. We typically think of monogenic disorders that may be cured by a single genomic change. Alternatively, it may be applied for "enhancement," such as improving traits like intelligence and physical capabilities. It could also be used to alter traits, such as hair or eye color. However, there is a general consensus that gene therapy is only relevant for disease prevention and treatment, whereas "enhancement" therapy should not be performed.

What are the risk factors of heritable germline editing? It is essential to realize that with this technology, we might be changing not only a single individual but a host of future individuals as well. In this manner, there might even be potential for harm to humanity as a whole. One of the most significant risk factors is that "off-target" effects are possible, meaning that genetic changes accidentally are introduced in genome locations that were not intended. There might be other types of undesired effects, even in the absence of off-target effects. This may not be likely in the event that we are replacing a pathogenic mutation with an allele common in the population, but introducing a less common genomic variant that is poorly documented could turn out to be harmful. A change in a reproductive cell may have undesired long-term consequences, and once these bad effects are discovered, the changes may have already been passed on to children of the patient. Long-term follow-up studies will certainly be important in germline editing experiments.

There is also a social justice issue associated with germline editing, as it is expected to be costly and for this reason may become available only to a few. This dilemma also applies to many other technologies, such as PGD. A closely related problem is the development and use of very expensive drugs to treat rare genetic disorders (Chapter 14). Should this money instead be used to relieve the suffering of millions of poor people with the help of already existing technologies? Another concern of germline editing is that if it became common among the wealthier, it could lead to a change in the prevalence of genetic diseases between wealthy and poor populations.

There are additional arguments that have been raised against heritable germline therapy. One is that persons with disabilities may be discriminated against. Widespread use of gene therapy could make society less accepting of people who are different. Furthermore, there are other ideas such as that artificial changes in the human genome are against a "natural order" and that we should not "play God" and influence natural evolution characteristic of the human genome. It has also been reasoned that germline editing is a version of eugenics, the idea of "improving the genetics of the human race" prevalent in the early part of the twentieth century in countries including the United States and Nazi Germany.

The National Academies of the United States presented a carefully prepared document in 2017 in which ethical aspects of gene therapy are thoroughly discussed. It is recommended that germline editing trials be permitted, but only if they follow the standard protocols for human clinical trials and if they are subject to rigorous oversight. Scientists should not yet perform germline editing on embryos intended for establishing a pregnancy—that is, "heritable germline editing." Furthermore, the research should be restricted to the treatment and prevention of disease. Experiments with genetic "enhancement," like improving on an individual's intellect or physical strength, should not be allowed.

CONCLUDING REMARKS

There is a dramatic development in gene technology concerned with the human genome. This technology generates a whole lot of information and opens up many possibilities with regard to gene therapy and gene editing. However, there are a number of decisions we need to make on how this information and technology should best be applied.

More and more individuals have their genomes analyzed. Such analysis may be of particular interest if you suspect that you are suffering from a genetic disorder, but members of this group of individuals do not necessarily want to know about their genomes. In any case, only in rare cases might there be a cure to the disorder although you know what mutation is causative.

We should be careful about heritable genome editing, as there are important safety concerns. But if the safety hurdles can be overcome, for what purposes should this editing be performed? In the case of PGD, what mutations should be selected against?

It should be kept in mind that for many of the ethical questions related to genomic information and to gene therapy, there are no simple answers. For some questions, there will be a different answer depending on the cultural background of the person you ask. Discussion of ethical topics should nevertheless be continued, and in that discussion it will be necessary to stay up to date with the latest scientific findings.

FURTHER READING

Gene-environment interplay

Berg J. 2016. Gene-environment interplay. *Science* 354(6308):15.

Ethical aspects on genetic information

Nuffield Council on Bioethics. 2015. *The collection, linking and use of data in biomedical research and health care: Ethical issues*. Nuffield Council on Bioethics, London.

Budimir D, Polasek O, Marusic A et al. 2011. Ethical aspects of human biobanks: A systematic review. *Croat Med J* 52(3):262–279.

Gymrek M, McGuire AL, Golan D et al. 2013. Identifying personal genomes by surname inference. *Science* 339(6117):321–324.

Zhang S. 2017. What happens when you put 500,000 people's DNA online. *The Atlantic*. www.theatlantic.com/science/archive/2017/11/what-happens-when-you-put-500000-peoples-dna-online/543747/?utm_source=fbb

Kaiser J. 2016. Baby genome screening needs more time to gestate. *Science* 354(6311):398–399.

Interview with Hank Greely on genetics ethics

Greely H, Gitschier J. 2014. The ethics of our inquiry: An interview with Hank Greely. *PLOS Genet* 10(11):e1004780.

BRCA1 and *BRCA2* mutations and breast cancer prognosis

van den Broek AJ, Schmidt MK, van't Veer LJ et al. 2015. Worse breast cancer prognosis of *BRCA1/BRCA2* mutation carriers: What's the evidence? A systematic review with meta-analysis. *PLOS ONE* 10(3):e0120189.

Ethical aspects on gene therapy and genome editing

Anonymous. 2017. Take stock of research ethics in human genome editing. *Nature* 549(7672):307.

Anonymous. 2017. The future of human genome editing. *Nat Genet* 49(5):653–653.

Fogarty NME, McCarthy A, Snijders KE et al. 2017. Genome editing reveals a role for OCT4 in human embryogenesis. 550:67.

National Academies of Sciences, Engineering, and Medicine. 2017. *Human genome editing: Science, ethics, and governance*. National Academies Press, Washington, DC.

Scheufele DA, Xenos MA, Howell EL et al. 2017. U.S. attitudes on human genome editing. *Science* 357(6351):553–554.

On the cost of gene therapy

Orkin SH, Reilly P. 2016. MEDICINE. Paying for future success in gene therapy. *Science* 352(6289):1059–1061.

Appendixes

APPENDIX 1

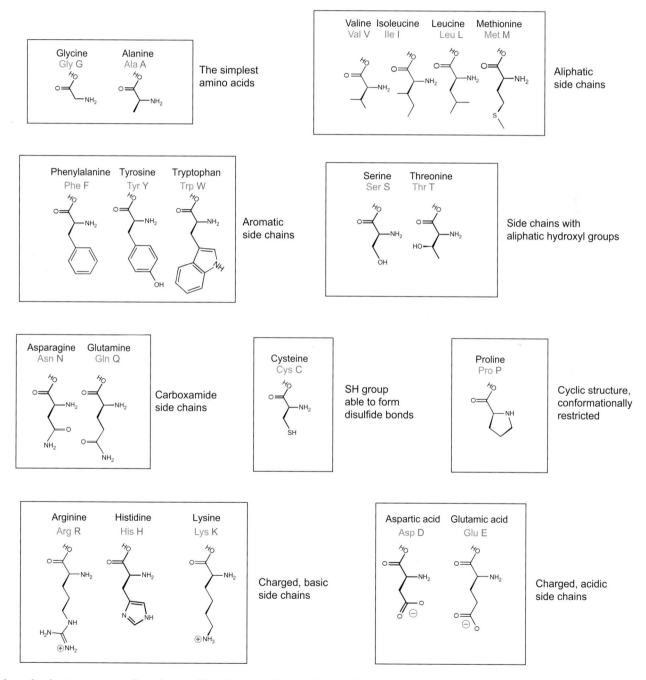

Chemical structures of amino acids. The smallest amino acids are glycine and valine; the side chain of glycine is just a hydrogen atom (H), and the side chain of alanine is a methyl group, CH_3. Cysteine has a reactive SH group that may form a disulfide bond with another cysteine located in the same protein, thereby stabilizing the protein structure. Proline is unique among amino acids as its side chain is joined to both the amino group and the α-carbon. Thereby, a cyclic structure is formed that makes proline conformationally restricted.

APPENDIX 2

Nucleotide ambiguity symbols		
Code	Nucleotides	Complement
A	A	T
G	G	C
C	C	G
T	T	A
U	U	A
R	A or G, purine	Y, pyrimidine
Y	C or T, pyrimidine	R, purine
S	C or G, strong pairing	S, unchanged
W	A or T, weak pairing	W, unchanged
K	G or T, keto	M, amino
M	A or C, amino	K, keto
B	C or G or T, not A	V, not T
V	A or C or G, not T	B, not A
D	A or G or T, not C	H, not G
H	A or C or T, not G	D, not C

Glossary

3′ and 5′ splice sites (acceptor and donor sites) Splicing occurs through a two-step mechanism. In the first reaction, the 2′ hydroxyl of a branch site adenosine close to the 3′ end of the intron attacks the phosphodiester bond that is at the 5′ end of the intron, the 5′ splice site, or donor site. In the second reaction, the free 3′ end of the exon generated in the first step attacks with its hydroxyl group the phosphodiester bond at the 3′ end of the intron, the 3′ splice site, or acceptor site.

α-helix A common helical structure in proteins in which every backbone N-H group donates a hydrogen bond to the backbone C = O group of the amino acid located three or four residues earlier along the protein sequence.

adenosine deaminase severe combined immunodeficiency (ADA-SCID) An inherited disease caused by a defect in the gene encoding the enzyme adenosine deaminase (ADA). The deficiency in this enzyme damages the immune system and causes severe combined immunodeficiency (SCID). Affected individuals lack immune protection from bacteria, viruses, and fungi.

alignment The comparison or fitting of two DNA or protein sequences that are evolutionary related.

allele A genetic variant of a specific genomic region, for instance, a region corresponding to a gene.

amino acid Building block of proteins. All amino acids consist of one amino group, one carboxyl group, and one side chain specific to the amino acid.

alternative splicing The regulated process whereby exons are joined in different combinations. In this manner, a number of different RNAs and proteins may be formed from a single gene.

annotation Genome annotation is the assignment of sites or units in a genome, such as genes, exons, transcription regulatory sites, and transposons.

anticipation The phenomenon that the symptoms of some genetic conditions tend to appear at an earlier age and/or become more severe as the disorder is passed from one generation to the next.

anticodon Three-nucleotide sequence in tRNA that is complementary to a codon in the mRNA.

apoptosis A highly regulated process of programmed cell death. Apoptosis plays an important role in maintaining the health of the body by eliminating unneeded or abnormal cells.

apurinic site A site in DNA where a purine is lacking

archaea Archaea are single-cell microorganisms. They lack the membrane-bound nucleus characteristic of eukaryotes. The evolutionary history is distinct from bacteria. Analyses of biochemical components show that archaea are more similar to eukaryotes than are bacteria.

AU-rich element (ARE) Region in the 3′ UTR in mRNA with a high frequency of adenine and uracil. AU-rich elements are common determinants of mRNA stability.

autosome A chromosome that is not a sex chromosome.

β-sheet A common secondary structure element in proteins. Consists of strands connected laterally by backbone hydrogen bonds. Also known as *β-pleated sheet*.

β-plus thalassemia Disorder where patients have one allele with the sickle variant and one allele with another mutation in the β-globin gene causing a reduced amount of normal β-globin protein.

bacteria Bacteria are single-cell microorganisms. They lack a membrane-bound nucleus. The evolutionary history is distinct from archaea.

basal (or general) transcription factor A transcription factor part of the core machinery of transcription initiation; needed for the transcription of every gene.

base excision repair A DNA repair mechanism responsible for excision of bases that have been damaged by deamination, oxidation, or methylation. One example of such a damaged base is 8-oxoG. After excision, the correct base is inserted as guided by the complementary strand.

bioinformatics Computational handling and analysis of information generated by current methods in molecular biology and biomedicine. An interdisciplinary science with contributions from mathematics, computer science, and biomedicine.

cancer driver gene/mutation A cancer cell is characterized by numerous mutations. A driver mutation is a mutation in a cell that drives cancer progression—it provides the cell with a selective growth advantage. In contrast, "passenger" mutations are mutations that confer no selective advantage.

cancer genome A genome obtained by DNA sequencing of a sample of tumor tissue. The genome may be derived from a single cell or from a homogeneous or heterogeneous group of cancer cells.

CAR-T cell therapy A form of therapy in which a patient's T cells (a type of immune system cell) are genetically changed so they will attack cancer cells.

ChIP-sequencing (ChIP-seq) Method to determine what proteins (like transcription factors) interact with what regions in DNA. The first step is to covalently cross-link complexes between DNA and protein. Then DNA is fragmented, so that to each protein molecule is attached only a smaller piece of DNA. Then, an antibody against a specific protein is used to immunoprecipitate the target protein. From this precipitate, DNA is finally isolated and subjected to DNA sequencing.

chloroplast The organelle in plants in which photosynthesis takes place. Photosynthesis is the production of carbohydrates using sunlight as the energy source.

chromatin A complex in the cell nucleus, formed with chromosomal DNA, proteins, and other molecules.

chromosomal translocation An exchange of larger chromosomal segments between two different (nonhomologous) chromosomes. One example is the translocation known as the Philadelphia chromosome, involving an interchange of material between chromosomes 9 and 22. This abnormality is associated with the cancer diseases chronic myelogenous leukemia (CML) and acute lymphoblastic leukemia (ALL).

chromosome DNA molecule that contains part of or the entire genome of an organism. In a human cell, the genome is divided into 22 autosomal chromosomes and the X and Y sex chromosome. A eukaryotic chromosome also includes proteins such as histones.

chromothripsis The occurrence of many rearrangements affecting one or a few chromosomes in a cancer cell. May be the result of a single event causing multiple double-strand breaks that are repaired using error-prone mechanisms.

cis In the context of gene regulation, a "cis"-acting element is a DNA sequence element that occurs at a genomic site close to the gene under control. A "trans"-acting factor is encoded by a gene located at a genomic site distinct from the gene whose expression is under control.

codon A three-nucleotide sequence specifying a distinct amino acid (or translational stop).

complex or multifactorial disorder A disorder with multiple causes, such as mutations/variants in several genes in combination with lifestyle and environmental factors.

compound heterozygote An individual is a compound heterozygote when there are two recessive alleles for the same gene, and these two alleles are mutated at different locations.

consensus sequence In an alignment of DNA or protein sequences, the consensus sequence is showing for every sequence position the most common nucleotide/amino acid.

contig A contig is a set of overlapping sequences that form a region of the genome from which they were derived. A contig does not contain gaps (regions of unknown sequence). Contigs may be joined to form larger scaffolds that do contain gaps.

copy number variation (CNV) A larger (>1,000 nucleotides) DNA segment that is present at a variable copy number in comparison with a reference genome. CNV may arise as a result of deletions, insertions, and duplications. A common case of CNV is when portions of the genome are repeated because of duplication events, and the number of repeats in the genome varies between individuals.

core promoter The minimal region of the promoter required for transcription initiation.

coverage Coverage, in the context of DNA sequencing, is the number of reads that include a given nucleotide position.

CpG site A site in the genome containing the nucleotide sequence 5'-CG-3'.

cryo-electron microscopy An electron microscopy technique to elucidate the three-dimensional structure of macromolecules in which the sample is cooled to very low temperatures. The technique allows studies of macromolecules in their native environments.

de novo mutation A mutation that came about during the production of a parental sperm or egg cell, or early in the fertilized egg. When considering an individual with a *de novo* mutation, the mutation is not identified when analyzing the DNA of the parents.

degenerate code In the context of the genetic code, "degenerate" refers to the fact that to many of the amino acids, there are more than one codon.

deletion mutation A mutation where one or more nucleotides are removed.

depurination Chemical reaction where a purine (adenine or guanine) is removed from DNA.

diploid Referring to the situation that a cell contains two copies of each chromosome. Compare "haploid."

distance matrix In the context of molecular phylogenetics, a distance matrix shows pairwise distances between sequences. The distance between two sequences is obtained by counting the number of nucleotide or amino acid differences between two aligned sequences.

DNA Deoxyribonucleic acid, polymer of nucleotide (phosphate-deoxyribose-base) units, and a carrier of genetic information.

DNA replication Enzymatic process where new copies of DNA are made.

dominant An allele is dominant if only one copy is sufficient to produce the phenotype.

downstream promoter element (DPE) DNA sequence element part of the core promoter, located downstream of the transcription start site.

dotplot A two-dimensional graph where two sequences are compared; they are on different axes and all instances of identity are indicated with a "dot."

enhancer DNA element that will recruit transcription factors and thereby stimulate transcription. Enhancers are often located at a large distance—several thousand nucleotides or more—from the transcription start site.

enhancer RNA (eRNA) A class of noncoding RNA molecules transcribed from the DNA sequence of enhancer regions.

epigenetic Epigenetic information is not encoded in the DNA nucleotide sequence but is nevertheless transmissible during cell division. Epigenetic mechanisms involve chemical or structural changes to chromatin, instead of changes in nucleotide sequence. Examples of such mechanisms are DNA methylation and histone modification.

epigenetics The study of heritable changes in gene expression that do not involve changes in the nucleotide sequence of DNA.

epigenome An account of all of the epigenetic modifications, such as DNA methylation and histone modifications, of a genome.

euchromatin A lightly packed form of chromatin, typically associated with active genes.

eukaryotes Organisms whose cells have a membrane-surrounded nucleus as well as organelles such as mitochondria.

exon A portion of a eukaryotic gene that corresponds to a part of the mature mRNA (or ncRNA) after splicing. An exon of a protein gene may contain 5′ UTR, coding sequence, or 3′ UTR.

exon skipping A mechanism during RNA splicing where one or more exons present in the primary RNA transcript are left out from the final mRNA.

frameshift mutation A mutation that disturbs the reading frame during protein synthesis.

fusion gene A hybrid gene formed from two genes that were previously separate. A fusion gene may occur as a result of chromosomal translocation.

gamma retrovirus Retroviruses are RNA viruses that replicate through an DNA intermediate. Gamma retroviruses is a subgroup of these viruses. Examples include the murine and feline leukemia viruses.

gene A portion of DNA in a genome that is responsible for the production of a specific protein or RNA transcript. By and large the elements of a gene are located close to each other on a chromosome, although regulatory elements like enhancers may be distant from the regions encoding the protein or RNA product.

gene ontology (GO) A unique vocabulary to describe genes and proteins where the terms are arranged in a graph. The root nodes of the graph are the three main GO categories "molecular function," "biological process," and "cellular component." The purpose of GO is to facilitate the analysis of data related to genes and proteins, such as gene expression data.

gene therapy Procedures where DNA or RNA is delivered to patients to treat a disease. Examples of gene therapy include the (1) introduction of a normal copy of defect gene, (2) destruction of a malfunctioning gene, and (3) administration of DNA or RNA to modify the expression of a gene.

genetic code Table showing the relationship of codons and the amino acids they specify.

genetic linkage analysis A method based on the fact that during meiosis two different loci on a chromosome have a certain probability of being separated through the process of chromosomal crossing-over. The closer the two loci are, the more likely they are to stay together—they are said to be genetically linked. Conversely, two loci that are distant are more likely to be separated into different chromatids during chromosomal crossing-over. The location of a disease gene may be inferred from an analysis of genotypes in a family affected by the disease.

genome The genetic material of an organism.

genome browser A web-based tool to examine the properties of a genome where the user typically is viewing a distinct region of a chromosome.

genome-wide association study (GWAS) A study of a genome-wide set of genetic variants in different individuals to see if any variant is associated with a trait. A GWAS often compares the DNA of individuals with a particular trait to the DNA of control individuals without the trait.

genotype The genetic makeup of an organism with reference to a single trait, set of traits, or entire complex of traits. See also *phenotype*.

germline Lineage of cells or inherited material associated with eggs and sperm.

graph In computer science, a graph is a data structure with a set of vertices (also known as *points* or *nodes*) and edges (lines connecting the vertices).

haploid Refering to the situation that a cell contains only one copy of each chromosome. Compare "diploid."

hemoglobin An oxygen transporting protein present in red blood cells.

hereditary hyperferritinemia cataract syndrome A disorder characterized by an excess of the iron storage protein ferritin in the blood (hyperferritinemia) and tissues of the body. A major symptom is a clouding of the eye lenses (cataract).

heterochromatin A condensed and compact form of chromatin. Heterochromatin is typically transcriptionally silent.

heteroplasmy Organellar genomes, like those of mitochondria, are often not genetically identical within the same cell. The presence of more than one type of organellar genome within a cell or within an individual is known as heteroplasmy.

heterozygous Having two nonidentical copies of an allele. A characteristic of diploid cells.

histone Histones are basic proteins in eukaryotes that package DNA into nucleosomes. They are the major protein components of chromatin and play a significant role in epigenetic control of gene expression.

homologous In the context of molecular biology, the word "homologous" can refer to an evolutionary relationship. Two DNA or protein sequences are homologous if they share a common ancestor during the evolution of life. Alternatively, "homologous" may refer to the relationship of chromosomes in a cell. In a diploid cell, all pairs of autosomal chromosomes have one paternal and one maternal copy—these two chromosomes are said to be "homologous."

homologous recombination An event where sequences are exchanged between two DNA molecules with similar sequences. It is used during meiosis to create genetic variation, as well as to repair double-strand breaks in DNA.

homozygous Having two identical copies of an allele. A characteristic of diploid cells.

huntingtin Protein encoded by the *HTT* gene, the gene responsible for Huntington's disease. Huntingtin contains a polyglutamine tract that is expanded in the disease.

imprinting For each of our autosomal genes, we inherit one copy from our father and one from our mother. Imprinting refers to the situation where only one of the two genes is expressed.

indel mutation A mutation that is either an insertion or deletion

initiator element, Inr DNA sequence element in the core promoter covering the transcription start site with the consensus sequence BBCABW.

insertion mutation A mutation where one or more nucleotides are inserted.

intron A portion of a eukaryotic gene in between two exons that is removed during splicing of the primary transcript. The intron is not part of the fully mature mRNA or ncRNA.

inverted repeat A situation where a sequence on one of the two DNA strands is repeated on the other strand. Same as *palindrome*.

iron responsive element (IRE) RNA hairpin structure important for translational control of selected mRNAs related to iron metabolism.

karyotype The number and appearance of chromosomes in the nucleus of a eukaryotic cell as viewed by light microscopy.

lagging strand The strand of newly formed DNA whose direction of synthesis is opposite to the direction of the growing replication fork.

lariat After splicing, the intron is in a lariat form, a branched structure involving the branch site

adenosine. "Lariat" is derived from the resemblance of the intron structure to a lasso.

leading strand The strand of newly formed DNA that is being synthesized in the same direction as the growing replication fork.

lentivirus Virus that belongs to the group of retroviruses, RNA viruses that replicate through a DNA intermediate. Members of the group lentivirus are characterized by a long incubation period. The most well-known lentivirus is human immunodeficiency virus (HIV), which causes AIDS.

long noncoding RNA (lncRNA) May be arbitrarily defined as an RNA larger than 200 nucleotides in length. Such an RNA is often mRNA-like in the sense that it is transcribed by RNA polymerase II, polyadenylated, and spliced.

macromolecule A large biopolymer molecule such as a protein or nucleic acid.

Manhattan plot In a genome-wide association study (GWAS), a Manhattan plot shows the genomic localization of a genetic variant on the x-axis and the negative logarithm of the association P-value on the y-axis.

meiosis Meiosis is the process where sperm and egg cells are generated. During meiosis, a single cell divides twice to produce four cells that are haploid—that is, they each contain half of the original amount of genetic information. During meiosis, crossing-over results in a genetic makeup of the haploid cells where each chromatid is genetically distinct from the parental chromatids.

Mendelian disorder A disorder showing Mendelian inheritance, a type of inheritance following the laws proposed by Gregor Mendel in the 1860s.

messenger RNA (mRNA) An RNA molecule acting as a template for the production of proteins.

microRNA (miRNA) Small, noncoding RNA regulating the stability and translational efficiency of endogenous mRNAs.

microsatellite DNA A tract of tandemly repeated DNA motifs that range in length from one to six or up to ten nucleotides and that are typically repeated 5–50 times. Microsatellites are distributed throughout the human genome. Same as *simple sequence repeat (SSR)*.

mismatch repair A DNA repair mechanism where incorrectly paired bases are removed.

missense mutation Mutation that changes one amino acid into another.

mitochondria Organelles in eukaryotes in which carbohydrates are oxidized to generate cellular energy in the form of ATP molecules.

molecular cloning A set of molecular biology methods that are used to construct recombinant DNA molecules that are allowed to replicate in a host cell.

molecular phylogenetics A branch of phylogenetics that makes use of molecular sequences, DNA, RNA, and protein sequences to examine evolutionary history.

molecular sequence Linear sequence of building blocks of biopolymers, such as amino acids in proteins and nucleotides or bases in nucleic acids.

multiple sequence alignment A comparison of two or more molecular sequences by alignment.

mutation A change in the sequence of nucleotides in DNA. Mutations can arise during DNA replication or as a result of DNA damage.

myotonic dystrophy Inherited disorder affecting muscle function. It is characterized by progressive muscle wasting and weakness.

nanopore sequencing DNA sequencing method that relies on the measurement of changes in electrical current density when a DNA molecule is transported across a small orifice, a nanopore.

neighbor joining A tree construction method in molecular phylogenetics. Relies on a distance matrix obtained by pairwise comparison of sequences. Neighbor joining is computationally fast as compared to many other methods in molecular phylogenetics.

noncoding RNA (ncRNA) An RNA molecule transcribed from DNA, which does not encode a protein.

nonsense codon A codon that does not encode an amino acid, but is a stop signal for protein synthesis.

nonsense-mediated mRNA decay (NMD) A mechanism to eliminate mRNAs with premature stop (nonsense) codons.

nonsense mutation A mutation where a codon encoding an amino acid is converted into a stop codon. Also known as a *stop-gain mutation*.

nonsynonymous mutation Mutation that affects the amino acid encoded by the codon. Nonsynonymous mutations include missense and nonsense mutations.

nuclear magnetic resonance (NMR) Method for analysis of macromolecular structure in which the macromolecule is typically in water solution. The method relies on the fact that certain atomic nuclei are intrinsically magnetic.

nucleosome A structure representing the first level of DNA packing in eukaryotic cells. In a nucleosome, a region of DNA is wound around a core of eight histone protein subunits.

nucleotide Building block of nucleic acids and consists of phosphate, a sugar residue, and a nitrogenous base.

nucleotide excision repair A DNA repair mechanism where bulky groups like pyrimidine dimers are removed in 12–24 nucleotide long strands.

oligonucleotide A single-stranded nucleic acid molecule (DNA or RNA) with a relatively small number of nucleotides. Oligonucleotides often appear in the context of biotechnology applications, in which case they are synthesized chemically.

oncogene Gene that will promote cancer development when it is mutated or aberrantly expressed.

one gene–one enzyme hypothesis The idea that each enzyme is encoded by a specific gene—an idea put forward by Beadle and Tatum in 1941.

one gene–one protein hypothesis The one gene–one enzyme hypothesis generalized to all proteins.

open reading frame (ORF) A continuous stretch of codons that contain a start codon at the 5′ end and a stop codon at the 3′ end.

orthology A category of evolutionary relationship. Orthology arises as a result of a speciation event. Orthologous proteins are proteins that carry out the same function in different species.

palindrome A situation where a sequence on one of the two DNA strands is repeated on the other strand. Same as *inverted repeat*.

paralogy A category of evolutionary relationship. Paralogy comes about by gene duplication events. Paralogous proteins have similar but not identical functions within an organism.

peptide bond Covalent bond linking two amino acids in a protein.

phenotype The observable characteristics or traits of an organism. The phenotype is determined by genotype and environmental conditions.

phylogenetic tree Tree to show the evolutionary relationship of entities such as biological species.

phylogenetics Phylogenetics or phylogeny is the study of the evolutionary history and relationships among entities such as individuals or species.

point mutation Synonymous with substitution mutation, the replacement in DNA of one nucleotide (base) with another.

poly(A) tail A sequence of "A"s that is post-transcriptionally added to the 3′ end of eukaryotic mRNAs.

polymerase chain reaction (PCR) A laboratory method to amplify a region of DNA, such that many copies are formed of this region. The method relies on multiple cycles of DNA replication and the use of primers defining the ends of the region amplified.

polymorphism Genetic polymorphism is the occurrence of variation in many genomic sites, such as single nucleotide polymorphism (SNP) and structural variation.

polypeptide A long chain of amino acids linked by peptide bonds.

premutation A gene variant that produces a normal individual but is predisposed to become a full mutation in subsequent generations.

primer A short strand of RNA or DNA (often about 18–22 nucleotides) that serves as a starting point for DNA synthesis. A primer is needed for DNA replication because DNA polymerase can only add nucleotides to an existing strand of nucleic acid.

primary structure Sequence of building blocks in a macromolecule, such as amino acids in proteins or nucleotides in DNA or RNA.

promoter A region of a gene that directs its transcription. A promoter is located close to the transcription start site and is about 100–1,000 nucleotides long, where the major part is upstream of the transcription start site. The promoter has binding sites for transcription factors.

protein subunit A protein (polypeptide) that interacts with other proteins to form a protein complex.

proximal promoter A region upstream of the core promoter that may be up to 1,000 nucleotides in size.

quaternary structure (protein) The positions and structures of all polypeptides that are present in a protein with more than one polypeptide.

recessive An allele is recessive if two copies are required for a phenotypic characteristic.

reference genome The human reference genome (or reference assembly) is a haploid genome reconstructed from a set of different donors. A genome of an individual may be compared to this reference genome to reveal what is genetically characteristic of the individual. The reference genome is not static but now and then updated, reflecting progress in assembly.

replication slippage A form of mutation that leads to either repeat expansion or repeat contraction during DNA replication.

restriction enzyme map A map to show the sites where one or more restriction enzymes cut in a DNA nucleotide sequence.

retrotransposon A transposon that moves around in a genome with a mechanism involving an RNA intermediate. The retrotransposon DNA is first transcribed to RNA, and then a DNA copy is produced of the RNA using reverse transcription. This DNA copy is finally inserted into a novel site in the genome.

retrovirus An RNA virus that replicates through a DNA intermediate. Uses its own reverse transcriptase to copy the RNA genome to DNA. The DNA copy of the viral genome is inserted into the host genome.

reverse transcriptase An enzyme responsible for the synthesis of DNA using an RNA molecule as a template.

reversible terminator sequencing A method of DNA sequencing that is based on DNA replication, and "reversible terminator" nucleotides are used that are blocked at their 3′ end to prevent further extension of the DNA chain. The nucleotide also carries a base-specific fluorophore which is detected. The nucleotide is then deblocked and the fluorophore is removed, thereby making possible the incorporation of the next reversible terminator nucleotide.

ribonucleic acid (RNA) A nucleic acid like DNA, but with ribose instead of deoxyribose as the sugar moiety, and with the base uracil instead of thymine.

ribose Pentose sugar moiety in RNA.

ribosome Large assembly of RNA and protein subunits where protein synthesis takes place.

RNA (ribonucleic acid) A nucleic acid like DNA, but with ribose instead of deoxyribose as the sugar moiety, and with the base uracil instead of thymine.

RNA interference A process where small noncoding RNAs, miRNAs, or siRNAs inactivate target mRNAs based on sequence complementarity. Same as *RNA silencing*.

RNA secondary structure A structure formed by internal base-pairing within an RNA molecule.

RNA silencing A process where small noncoding RNAs, miRNAs, or siRNAs inactivate target mRNAs based on sequence complementarity. Same as *RNA interference*.

RNA world hypothesis The hypothesis that RNA had an essential role at an early stage in the development of life on our planet and predated proteins and DNA.

satellite DNA Tandemly repeating DNA found in centromeres and in heterochromatin.

selenocysteine insertion sequence (SECIS) A sequence that is a signal that the codon UGA is to be decoded as selenocysteine.

second genetic code Amino-acyl tRNA synthetases make sure an amino acid is covalently joined to a specific tRNA with an anticodon corresponding to that tRNA. In the enzymatic reaction, there are tRNA structural elements that are important for amino acid identity. These elements have been referred to as a second genetic code.

sense codon A sense codon specifies a distinct amino acid.

secondary structure Local folded structures in proteins or nucleic acids. In proteins the most common secondary structures are α-helix and β-sheet. In nucleic acids secondary structure refers to internal base-pairing within the molecule.

sequence assembly To determine the sequence of a larger genome, like that of a human, shorter DNA fragments are first sequenced. Sequence assembly refers to the computational procedure to reconstruct the complete genome from these sequences. In this process, the shorter sequences are merged based on sequence overlap.

sequence logo A graphical representation of the sequence conservation of nucleotides in nucleic acids or amino acids in protein.

shotgun sequencing A larger DNA to be sequenced is broken up into shorter fragments using a procedure of random fragmentation. Each of the resulting fragments is sequenced individually. The sequence of the larger DNA is finally reconstructed by computational assembly of the shorter sequences on the basis of sequence overlap.

sickle cell anemia Inherited disorder where the affected individual has two copies of a β-globin gene in which a point mutation causes a glutamic acid to valine substitution. The mutation results in an aggregation process where globin molecules form fibers such that red blood cells are distorted and normal blood flow is prevented. There is a lowered amount of red blood cells and poor oxygen transport.

sickle cell trait Phenotype of individual with one copy of normal *HBB* gene and one copy of sickle cell variant of *HBB* gene.

simple sequence repeat (SSR) A tract of tandemly repeated DNA motifs that range in length from one to six or up to ten nucleotides and are typically repeated 5–50 times. Microsatellites are distributed throughout the human genome. Same as *microsatellite DNA*.

single gene or monogenic disorder A genetic disorder resulting from mutation(s) in one specific gene only.

single molecule real-time sequencing (SMRT) A method of DNA sequencing. A single DNA polymerase is fixed on the bottom of a very small reaction chamber. The nucleotides used contain a fluorophore like in reversible terminator sequencing. When a nucleotide enters the catalytic site of the polymerase, it stays there for a short while, and a light pulse is produced and recorded. When the nucleotide is incorporated into DNA, the terminal phosphates and the fluorophore are cleaved off, and a new nucleotide may enter into the polymerase active site.

single nucleotide polymorphism (SNP) A genetic variation in a single nucleotide position in the genome. SNPs typically reflect substitution mutations, but short insertions and deletions are also included in the concept of a SNP. For a variation in a genomic position to be classified as a SNP, more than 1% of a population does not carry the same nucleotide in that position.

single nucleotide variant (SNV) Same as *single nucleotide polymorphism (SNP)*, but a SNV is a variation disregarding frequency of alleles in the population.

small interfering RNA (siRNA) Similar to microRNA, but the function of siRNA is to downregulate exogenous sequences, such as those that enter the cell during a viral infection.

small noncoding RNA (ncRNA) May be arbitrarily defined as a noncoding RNA less than 200 nucleotides in length.

SNP array SNP array technology makes use of a chip with immobilized short DNA sequences, each corresponding to a certain SNP. When this chip is subjected to a DNA sample from an individual, all perfectly matching sample sequences will give rise to a signal based on base-pairing to the chip sequences.

somatic Referring to cells that are not germline.

specific transcription factor A transcription factor regulating transcription of only one gene or a set of genes (as opposed to the general or basal transcription factors that are involved in the transcription of all genes).

spinal muscular atrophy A genetic disorder that affects the control of muscle movement. It is caused by a loss of motor neurons in the spinal cord and the part of the brain that is connected to the spinal cord (the brainstem). The loss of motor neurons leads to weakness and wasting (atrophy) of muscles used for activities such as crawling, walking, sitting up, and controlling head movement. It is a fatal disorder being the most common genetic cause of infant death. It most often leads to death before age 2 years.

spinocerebellar ataxia (SCA) A group of inherited disorders characterized by degenerative changes in the part of the brain related to movement control and sometimes in the spinal cord.

spliceosome Large complex of proteins and RNA-protein complexes responsible for splicing.

splicing Processing of a precursor mRNA (or ncRNA), where exons are joined and introns are removed.

stem cell Stem cells are undifferentiated cells that can differentiate into specialized cells or divide to produce more stem cells.

stop-gain mutation A mutation where a codon encoding an amino acid is converted into a stop codon. Also known as a *nonsense mutation*.

stop-loss mutation A mutation converting a stop codon to a codon encoding an amino acid. Also known as a *sense mutation*.

structural variation A form of individual genetic variation involving larger pieces (>50 nucleotides) of DNA sequence. Structural variation includes deletions, duplications, copy-number variants, insertions, inversions, and translocations.

substitution In the context of DNA the replacement of one nucleotide (base) with another. A "substitution mutation" is synonymous with a "point mutation."

substitution matrix A table of scores used in the alignment of protein or DNA sequences. Each score reflects the probability of a specific amino acid or nucleotide substitution.

synonymous mutation A mutation that does not affect the amino acid encoded by the codon.

TATA box A sequence of DNA (with consensus sequence TATA) found in the core promoter region of eukaryotic cells. The TATA box is a binding site for the TATA-binding protein (TBP).

tautomer One form of two or more isomeric organic compounds. Tautomers may differ only in the position of the protons and electrons and may easily be interconverted.

telomere Region at each end of a linear chromosome that is responsible for maintaining an intact end as well as part of a machinery to avoid the end being recognized as a double-strand break. A telomere consists of repeats of a sequence which in vertebrates is TTAGGG.

tertiary structure Overall three-dimensional structure of a macromolecule such as a protein or nucleic acid.

topologically associating domains (TADs) Chromosomal regions that show high levels of interaction within the regions but little or no interaction with neighboring regions.

trait A distinct variant of a phenotypic characteristic. Examples include eye and hair color.

trans In the context of gene regulation, a "cis"-acting element is a DNA sequence element that occurs at a genomic site close to the gene under control. A "trans"-acting factor is encoded by a gene located at a genomic site completely distinct from the gene whose expression is under control.

transcription The process by which RNA is produced using a DNA molecule as template.

transfer RNA (tRNA) An RNA molecule acting as an adaptor during protein synthesis between the languages of nucleic acids and proteins. Each tRNA carries an anticodon that is able to read one or more codons in the mRNA and is at the same time carrying the amino acid corresponding to those codons.

translation The process by which protein is synthesized using mRNA as template.

translesion synthesis A mechanism of DNA repair that allows synthesis past sites of DNA damage such as pyrimidine dimers or sites where a purine or pyrimidine has been lost (apurinic or apyrimidinic sites, respectively).

transposon A sequence of DNA that can move to new positions within the genome.

trinucleotide tandem repeat Trinucleotide sequences that are repeated multiple times and where the trinucleotide sequences are directly adjacent to each other.

tumor suppressor gene A gene that normally prevents cells from taking critical steps toward cancer. However, when these genes are mutated, cells may more easily progress toward cancer.

untranslated region (UTR, 5′ or 3′) A region of mRNA that is not part of the actual protein-coding sequence. A 5′ UTR is on the 5′ side of the coding sequence, and a 3′ UTR is on the 3′ side.

upstream open reading frame (uORF) An open reading frame positioned upstream of the main coding sequence of a mRNA.

uracil A nitrogenous base in RNA that corresponds to thymine in DNA.

variant (genetic) Sequence at a site in the genome reflecting or characterized by individual variation. Genetic variation is caused by mutations, and "variant" may be used as synonymous with "mutation."

variant calling The process of identifying variants in a genome sequence by comparison to a reference genome.

whole exome sequencing (WES) The process of determining the DNA sequence of the exonic parts of a genome.

whole genome sequencing (WGS) The process of determining the complete DNA sequence of the whole genome of an organism.

whole genome shotgun sequencing A procedure of whole genome sequencing when the whole genome is reconstructed from shorter sequencing reads directly, without first identifying and sequencing fragments of the genome obtained by restriction enzyme cleavage.

wobble In the pairing of a codon to an anticodon, the Watson-Crick base-pairing rules apply, except in the case of the third position of the codon (first position of the RNA anticodon) the pairing is less strict and referred to as a wobble pair. One example is G in the third position of the codon that may pair with either C or U.

X-linked recessive inheritance Inheritance reflected by a mutation in a gene on the X chromosome that causes the trait or disease to be expressed specifically in males. Males have only one X chromosome, and for this reason they are dependent on having functional copies of genes located on that chromosome.

X-ray crystallography Method for analysis of macromolecular structure. The molecule is in a crystalline form and subjected to x-ray irradiation. By analyzing the diffraction pattern of x-rays, the structure may be elucidated.

zinc finger A finger-shaped protein structural motif where a zinc ion is coordinated to amino acids. Many proteins with zinc finger motifs are transcription factors where the zinc fingers are responsible for sequence-specific recognition of DNA.

Recommended Textbooks for Further Reading

Berg JM, Stryer L, Tymoczko JL, Gatto GJ. 2018. *Biochemistry.* Macmillan Higher Education, New York.

Alberts B. 2015. *Molecular biology of the cell.* Garland Science, New York.

Brändén C-I, Tooze J. 2009. *Introduction to protein structure.* Garland, New York.

Klug WS. 2017. *Essentials of genetics.* Pearson, Boston.

Jorde LB, Carey JC, Bamshad MJ. 2016. *Medical genetics.* Elsevier, Philadelphia.

Nussbaum RL, McInnes RR, Willard HF, Hamosh A. 2016. *Thompson & Thompson genetics in medicine.* Elsevier, Philadelphia.

Richards JE, Hawley RS. 2011. *The human genome: A user's guide.* Elsevier, Amsterdam.

Gillham NW. 2011. *Genes, chromosomes, and disease from simple traits, to complex traits, to personalized medicine.* FT Press Science, Upper Saddle River, NJ.

Watson J. 2017. *DNA: The secret of life.* Arrow Books, [S.l.].

Snyder M. 2016. *Genomics and personalized medicine: What everyone needs to know.* Oxford University Press, New York.

Bjorklund R. 2011. *Sickle cell anemia.* Marshall Cavendish Benchmark, New York.

Shreeve J. 2004. *The genome war: How Craig Venter tried to capture the code of life and save the world.* Alfred A. Knopf, New York.

Chakravarti A. 2014. *Human variation: A genetic perspective on diversity, race, and medicine.* Cold Spring Harbor Laboratory Press, New York.

Chiu LS, Seachrist JA. 2007. *When a gene makes you smell like a fish—And other tales about the genes in your body.* Oxford University Press, Oxford; New York.

Reilly P. 2015. *Orphan: The quest to save children with rare genetic disorders.* Cold Spring Harbor Laboratory Press, New York.

Dudley JT, Karczewski KJ. 2014. *Exploring personal genomics.* Oxford University Press, Oxford.

Quackenbush J. 2011. *The human genome: The book of essential knowledge.* Charlesbridge, Watertown, MA.

Sulston J, Ferry G. 2003. *The common thread: Science, politics, ethics and the human genome.* Corgi, London.

Lesk AM. 2017. *Introduction to genomics.* Oxford University Press.

Read AP, Donnai D. 2015. *New clinical genetics.* Scion Publishing Ltd, Banbury, England.

Dawkins R, Wong Y. 2017. *The ancestor's tale: A pilgrimage to the dawn of life.* Mariner Books.

St. Clair C, Visick J. 2015. *Exploring bioinformatics: A project-based approach.* Jones & Bartlett Learning, Burlington, MA.

Zvelebil M, Baum JO. 2008. *Understanding bioinformatics.* Garland, New York.

Lemey P, Salemi M, Vandamme A-M. 2017. *The phylogenetic handbook: A practical approach to phylogenetic analysis and hypothesis testing.* Cambridge University Press, Cambridge, UK.

Haddock SHD, Dunn CW. 2011. *Practical computing for biologists.* Sinauer Associates, Sunderland, MA.

Pevsner J. 2016. *Bioinformatics and functional genomics.* Wiley Blackwell, Chichester, West Sussex.

Index